# CARBON DYNAMICS
# IN
# EUTROPHIC, TEMPERATE
# LAKES

# CARBON DYNAMICS IN EUTROPHIC, TEMPERATE LAKES

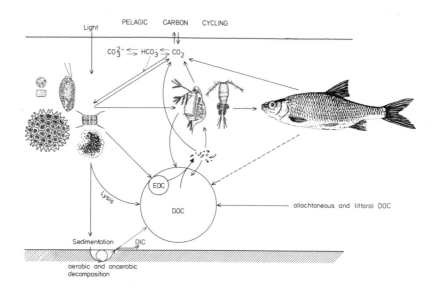

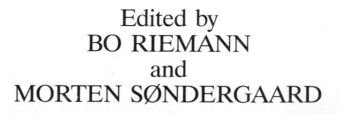

Edited by
**BO RIEMANN**
and
**MORTEN SØNDERGAARD**

ELSEVIER SCIENCE PUBLISHERS B.V.
Sara Burgerhartstraat 25
P.O. Box 211, 1000 AE Amsterdam, The Netherlands

*Distributors for the United States and Canada:*

ELSEVIER SCIENCE PUBLISHING COMPANY INC.
52, Vanderbilt Avenue
New York, NY 10017, U.S.A.

ISBN 0-444-42736-8

Printed in The Netherlands

# Contents

**Chapter 1**

## Preface
by Bo Riemann and Morten Søndergaard

**Chapter 2**

## Dissolved organic carbon (DOC) in lakes
by Niels O. G. Jørgensen

**Chapter 3**

# Phytoplankton
by Morten Søndergaard and Lars Møller Jensen with contributions by
Jørgen Kristiansen (section 3.2.) and Wayne H. Bell (section 3.3.5.)

## Chapter 4

## Bacteria

by Bo Riemann and Morten Søndergaard with contributions by Niels O. G. Jørgensen (Section 4.5.)

## Chapter 5

# Zooplankton

by Suzanne Bosselmann and Bo Riemann

## Chapter 6

# Planktivorous fish

by Lars Johansson and Lennart Persson

**Chapter 7**

# Carbon metabolism and community regulation in eutrophic, temperate lakes

by Bo Riemann, Morten Søndergaard, Lennart Persson and Lars Johansson

**Chapter 8**

# Epilogue

# Chapter 1. PREFACE

## 1.1. Introduction

This book contains papers dealing with pools, pathways, fluxes and transformation of organic matter in the pelagic zone of eutrophic, temperate lakes. It is a direct result of the work of a research group during the period 1981-85.

In 1981 the Danish Natural Science Research Council initiated an arrangement with "temporary research groups". One of the groups entitled "The carbon project" studied "Carbon metabolism in temperate, eutrophic lakes with special reference to quantificaiton of bacterial production in relation to phytoplannkton growth and nutrients". The staff in this project were Bo Riemann, Morten Søndergaard, Niels O. G. Jørgensen, Suzanne Bosselmann and Lars Møller Jensen. However, including technicians, students and guest scientsits, more than 25 persons were in periods working actively with the group. Over the period of four years the group has been working with a large number of lakes and some coastal marine environments in various parts of Denmark. Diel and seasonal field studies alternated with periods of intensive laboratory experiments.

Although priority was given to interactions between planktonic microorganisms and their surrounding environments, some attention was also paid to changes in biological structure and carbon pathways induced from higher trophic levels, e.g. by planktivorous fish. The results obtained and the many theoretical discussions within the group and with many friends and colleagues prompted us to summarize current understanding and specify gaps in our knowledge of pelagic carbon metabolism. This book is the result of our efforts.

As it turned out that we were not able to cover all important aspects of the pelagic carbon metabolism, a few people were invited to contribute in attempts to reach a more comprehensive treatment of the subject.

Throughout the 1981-85 period, a number of persons have participated in the project and contributed to the completion of the book. First of all, we wish to record our most sincere thanks to our technicians Winnie Martinsen, Bodil Pihlkjær and Jette Bargholtz, who with a great sense of responsibility contributed to the overall planning and management of the program. Without their flexible attitude towards the demands from many people and patience during days with long working hours, the successful completion of our objectives would not have been reached. Moreover, we wish to thank Svend H. Nielsen, who completed all the drawings to the book and Hanne Møller who typed most of the manuscript.

The major part of the project was financed by the Danish Natural Science Research Council. The Environmental Protection Agency, Freshwater Laboratory, Silkeborg gave the pontoon bridge contruction and helped continuously with nutrient analyses in particular and other resources. Finally, The Freshwater Biological Laboratory, University of Copenhagen and The Botanical Institute and Institute of Genetics and Ecology, both University of Aarhus all supported the project with supplies, equipment and some basic expenses. We thank the scientific staff for numerous discussions and encouragement during the entire program.

A number of people have helped with the completion of specific sections of the book. We have specified our gratitude in acknowledgements in each chapter.

## 1.2. Autecology versus synecology

Autecological studies make an important contribution to synecology. However, it has always been a problem to relate results on single species to environmental fluctuations and the biological complexity in nature. For example, zooplankton grazing of phytoplankton can be measured in the laboratory using cultures of algae and animals. Several techniques are available and useful in estimating grazing with various concentrations of prey and predator. But under natural conditions it is extremely difficult to measure *in situ* grazing. The results from the laboratory experiments cannot be directly applied, because e.g. concentrations, size, morphology and nutrient regime of the phytoplankton are factors influencing the grazing rates of the zooplankton. Thus, methods developed to study species often fall short in an overall ecological context, although valueable qualitative information and general ecological concepts have been gained over the years.

Some of the ecological methods developed recently fulfil the key assumptions for synecological studies: they are specific to a well-defined proportion of organisms, and they are quantitative. Direct enumeration of bacteria by epiflourescence microscopy, $^3$H-thymidine incorporation into bacterial DNA as a measure of their net production and chlorophyll a analyses as a tool to estimate the phytoplankton biomass are three examples of synecological methodological approaches; either designed for or extensively used in natural environments. The results obtained by these methods do not intend to be particularly precise and/or accurate. In fact, a number of precautions should always be verified in practice before the results can be used. But the results do under optimal conditions represent a pool or a process of a community of specific organisms.

## 1.3. Modeling the pelagic environment

Ecological models are indeed the scientific target for most of the aquatic research, but there are certainly large gaps between the small process-modelers who like "the mariners listened to the voices of the Sirens" and to the total ecosystem modelers "who stuffed their ears, if not with wax, like Odysseus, then with paper, and pressed on" (Shapiro 1979. Ecosystem modeling - a personal perspective. In Scavia, D. & Robertson, A. (Eds.), Perspectives on lake ecosystem modeling. Ann Arbor Science Publ. Inc.).

An example of a total ecosystem conception, which was designed and extensively used during the past two decades, is the "phosphorus syndrome". The general perception included that phosphorus caused excell algal growth and the subsequent changes in the densities and sizes of phyto- and zooplankton were side effects of the phosphorus input. Moreover, such changes were suspected to reverse, when phosphorus was removed from the influents.

Contrary to expectations such reversals were not shown, and results from laboratory and field experiments demonstrated that a number of fundamental assump tions for the phosphorus models were not fulfilled. As an example, zooplankton communities in eutrophic, temperate lakes are in periods dominated by small-sized animals like *Chydorus* and *Bosmina*. This observation was interpreted as succession towards small-sized animals, because the excess nutrient-additions allowed the development of large, colonial phytoplankters, which are difficult for the animals to consume. However, changes on higher trophic levels do also occur as a result of increased phosphorus addition. Some of these changes cause an increased fish predation on the macrozooplankton (e.g. *Daphnia* and *Eudiaptomus*).

We do not refute the importance of phosphorus and other nutrients in lake eutrophication. However, when excess algal growth is initiated, we suggest that the biological changes occurring on all trophic levels influence the eutrophication process. So, eutrophication is not only a case of increased input. A number of these biological changes may be important when lake restorations are planned, and none of the existing ecological models include these processes. Moreover, the reversibility of some of the biological changes that occur upon eutrophication are not known.

## 1.4. Food chain studies

During the past decade an increasing number of ecological studies have considered the complexity of the natural environments; more and more of the new methods include *in situ* procedures involving large parts of the whole trophic structure. The results from many of these studies have allowed the establishment of total carbon balances. Thus, it has been possible to evaluate the carbon contribution from algae and zooplankton to natural bacterial assemblages from measurements of the bacterial net assimilation and growth yield. It has also been possible to evaluate the potential role of bacteria and phytoplankton as food for zooplankton, and the effects of predation by planktivorous fish on the routes and rates of the detrital pathway.

We are still only beginning to understand the carbon metabolism in eutrophic, temperate lakes in details. However, the past years international research has created a solid foundation, and we hope that this book can be used as such in the construction of whole lake carbon metabolism.

This book is aimed at graduate students with a practical or theoretical interest in carbon metabolism. Our research colleagues can hopefully also find interest in the presented synthesis. We also believe that people at Environmental Protection Agencies might find sections of interest.

Bo Riemann                                            Morten Søndergaard

# Chapter 2. DISSOLVED ORGANIC CARBON (DOC) IN LAKES

by Niels O. G. Jørgensen

## 2.1. Location of DOC in aquatic ecosystems

The principal producers of organic matter in pelagic ecosystems are microscopic phytoplankton organisms. During their photosynthesis inorganic carbon is reduced and organic carbon compounds are produced. In the algae the organic compounds are allocated to biosynthetic pathways or used as fuel for metabolic processes; but in addition, a varying portion of the organic compounds can be released to the ambient water. As a result, phytoplankton photosynthesis leads to production of both particulate organic matter (POM, the algae) and dissolved organic matter (DOM, the released compounds). If the algae are grazed by zooplankton, organic compounds may also be released from spillage during feeding and from fecal pellets produced by the zooplankton (Lampert 1978). When the phyto- and zooplankton organisms die, another contribution of DOC occurs during autolysis of cellular components (Marshall and Orr 1961, Zygmuntowa 1981, Cole et al. 1984a, b, Krog et al. 1986, see also Chapter 4). The organic substances released from the plankton do not accumulate in the water, but serve as nourishment for heterotrophic bacteria. As a result of the bacterial metabolism, organic compounds are mineralized to $CO_2$, $NH_4$, $PO_4$ and other inorganic compounds which then become available for new primary production. This very simplified description of the fate of phytoplankton-produced organic compounds illustrates that DOM is an important part in cycling of matter in aquatic environments.

DOM makes up a portion of the pool of detritus in natural waters. Detritus includes all types of dead, undecomposed organic matter. In addition to DOM, dead POM is another important fraction of detritus. The bulk of particulate detritus in pelagic waters consists of debris of phyto- and zooplankton origin. Like DOM, the particulate detritus is degraded and metabolized by bacteria and/or grazed by zooplankton. During activity of the bacteria not only new bacterial biomass, but also DOM is produced, as organic substances are released during microbial degradation of the detritus (Amano et al. 1982).

Traditionally, membrane or glass-fibre filters are used to separate POM and DOM (Strickland and Parsons 1977). However, these filters do not retain the smallest bacteria and viruses (see e.g. Brock et al. (1984) for sizes of procaryotes and viruses). Quantitatively, this is not important, since these organisms consitute a minor proportion of the total content of organic matter. The dry weight of material retained by the membrane or glass-fibre filters represents as assessment of the particulate matter. It does not, however, give an exact indication of the organic material as many organisms have a high content of inorganic substances. For instance, the silica frustule of diatoms may constitute up to 40% of the dry weight. Similarly, inorganic matter makes up more than the weight of the organic biomass of some zooplankton species, (Parsons et al. 1977).

The organic content of particulate matter may be determined directly by measuring dominant elements such as nitrogen, phosphorus and carbon. Among these constituents the total content of carbon (POC) is the most satisfactory. All organic matter contains carbon, but not always nitrogen or phosphorus, and photosynthetic production of organic matter (reduction of $(CO_2)_n$ to $(CH_2O_n)$) is commonly measured in units of carbon. Also, chemical analyses of carbon are relatively simple and can be performed with great accuracy. Organic carbon was usually measured by a "wet" combustion with dichromic-sulphuric acid (Strickland and Parsons 1977), but today organic carbon is commonly combusted (oxidized) to $CO_2$ which is measured by infrared spectrophometry (Menzel and Vaccaro 1964). Correspondingly, the content of dissolved organic matter is typically determined as the total amount of dissolved organic carbon (DOC), allowing a direct comparison of POC and DOC. However, in many studies of dissolved organic matter, specific compounds rather than the total pool of DOC are measured to get insight into dynamic relations of specific DOC compounds.

In living organisms, carbon is the most abundant individual element. In terms of weight, carbon constitutes 47% of all protein amino acids, 40% of carbohydrates and about 75% of common lipids. In phyto- and zooplankton, protein constitutes 35-68% and 53-64%, carbohydrates 20-42% and 1-3% and lipids 4-23% and 10-23%, respectively, of the dry weight of the organisms (Parsons et al. 1961, Raymont et al. 1969). Consequently, the average carbon content of most pelagic organisms is about 40% of the dry weight. The ratio between carbon and another important element, nitrogen, in marine plankton is 3 to 8 (Holm-Hansen et al. 1966, Omori 1969). The C:N ratio of DOM in the sea has been found to vary from 2.5 (deep sea) to 30 (warm environments), with values close to 3 in temperate waters (Duursma 1961, Degens 1970). This nitrogen enrichment of DOM relative to the organic matter of the plankton suggests that nitrogen compounds are abundant elements of the DOM. Most of the nitrogen compounds are probably polymers or high molecular-weight complexes (Degens 1970).

C:N ratios of DOM in lakes appear to be higher than in the sea. Unfortunately, few data have been published. In different lakes in Wisconsin, USA, Birge and Juday (1934) found that the C:N ratio increased from 12 to 19 with increasing loads of POM and DOM. Freshwater plankton appear to have slightly larger C:N ratios than marine plankton (Reynolds 1984), but the increased C:N ratios in lakes suggest that the organic matter in part has lost its content of nitrogen, probably during microbial degradation. Alternatively, organic compounds with a low nitrogen content, such as allochthoneous matter of terrestrial origin, can be a significant source of DOM in lakes (Wetzel 1983). More analyses of DOM in freshwater habitats are required to confirm that higher C:N ratios are more common in lakes than in the sea.

## 2.2. DOC pools in lakes

The total amount of dissolved organic matter, determined as DOC, has been measured in a number of lakes of different primary productivities. In meso- and eutrophic lakes, concentrations of DOC typically range from 4 to 10 mg l⁻¹ (Wetzel et al. 1972, Allen 1978, Ochiai and Handa 1980, Søndergaard and Schierup 1982, Søndergaard 1984). In more productive waters like hypereutrophic lakes, DOC concentrations up to 25 mg l⁻¹ may occur (Seki and Nakano 1981). Correspondingly, the lower productivity in oligotrophic waters usually reflected in lower DOC pools do typically not exceed 3 mg l⁻¹ (Kaplan and Bott 1982, Søndergaard 1984). This positive relationship between productivity and pools of DOC suggests that levels of DOC may be estimated if the primary production is known. Søndergaard (1984) compared concentrations of DOC with annual phytoplankton production in 14 Danish lakes and found a positive correlation between growth of phytoplankton and pools of DOC (Fig. 2.1). Apparently this relationship is not always linear, as a logarithmic curve fit gave a better correlation at low productivities than did the linear regression. A similar connection between DOC and productivity was observed in five American lakes by Allen (1978).

A number of individual compounds of the DOC pool are known to be removed by assimilation of pelagic bacteria (see Chapter 4). Hence, the relationship between DOC levels and phytoplankton growth may in fact indicate a balance between bacterial activity and phytoplankton production. In most aquatic environments, such a coupling between phytoplankton growth and bacterial heterotrophy actually seems to exist (Hobbie and Rublee 1977, Blaauboer et al. 1981, Søndergaard et al. 1985). However, if most DOC has a non-phytoplankton origin (e.g. due to input of terrestrial organic matter, inlet of humus-rich water or an intense production of littoral plants) levels of DOC would not be linked to the phytoplankton primary production.

In most lakes pools of DOC appear to be rather constant. In Lake Lawrence, Michigan, USA, Wetzel et al. (1972) found seasonal and vertical changes of DOC most often ranging between 5 and 7 mg l⁻¹. Contrary, the pool of POC varied more than 10-fold in the same period, and apparently this variation was not correlated with the changes of DOC. The stability of the DOC pool in Lake Lawrence agrees with observations from the Danish Lake Mossø (Fig. 2.2). During 30 days in spring, more than a 4-fold increase of the POC pool occurred. This was mainly due to an active phytoplankton growth as indicated by the chlorophyll *a* concentration. Within that period, only minor changes of the DOC pool were measured. The reason for this stability of DOC pools in lakes, despite widely fluctuating POC pools, probably should be sought in both the composition of the DOC and in the microbial assimilation of specific DOC species. Although a large portion of DOC in most freshwaters consists of low molecular weight compounds, they may be resistant to degradation by microorganisms. It is likely that the readily degradable products of POM origin (simple carbohydrates, organic acids and amino acids) only contributes insignificantly to the DOM pool, due to their fast uptake by bacteria. This is true, certainly, of free amino acids that are commonly released during

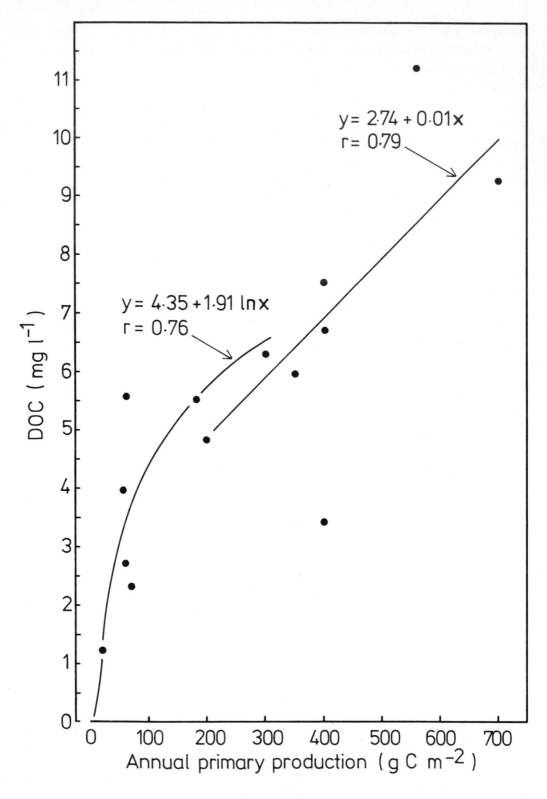

Fig. 2.1.
Concentrations of DOC versus annual pelagic primary production in 14 Danish lakes. Redrawn from Søndergaard (1984).

photosynthesis in natural phytoplankton (Mague et al. 1980, Jørgensen 1986), and of most DOC leaching from dead algae (Hansen et al. 1986).

Most freshly released DOM is immediately utilized, but presumably some portion must be (or become) recalcitrant to decomposition, thus contributing to the main DOM pool. Since all DOM is derived from POM, some relations between pools of DOM and POM may be expected to exist. Seki and Nakano (1981) compared the POM:DOM ratio with the total content of organic carbon (TOC) in waters of increasing productivity, and they ranked the most important components of the POM (Fig. 2.3). TOC (POC + DOC) varied from 0.5 mg l$^{-1}$ in oligotrophic water to more than 400 mg l$^{-1}$ in a hypereutrophic lake. When the productivity increased, the proportion of POM to DOM increased also. Under oligotrophic

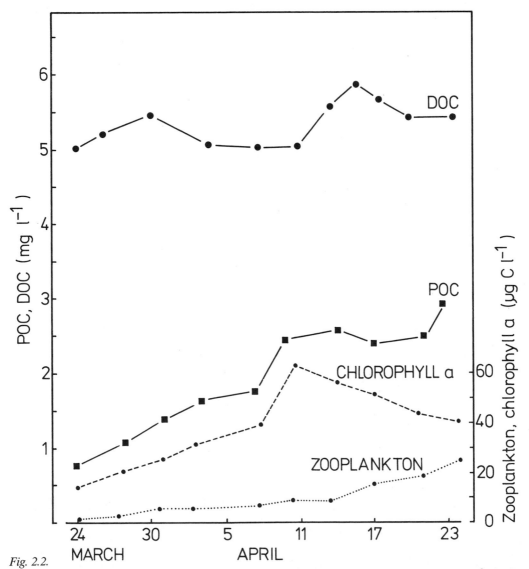

*Fig. 2.2.*
*Concentrations of DOC and POC (in mg l.$_1$) and chlorophyll a and zooplankton (in µg C l$^{-1}$) in the Danish Lake Mossø during a phytoplankton bloom in March- April 1980. Redrawn from Riemann et al. (1982) and Søndergaard and Schierup (1982).*

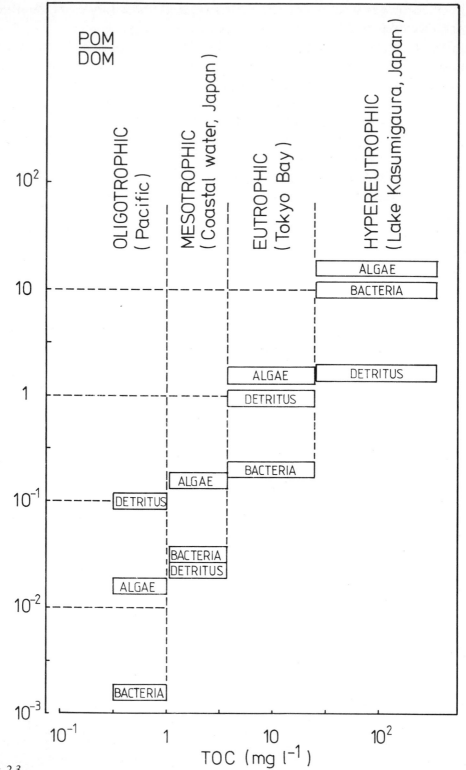

Fig. 2.3.

*POM/DOM ratios in relation to total organic carbon (TOC) in pelagic environments of increasing phytoplankton production. Detritus includes particulate matter only. See text for details. Redrawn from Seki and Nakano (1981).*

conditions, there was 10-fold more DOM than POM, whereas POM in the hyper-eutrophic lake exceeded the concentration of DOM by a factor of 20. In eutrophic waters, rather similar concentrations of POM and DOM may occur during periods of high phytoplankton biomass. The 200-fold range of POM in Fig. 2.3 does not indicate a similar difference in total productivity of the ecosystem. In oligotrophic waters, the photic zone usually extends to a depth of several meters, or more than 100 m in some oceans (Tait 1972), whereas the photosynthetic zone in hypereutrophic lakes may be only few cm. But the primary production per surface area, e.g. g C $m^{-2}$ $year^{-1}$, has generally no more than a 20-fold variation from oligotrophic to hypereutrophic conditions (Riemann and Mathiesen 1977, Søndergaard 1984). In their study, Seki and Nakano (1981) found that algae (phytoplankton) were the most important component of the POM, except in oligotrophic water, in which particulate detritus was more abundant than algae. In meso- and hypereutrophic waters, either bacteria or detritus may dominate the POM next to phytoplankton.

## 2.3. Composition of DOC

The occurrence of dissolved organic matter in natural waters has been known for at least a century. Among the first scientists to recognize the existence of DOM was Thomson who in 1874 determined concentrations of DOM in the sea and assumed that it was an important nutrient for certain sea-living organisms. Later, not only concentrations, but also major components of DOM, e.g. content of nitrogen (Krogh 1933) were determined in different environments (reviewed by Duursma 1961). Within the last two decades the knowledge on qualitative aspects has expanded due to development of new analytical techniques. Studies on composition of DOM and DOC have dealt both with different molecular-weight classes of the DOC, and with the identification of specific substances.

### 2.3.1. Molecular weight composition

It is generally believed that low molecular-weight compounds such as mono- and disaccharides, organic acids and amino acids are the preferred organic substrates for heterotrophic bacteria, as these molecules can readily enter metabolic pathways in bacteria (Wright and Hobbie 1966, Vaccaro et al. 1968). In support of that, Ogura (1974) showed that organic compounds < 500 Daltons ( = D) are generally decomposed faster than larger molecules. Organic compounds that have a fast turnover, because they are assimilated by bacteria immediately after release are denoted *labile*, while compounds with a slow turnover are denoted *refractory*.

Size fractionation of molecules can be carried out by ultrafiltration, i.e. filtration under pressure through membranes with specific pore sizes, or by gel permeation chromatography in which the molecules are separated on a column packed with polysaccharides, typically dextran. The organic compounds are commonly divided into 3 or 4 weight fractions, e.g. < 500 D, 500-10,000 D, 10,000-100,000 D and > 100,000 D.

Separation of DOC in natural waters demonstrates that the low molecular-

Table 2.1. Molecular weight composition of DOC in natural waters. D = Daltons

| Locality | Method | Low-molicular fraction | High-molecular fraction | Reference |
|---|---|---|---|---|
| Star Lake, Vermont, USA | Ultramembrane filtration | 27% < 500 D | 30% > 100,000 D | Allen (1976) |
| 5 different lakes, Vermont and New Hampshire, USA | Ultramembrane filtration | 30-64% < 500 D | - | Allen (1978) |
| 3 different lakes, Japan | Gel-filtration (Sephadex G-25) | 35-68% < 200 D | (20-33% > 1,400 D) | Hama and Handa (1980) |
| Lake Mossø, Denmark | Gel-filtration (Sephadex G-50) | 64-94% < 700 D* | 1-13% > 10,000 D | Søndergaard and Schierup (1982) |
| 11 different lakes, Denmark | Gel-filtration (Sephadex G-50) | 71 ± 11% < 700 D | 16 ± 5 > 10,000 D | Søndergaard (1984) |
| Mirror Lake | Ultramembrane filtration | 16 ± 14% < 100 D | 31% > 100,000 D | Cole et al. (1984a) |
| Tokyo Bay surface water, Japan | Ultramembrane filtration | 24-42% < 500 D | 8-23% > 100,000 D | Ogura (1974) |

*70% was < D. Water samples collected regularly over a 6-week period.

weight fraction (< 200-700 D) constitutes a significant portion of the total DOC pool (Table 2.1). In some lakes, this fraction was more important than all larger molecules. The fraction between 500 and 10,000 D (not included in Table 2.1) is generally of least importance (Ogura 1974, Allen 1978). Somewhat surprisingly large molecules (> 100,000 D) can make up 30% of the DOC in lake water, but apparently less in seawater.

Various organic molecules have been determined in the different size classes. Carbohydrates appear mainly to occur in the > 1,400 D fraction, while amino acids are abundant in both the > 1,400 D and the < 200 D fraction (Burnison 1978, Hama and Handa 1980). The latter fraction has also been found to include several organic acids like acetic and lactic acid (Hama and Handa 1980). The > 100,000 D fraction probably consists of protein-like compounds and condensates with a complex chemical structure such as humic substances (Ogura 1974).

Labile compounds of the DOC mainly consist of low molecular-weight substances, but in addition, proteins may have a high turnover rate due to their degradation by bacteria (Hollibaugh and Azam 1983). They observed that proteins > 100,000 D were utilized by bacteria at rates similar to those of free amino acids. The occurrence of large, easily degradable molecules, possibly proteins, was also observed by Allen (1978). He used ultra-violet oxidation as a measure of lability of DOC in lake water and found a similar decomposition of molecules < 2,000 D and > 100,000 D.

The refractory compounds are probably all humus-like substances, i.e. complex molecules consisting of aromatic compounds like phenols, sugars, organic acids and peptides, with molecular weights from few hundreds to >100,000 D. Humic compounds are probably rich in nitrogen, as most dissolved organic nitrogen in lake water occurs in the 10,000-50,000 D fraction (Tuschall and Brezonik 1980). This fraction can make up the bulk of natural humus (Ogura 1974). Indirect evidence that humic substances are refractory was given by Allen (1978) who observed the smallest UV decomposition in the 2,000-50,000 D fraction.

Bacterial degradation of different molecular weight fractions in lakes of increasing productivities has indicated that in eutrophic lakes a larger proportion of DOM, relative to oligotrophic lakes, is labile and can be degradated by the bacteria (Allen 1978, Søndergaard 1984). Søndergaard reports that in Lake Almind and Lake Ørn with annual productivities of 60 and 400 g C m$^{-2}$, respectively, 17% (Lake Almind) and 53% (Lake Ørn) of the DOC could be decomposed by pelagic bacteria within 10 days. If the bacterial populations in eutrophic lakes have the potential of degrading half the DOC pool, why does a large pool of low molecular-weight (<700 D) organic matter usually occur in the lakes, e.g. in Lake Mossø (Table 2.1). No exact answers can be given. Speculatively, Liebigs law of minimum may explain this paradox. Thus, if organic compounds required for growth of bacteria such as vitamins and essential amino acids (Morita 1984) are not present, the remaining organic compounds may not be degradable to the bacteria.

### 2.3.2. Specific DOC compounds

Natural concentrations of labile DOC compounds, such as free saccharides, free amino acids and organic acids, are generally very low. In contrast humic substances, certain polysaccharides and other refractive compounds may occur in significantly larger amounts. The low concentrations of most labile compounds make their detection difficult. Often it is necessary to increase the concentration of the compounds before analysis by removing most of the water by evaporation or freeze-drying. For instance, before gas chromatographic analysis of free saccharides, natural water samples must be concentrated up to 20 times (Münster 1984). Alternatively, specific labeling, especially by reagents producing fluorescent products, may increase the detection limit of DOC species substantially. Thus, using fluorescent derivatization of free amino acids, concentrations below $10^{-14}$ M can be measured directly with high pressure liquid chromatography (Lindroth and Mopper 1979). Less sensitive procedures are required for analysis of the relatively large concentrations of polysaccharides commonly found in lake water. The polysaccharides can be quantified by standard colour reactions after a single concentration step (Strickland and Parsons 1972).

Most chemical measurements of specific DOC species have unfortunately been performed in marine environments; few data on DOC compounds in freshwater are available. Examples of organic compounds identified in lakes are given in Table 2.2. Among free, dissolved molecules, mono- and disaccharides, organic acids and amino acids are commonly found. Glucose and fructose (fructose is not included in Table 2.2) appear to be the dominant free saccharides in natural waters (Liebezeit 1980, Ittekkot et al. 1981). The abundance of glucose is expected,

Table 2.2. Specific dissolved organic molecules in lake water

| Locality | Compound | Concentration mg l⁻¹ | Analytical procedure | Reference |
|---|---|---|---|---|
| **A. Dissolved free compounds** | | | | |
| Lake plussee, FRG | Glucose, cellobiose, sucrose, galactose | 0.4-145 (single saccharides) | Gas chromatography | Münster (1984) |
| Lake Stocksee | Glucose | 18.0 | Liquid chromatography | Gocke et al. (1981) |
| Lake Mossø, Denmark | Glucose | 0.5-100 | Enzymatic | Riemann et al. (1982) |
| 3 different lakes, Japan | Formic, acetic, lactic, propionic and butyric acid | 3-125 (total pool) | Gas chromatography + mass spectrometry | Hama and Handa (1980) |
| Upper Klamath Lake Oregon, USA | 16 amino acids (serine, lysine and glycine dominant) | 82-197 (total pool) | Cation exchange concentration + automatic amino acid analyzer | Burnison and Morita (1974) |
| 3 different lakes, Denmark | 25 amino acids (serine, glycine, alanine and ornithine dominant) | 38-240 (total pool) | Fluorescent derivatization + high pressure liquid chromatography | Jørgensen (1986) |
| Lake Plussee, FRG | Citric, oxalic, glycollic and malic acid Short-chain fatty acids | up to 100 (single compounds) | Paper chromatography | Weinmann (1970) |
| **B. Dissolved, hydrolysable compounds** | | | | |
| 3 different lakes, Japan | Carbohydrates | 950-14300 340-4300 | Anthrone sulphuric acid method Gas chromatography | Ochiai and Hanay (1980) |
| Lake Plussee, FRG | Carabohydrates | 35-1670 | Paper chromatography | Weinmann (1970) |
| 3 different lakes, Japan | Carbohydrates Proteins | 240-910 144-320 | Phenol sulphuric acid method Fluorescent derivatization | Hama and Hanya (1980) |
| Lake Nakanauma, Japan | Carbohydrates Proteins | 1000 1400 | Automatic amino acid analyzer Anthrone sulphuric acid method | Ochiai et (1980) |
| Lake Schöhsee, FRG | Carbohydrates | 155* | Gas-liquid chromatography | Stabel (1977) |

*(annual mean)

as it is the major component of many carbohydrates (Lehninger 1970); but it may actually also be the source of fructose, as isomerization of glucose to fructose occurs spontaneously under the weak alkaline conditions that occur during high rates of phytoplankton photosynthesis (Mopper 1980). The occurrence of serine, glycine, alanine, lysine and ornithine as dominant free amino acids does not correspond to the amino acid composition of proteins in plankton (Parsons et al. 1977). This suggests that protein amino acids are selectively metabolized by bacteria. Alternatively, other sources of amino acids are important, such as bacterial transformation of different organic nitrogen compounds into amino acids (Lee and Cronin 1982), or photosynthesizing phytoplankton which have been found to release free amino acids into the ambient water (Jørgensen 1986). Natural concentrations of free saccharides and amino acids in lakes may be very low. For instance, when minimum concentrations of glucose and amino acids occur, the average amount of the single compounds corresponds to dissolving one mole of each in 50,000,000 l of water.

The low molecular-weight compounds all demonstrate significant changes in concentrations in lakes, but glucose in particular appears to fluctuate (Bölter 1981, Münster 1984). The mechanisms causing these variations are unknown, as changes of low molecular-weight organic molecules do generally not correlate with changes of the biological activity such as the production of phytoplankton and bacteria (Meyer-Reil et al. 1979, Münster 1984, Jørgensen 1986).

Amino acids and saccharides of hydrolysed proteins and carbohydrates, respectively, indicate that from 2 to 10-fold more amino acids and saccharides are present as protein and carbohydrate, than as free monomers. The concentrations of proteins and carbohydrates shown in Table 2.2 suggest that these compounds vary less than their monomers, but significant changes of both protein and carbohydrate pools obviously can occur in lakes. The data by Ochiai and Hanya (1980) illustrate that different analytical procedures can produce different concentrations of an organic compound in similar water samples. This aspect has rarely been examined in DOC studies, but the observation suggests that measured concentrations of organic substances do not always reflect the actual concentrations in the water.

The different organic compounds listed in Table 2.2 suggest that monomeric and polymeric molecules constitute about 250 and 1500 mg C $l^{-1}$, respectively. Assuming a carbon content of 50% of the dissolved organic compounds and a total DOC content of 5 mg $l^{-1}$, then free and combined organic molecules make up 2.5 and 15% of the DOC. Substances other than those in Table 2.2 are known to be present in lake water, e.g. small amounts of hydrocarbons and vitamins, but the estimate suggests that only about 20% of the DOC has been identified. Hama and Handa (1980) came to a similar conclusion after analysing DOC species in several Japanese lakes.

A large portion of the residual 80% DOC probably consists of various humic compounds (humic acids, fulvic acids and humins). This may in particular be true of lakes with a large input of terrestrial organic matter (Wetzel 1983). Due to its colour, the humic fraction of DOC in sea water has been named "Gelbstoff" (Kalle 1966). Humic substances are believed to be very resistant to both chemical and biological attacks (Gagosian and Stuermer 1977). However, some strains of bacteria have proven capable of decomposing humic substances isolated from freshwater (Haan 1972, 1974); and recently, Hessen (1985) reported that humic compounds in some oligotrophic lakes may even be important organic substrates for bacteria. The humic compounds that are preferentially degraded, may belong to the low molecular-weight fraction ($<700$ D). The molecular weight composition in Table 2.1 indicates that this fraction may constitute from 24 to 94% of the DOC. This is a larger portion of the DOC than the identified about 20%. Hence, labile humic compounds may be contained in the low molecular-weight fraction.

Lipids may be another important group of compounds of the residual DOC fraction. In sea water, dissolved and chloroform-extractable lipids make up from 8 to 16% of the DOC (Jeffrey 1970). He identified a wide spectrum of lipids in sea water, such as monoglycerides, fatty acids, sterols (among these cholosterol), phospholipids and vitamin A.

The residual fraction may also include colloids. Colloids are particle-like, high-molecular polymer compounds of sizes measured in the nm range. The content of amino acids in freshwater colloidal matter in the Patuxent River, Maryland, USA, varied from 100 to 500 mg l$^{-1}$ (Sigleo et al. 1983). In the Mackenzie River, Canada, colloidal amino acids made up from 11 to 80 mg l$^{-1}$, while about 335 mg l$^{-1}$ was measured in lakes in Florida (Peake et al. 1977, Tucshall and Brezonik 1980). The total amounts of colloids were not determined. In a Danish lake, colloids made up 8% of the total DOC (Krogh and Lange 1932); but otherwise knowledge on the occurrence of colloidal matter in natural waters is scarce. Some reports suggest that colloids in both fresh and sea water may be more common than generally assumed (Lock et al. 1977, Breger 1970).

The different organic compounds in freshwaters have different turnover rates, leading to a wide range of residence times. The residence time of most labile and low molecular-weight compounds vary from less than one hour during summer to several days during the winter in temperate lakes (Wright 1970, Overbeck 1979, Bölter 1981, Jørgensen and Søndergaard 1984). In contrast, most high molecular-weight substances are refractory, and have very long residence times, with the exception of some proteins (Hollibaugh and Azam 1983). Examination of naturally radioactive carbon compounds in the sea indicates an average age of 3400 years of marine DOC (Williams et al. 1969). Similar, or even longer, residence times of refractive DOC compounds may occur in freshwater, in particular when a large amount of high molecular-weight humic substances is present. In bogs, intact plants seeds and pollen thousands of years old witness that humic compounds in combination with low pH-values are efficient preservatives. An important function of humic substances is their ability to chelate with metal ions. The chelation maintains some ions in solution and prevents a precipitation, e.g. of manganese and iron, which become available to uptake by algae and bacteria (Shapiro 1969). Other ions like mercury and cobalt apparently chelate so strongly with humic compounds that they cannot be assimilated by pelagic organisms (Huljev and Strohal 1983). In this way humic compounds may serve as detoxifying agents.

## 2.4. Sources of DOC

All aquatic organisms contribute dissolved organic matter to the ambient water at some stages of their lives. In healthy, living organisms, release of organic compounds can be due to excretion of metabolic waste products, such as urea from zooplankton (Butler et al. 1979), or it may be a reaction to sudden and unwanted changes of chemical and physical properties of the environment. Thus, changes in temperature and osmotic pressure can induce release of free amino acids in bacteria (Kay and Gronlund 1969). In dead organisms, autolytic enzymes (Brock et al. 1984) in combination with exoenzymes released by bacteria (Hoppe 1983, Rego et al. 1985) immediately start degrading organic compounds of the cells, to smaller, free molecules which can be assimilated by the bacteria. A number of organic compounds have been identified as excretion or release products of living and dead organisms. A brief survey of selected compounds and their sources are listed below. In some cases are observations from marine studies referred to, as most work on release of DOC species has been conducted in sea water.

### 2.4.1. Phytoplankton

Products of release from phytoplankton include organic acids, nucleic acids and their derivatives, mono- and disaccharides, carbohydrates, sugar alcohols, amino acids, peptides, proteins, vitamins and lipids (Hellebust\1965, Fogg 1966, Aaronsen et al. 1971). Obviously, most organic molecules of living algal cells can be detected as extracellular products. In most cases has the release been studied in laboratory cultures of algae in logarithmic growth phases. However, in these experiments, dead, injured or unhealthy algal cells incidently may have contributed organic compounds which may explain how vitamins and nucleic acids that are vital to living cells occur in the growth medium of the algae.

Do the released, extracellular compounds affect natural pools of DOC? In several field studies, changes of labile DOC species have been related to the growth rate of phytoplankton. In marine enclosures, Brockmann et al. (1979) observed a positive correlation between algal biomass and ambient pools of monosaccharides and some amino acids. Hammer and Eberlein (1981) measured increasing concentrations of free amino acids during a bloom of a marine diatom, but decreasing concentrations at the stationary phase of the algae. However, in several studies increasing concentrations of free amino acids have been found after phytoplankton blooms, both in lakes (Gardner and Lee 1975, Haan and Boer 1979, Jørgensen in press) and in the sea (Bohling 1970, Riley and Segar 1970). A large release of amino acids from dead phytoplankton also has been observed in laboratory cultures (Zygmuntowa 1981).

Quantitative information on specific, extracellular compounds has been determined in few situations. Mague et al. (1980) and Søndergaard et al. (1985) found that free amino acids made up 7 to 39% of extracellular organic compounds released by natural phytoplankton populations. In lakes, released free amino acids can make up from 6 to 25% of the pool of free amino acids in the water (Jørgensen 1986). Consequently, release of organic compounds from phytoplankton may contribute significantly to the flux of DOC compounds in lakes. The importance of released extracellular organic compounds relative to the phytoplankton production and algal lysis is discussed in Chapter 3.

## 2.4.2. Zooplankton

In experiments using phytoplankton as the food source it has been demonstrated that the cladoceran *Daphnia pulex* loses up to 17% of the ingested carbon as DOC (Lampert 1978). Most of the DOC originated from damaged algal cells, but DOC was also released during digestion and from fecal pellets. Most compounds released by zooplankton, mainly cladocerans, appear to be organic nitrogen compounds, in particular amino acids and other primary amines (Johannes and Webb 1970, Benson and Aldrich 1981, Gardner and Scavia 1981), though organic phosphorus compounds may also occur (Satomi and Pomeroy 1965). Organic phosphorus of copepods is apparently released very fast when the animals die. According to Marshall and Orr (1961), 75% of the phosphorus content of the copepod *Calanus* may be lost within 2 h after death.

Nitrogen compounds released by zooplankton have in some cases been suggested to increase natural DOC pools. Higher concentrations of free primary amines have been measured in sea water with zooplankton than in their absence (Eppley et al. 1981). Riemann et al. (1986) observed enhanced concentrations of free amino acids in lake water in which the zooplankton density had been increased, relative to untreated water samples. Both studies suggested that an increase in the pool of dissolved free amino acids not necessarily leads to an increased bacterial production.

## 2.4.3. Bacteria

DOC serves both as a carbon and energy source to heterotrophic bacteria, but the bacteria themselves can produce dissolved organic compounds during their metabolism. Iturriaga and Zsolnay (1981) found that marine bacteria released unidentified organic molecules of molecular weights both larger and smaller than various low molecular-weight organic compounds assimilated 72 h earlier. During degradation of dissolved proteins, bacteria have been observed to release free amino acids (Amano et al. 1982). In natural waters, decomposition of particulate organic matter by bacteria may be another important source of DOC. Exoenzymatic activity of bacteria may give rise to a loss of organic substances (Hoppe 1982, Rego et al. 1985). Organic compounds may also be released from bacteria as a reaction to chemical and physical stress (Kay and Gronlund 1969), as mentioned previously.

Indirectly, bacteria may be involved in changing concentrations of labile DOC compounds. In the sea, Andersson et al. (1985) observed similar diel fluctuations of the production of heterotrophic microflagellates and concentrations of free amino acids. They concluded that a significant portion of the released amino acids could be related to the flagellates grazing on bacteria. Heterotrophic microflagellates also seem to be common in freshwater (Sorokin and Pavaljeva 1972, Sheer et al. 1982, Riemann 1985), and hence, they might also influence pools of amino acids in lakes.

### 2.4.4. Production of humic compounds

In general, the exact sources of individual organic molecules cannot be determined in natural waters. This is particularly true of humic substances. There are indications that humic compounds may originate from abiotic as well as from biological processes. When heating up a mixture of aqueous solutions of glucose and different amino acids, Abelson and Hare (1971) obtained compounds with chemical properties resembling humic substances, including their brown colour, and with electron spin resonance patterns similar to those of natural humus. Supporting an abiotic formation of humus-like compounds, Degens (1970) suggested that mineral surfaces may act as templates for polymerization of organic molecules. Among the biological processes, products from lysis of microorganisms can be involved in the formation of humic substances (Wetzel 1983). Polymers (melanins) synthesized by natural tropical fungae have amino acid spectra similar to those of humic acids, probably indicating that melanins participate in the formation of humic compounds (Coelho et al. 1985). In the sea, a substantial part of the humic substances may be derived from exudates of kelps (Sieburth and Jensen 1970). Yet more detailed chemical analyses of both humic compounds and their possible precursors are necessary to draw conclusions on the sources of humic substances.

The excessive use of "can" and "may" in this section on chemical composition of DOC clearly emphasizes that the insight into actual molecular composition of dissolved organic matter in natural waters is limited. Within recent years, however, valuable informations on sources and fluxes of common organic molecules have emerged due to new, rapid and sensitive analytical procedures. So far these results demonstrate that concentrations of specific organic compounds generally cannot be related to single biological events, but that several pools and processes must be considered to understand the dynamics of dissolved organic compounds.

## Acknowledgements

I wish to thank Dr. T. Henry Blackburn, Department of Ecology and Genetics, University of Aarhus, Denmark, for his constructive criticism and suggestions to an earlier draft of the manuscript.

## 2.5. References

Aaronson, S., DeAngelis, B., Frank, O., & Baker, H. (1971). Secretion of vitamins and amino acids into the environment by *Ochromonas danica*. J. Phycol. 7: 215-218.

Abelson, P. H. & Hare, P. E. 1971. Reactions of amino acids with natural and artificial humus and kerogens. Carnegie Inst. Yearbook 69: 327-337.

Allen, H. L. 1976. Dissolved organic matter in lakewater: characteristics of molecular weight size-fractions and ecological implications. Oikos 27: 64-70.

Allen, H. L. 1978. Low molecular weight dissolved organic matter in five softwater ecosystems: a preliminary study and ecological implications. Verh. Int. Verein. Limnol. 20: 514-524.

Amano, M., Hara, S. & Taga, N. 1982. Utilization of dissolved amino acids in sea water by marine bacteria. Mar. Biol. 68: 31- 36.

Andersson, A., Lee, C., Azam, F. and Hagström, Å. 1985. Release of aminoacids and inorganic nutrients by heterotrophic marine microflagellates. Mar. Ecol. Prog. Ser. 23: 99-106.

Benzon, F. W. & Aldrich, J. C. 1981. A study of nitrogen excretion in the marine copepod *Temora longicornis.* Kieler Meeresforsch., Sonderh. 5: 186-190.

Birge, E. A. & Juday, C. 1934. Particulate and dissolved organic matter in inland lakes. Ecol. Monogr. 4: 440-474.

Blaauboer, M. C. T., Van Kuelen, R., & Kappenberg, Th. E. 1982. Extracellular release of photosynthetic products by freshwater phytoplankton populations, with special reference to the algal species involved. Freshwat. Biol. 12: 559-572.

Bohling, H. 1970. Untersuchungen über freie gelöste Aminosäuren in Meerwasser. Mar. Biol. 6: 213-225.

Breger, I. A. 1970. What you don't know can hurt you: organic colloids and natural waters, p. 563-574. In: Hood, D. W. (ed.), Organic matter in natural waters. Institute of Marine Science, Occasional Publications No. 1, College, Alaska.

Brock, T. D., Smith, D. W. & Madigan, M. T. 1984. Biology of microorganisms. Prentice/Hall International, Inc., London.

Brockmann, U. H., Eberlin, K., Junge, H. D., Maier-Reimer, E. & Siebers, D. 1979. The development of a natural plankton population in an outdoor tank with nutrient-poor seawater. II. Changes in dissolved carbohydrates and amino acids. Mar. Ecol. Prog. Ser. 1: 283-291.

Burnison, B. K. 1978. High molecular weight polysaccharides isolated from lake water. Ver. Int. Verein. Limnol. 20: 353-355.

Burnison, B. K. & Morita, T. Y. 1974. Heterotrophic potential for amino acids uptake in a naturally eutrophic lake. Appl. Environ. Microbiol. 27: 488-495.

Butler, E. I., Corner, E. D. S., & Marshall, S. M. 1970. On the nutrition and metabolism of zooplankton. VII. Seasonal survey of nitrogen and phosphorus excretion by *Calanus* in the Clyde Sea-area. J. mar. biol. Ass. U.K. 50: 525-560.

Bölter, M. 1981. DOC turnover and microbial biomass production. Kieler Meeresforsch. Sonderh. 5: 304-320.

Coelho, R. R. R., Linhares, L. F. & Martin, J. P. 1985. Amino acid distribution in some fungal melanins and of soil humic acids from Brazil. Plant and Soil 87: 337-346.

Cole, J. J., McDowell, W. H. & Likens, G. E. 1984a. Sources and molecular weight of "dissolved" organic carbon in an oligotrophic lake. Oikos 42: 1-9.

Cole, J. J., Likens, G. E. & Hobbie, J. E. 1984b. Decomposition of planktonic algae in an oligotrophic lake. Oikos 42: 257-266.

Degens, E. T. 1970. Molecular structure of nitrogenous compounds in sea water and recent marine sediments, p. 77- 106. In: Hood, D. W. (ed.), Organic matter in natural waters. Institute of Marine science, Occasional Publication No. 1, College, Alaska.

Duursma, E. K. 1961. Dissolved organic carbon, nitrogen and phosphorus in the sea. Netherlands J. Sea Res. 1: 1-148.

Eppley, R. W., Horrigan, S. G., Fuhrman, J. A., Brooks, E. R., Price, C. C. & Sellner, K. 1981. Origins of dissolved organic matter in southern California coastal waters: Experiments on the role of zooplankton. Mar. Ecol. Prog. Ser. 6: 149-159.

Fogg, G. E. 1965. The extracellular products of algae. Oceanogr. Mar. Biol. Ann. Rev. 4: 195-212.

Gagosian, R. B. & Stuermer, D. H. 1977. The cycling of biogenic compounds and their diagenetically transformed products in sea water. Mar. Chem. 5: 605-632.

Gardner, W. S. & Lee, G. F. 1975. The role of amino acids in the nitrogen cycle in Lake Mendota. Limnol. Oceanogr. 20: 379-388.

Gardner, W. S. & Scavia, D. 1981. Kinetic examination of nitrogen release by zooplankters. Limnol. Oceanogr. 26: 801- 810.

Gocke, K., Dawson, R. & Liebezeit, G. 1981. Availability of dissolved free glucose to heterotrophic microorganisms. Mar. Biol. 62: 209-216.

Haan, H. de 1972. Some structural and ecological studies on soluble humic compounds from Tjeukemeer. Verh. Int. Verein. Limnol. 18: 685-695.

Haan, H. de 1974. Effect of a fulvic acid fraction on the growth of a *Pseudomonas* from Tjeukemeer (The Netherlands). Freshwat. Biol. 4: 301-309.

Haan, H. & de Boer, T. 1979. Seasonal variations of fulvic acids, amino acids, and sugars in Tjeukemeer, The Netherlands. Arch. Hydrobiol. 85: 30-40.

Hama, T. & Handa, N. 1980. Molecular weight distribution and characterization of dissolved organic matter from lake waters. Arch. Hydrobiol. 90: 106-120.

Hammer, K. D. & Eberlein, K. 1981. Parallel experiments with *Thalassiosira rotula* in outdoor plastic tanks: Development of dissolved free amino acids during an algae bloom. Mar. Chem. 10: 533-544.

Hansen, L., Krog, G. F. & Søndergaard, M. 1986. Decomposition of lake phytoplankton. 1. Dynamics of short-term decomposition. Oikos, 46: 37-44.

Hellebust, J. A. 1965. Excretion of some organic compounds by marine phytoplankton. Limnol. Oceanogr. 10: 192-206.

Hellebust, J. A. 1974. Extracellular products, pp. 838- 863. In: Stewart, W. D. (ed.), Algal physiology and biochemistry. Bot. Monogr., vol. 10. Blackwell, Oxford.

Hessen, D. O. 1985. The relation between bacterial carbon and dissolved humic compounds in oligotrophic lakes. FEMS Microbiol. Ecol. 31: 215-223.

Hobbie, J. E. & Rublee, P. 1977. Radioisotope studies of heterotrophic bacteria in aquatic ecosystems, pp. 441-471. In: Cairns, J. Jr. (ed.), Aquatic microbial communities. Garland, New York.

Hollibaugh, J. T. & Azam, F. 1983. Microbial degradation of dissolved proteins in seawater. Limnol. Oceanogr. 28: 1104-1116.

Holm-Hansen, O., Strickland, J. D. H. & Williams, P. M. 1966. A detailed analysis of biologically important substances in a profile off southern California. Limnol. Oceanogr. 11: 548-561.

Hoppe, H.-G. 1983. Significance of exoenzymatic activities in the ecology of brackish water: Measurements by means of methyl- umbelliferyl-substrates. Mar. Ecol. Prog. Ser. 11: 299-308.

Huljev, D. & Strohal, P. 1983. Physico-chemical processes of humic acids - trace element interactions. Mar. Biol. 73: 243-246.

Ittekkot, V., Brockmann, U., Michaelis, W. & Degens, E. T. 1981. Dissolved free and combined carbohydrates during a phytoplankton bloom in the northern North Sea. Mar. Ecol. Prog. Ser. 4: 299-305.

Iturriaga, R. & Zsolnay, A. 1981. Transformation of some dissolved organic compounds by a natural heterotrophic population. Mar. Biol. 62: 125-129.

Jeffrey, L. M. 1970. Lipids of marine waters, p. 55-76. In: Hood, D. W. (ed.), Organic matter in natural waters. Institute of Marine Science, Occasional Publication No. 1, College, Alaska.

Johannes, R. E. & Webb, K. L. 1970. Release of dissolved organic compounds by marine and freshwater invertebrates, p. 257-274. In: Hood, D. W. (ed.), Organic matter in natural waters. Institute of Marine Science. Occasional Publication No. 1, College, Alaska.

Jørgensen, N. O. G. 1986. Fluxes of free amino acids in three Danish lakes. Freshwat. Biol. 16: 255-268.

Jørgensen, N. O. G. in press. Free amino acids in lakes: Concentrations and assimilation rates in relation to phytoplankton and bacterial production. Limnol. Oceanogr.

Jørgensen, N. O. G. & Søndergaard, M. 1984. Are dissolved free amino acids free? Microb. Ecol. 10: 301-316.

Kalle, K. 1966. The problem of the Gelbstoff in the sea. Oceanogr. Mar. Biol. Ann. Rev. 4: 91-104.

Kaplan, L. A. & Bott, T. L. 1982. Diel fluctuations of DOC generated by algae in a piedmont stream. Limnol. Oceanogr. 27: 1091-1100.

Kay, W. W. & Gronlund, A. F. 1969. Amino acid pool formation in *Pseudomonas aeruginosa*. J. Bacteriol. 97: 282-291.

Krog, G. F., Hansen, L. & Søndergaard, M. 1986. Decomposition of lake phytoplankton. 2. Composition and lability of lysis products. Oikos, 46: 45-50.

Krogh, A. 1933. Conditions of life in depth of the ocean. The Collecting Net 8: 153-156.

Krogh, A. & Lange, E. 1931. Quantitative Untersuchungen über Plankton, Kolloide und gelöste organische und anorganische Substanzen in dem Furesee. Int. Rev. ges. Hydrobiol. 26: 20-53.

Lampert, W. 1978. Release of dissolved oganic carbon by grazing zooplankton. Limnol. Oceanogr. 23: 831-834.

Lee, C. & Cronin, C. 1982. The vertical flux of particulate organic nitrogen in the sea: Decomposition of amino acids in the Peru upwelling area and the equatorial Atlantic. J. Mar. Res. 40: 227-251.

Lehninger, A. L. 1970. Biochemistry. Worth Publishers, Inc., New York.

Liebezeit, G. 1980. Aminosäuren und Zucker in marinen Milieu - neuere analytische Methoden und ihre Anwendung. Thesis, University of Kiel, FRG.

Lindroth, P. & Mopper, K. 1979. High performance liquid chromatography of subpicomole amounts of amino acids by precolumn fluorescence derivatizaiton with o-phthaldialdehyde. Anal. Chem. 51: 1667-1674.

Lock, M. A., Wallis, P. M. & Hynes, H. B. N.1977. Colloidal organic carbon in running waters. Oikos 29: 1-4.

Mague, T. H., Friberg, E., Hudges, D. J. & Morris, I. 1980. Extracellular release of carbon by marine phytoplankton; a physiological approach. Limnol. Oceanogr. 25: 262-279.

Marshall, S. M. & Orr, A. P. 1961. On the biology of *Calanus finmarchius.* XII. The phosphorus cycle: excretion, egg production, autolysis. J. mar. biol. Ass. U.K. 41: 463-488.

Menzel, D. W. & Vaccaro, R. F. 1964. The measurements of dissolved organic and particulate carbon in sea water. Limnol. Oceanogr. 9: 138-142.

Meyer-Reil, L.-A., Bölter, M., Liebezeit, G. & Schramm, W. 1979. Short-term variations in microbiological and chemical parameters. Mar. Ecol. Prog. Ser. 1: 1-6.

Mopper, K. 1980. Carbohydrates in the marine environment: Recent developments. In Biogeochimie de la materière organique à l'interface eau-sédiment marin. Colloques Internationaux du C.N.R.S. No. 293.

Morita, R. Y, 1984. Substrates capture by marine heterotrophic bacteria in low nutrient waters, p. 83-100. In: Hobbie, J. E.,Williams, P. J. LeB (eds.), Heterotrophic activity in the sea. Plenum Publishing Corporation, New York.

Münster, U. 1984. Distribution, dynamic and structure of free dissolved carbohydrats in the Plussee, a North German eutrophic lake. Verh. Int. Verein. Limnol. 22: 929-935.

Ochiai, M. & Handa, T. 1980. Change in monosaccharide composition in the course of decomposition of dissolved carbohydrates in lake water. Arch. Hydrobiol. 90: 257-264.

Ochiai, M. & Haney, T. 1980. Vertical distribution of monosaccharides in lake water. Hydrobiologia, 70: 165-169.

Ochiai, M., Nakajima, T. & Hanya, T. 1980. Chemical composition of labile fractions in DOM. Hydrobiologia 71: 95- 97.

Ogura, N. 1974. Molecular weight fractionation of dissolved organic matter in coastal seawater by ultrafiltration. Mar. Biol. 24: 305-312.

Ogura, N. 1975. Further studies on decomposition of dissolved organic matter in coastal seawater. Mar. Biol. 31: 101-111.

Omori, M. 1969. Weight and chemical composition of some important oceanic zooplankton in the North Pacific Ocean. Mar. Biol. 3: 4-10.

Overbeck, J. 1979. Studies on heterotrophic functions and glucose metabolism of microplankton in Plussee. Arch. Hydrobiol. Beih. Ergebn. Limnol. 13: 56-76.

Parsons, T. R., Stephens, K. & Strickland, J. D. H. 1961. On the chemical composition of eleven species of marine phytoplankters. J. Fish. Res. Bd. Can. 18: 1001-1016.

Parsons, T. R., Takahashi, M. & Hargrave, B. 1977. Biological oceanographic processes. Pergamon Press, Oxford.

Peake, E., Baker, B. L. & Hodgson, G. W. 1972. Hydrogeochemistry of the surface waters of the Mackenzie River drainage basin, Canada. II. The contribution of amino acids, hydrocarbons and chlorins to the Beaufort Sea by the Mackenzie River system. Geochim. Cosmochim. Acta 36: 867-883.

Raymont, J. E. G., Srinivasagam, R. T. & Raymont, J. K. B. 1969. Biochemical studies on marine zooplankton. VII. Observations on certain deep sea zooplankton. Int. Rev. ges. Hydrobiol. 54: 357-365.

Rego, J. V., Billen, G., Fontigny, A. & Somville, M. 1985. Free and attached proteolytic activity in water environments. Mar. Ecol. Prog. Ser. 21: 245-249.

Reynolds, C. S. 1984. The ecology of freshwater phytoplankton. Cambridge Univ. Press.

Riemann, B. & Mathiesen, H. 1977. Danish research into phytoplankton primary production. Folia Limnol. Scand. 17: 49-54.

Riemann, B., Søndergaard, M., Schierup, H.-H., Bosselmann, S. Christensen, G., Hansen, J. & Nielsen, B. 1982. Carbon metabolism during a spring diatom bloom in the eutrophic Lake Mossø. Int. Revue ges. Hydrobiol. 67: 145-185.

Riemann, B., Jørgensen, N. O. G., Lampert, W. & Fuhrman, F. A. 1986. Zooplankton induced changes in dissolved free amino acids and in production rates of freshwater bacteria. Microb. Ecol. 12: 247-258.

Riley, J. P. & Segar, D. A. 1970. The seasonal variation of the free and combined dissolved amino acids in the Irish Sea. J. mar. biol. Ass. U.K. 50: 713-720.

Salomi, M. & Pomeroy, L. R. 1965. Respiration and phosphorus excretion in some marine populations. Ecology 46: 877-881.

Seki, H. & Nakano H. 1981. Production of bacterioplankton with special reference to dynamics of dissolved organic matter in a hypereutrophic lake. Kieler Meeresforsch., Sonderh. 5: 408-415.

Shapiro, J. 1969. Iron in natural waters - its characteristics and biological availability as determined with the ferrigram. Verh. Int. Verein. Limnol. 17: 456-466.

Sheer, B. F., Sheer, E. B. & Berman, T. 1982. Decomposition of organic detritus: A selective role for microflagellate protozoa. Limnol. Oceanogr. 27: 765-769.

Sieburth, J. McN. & Jensen, A. 1970. Production and transformation of extracellular organic mater from littoral marine algae: a résumé, p. 203-224. In: Hood, D. W. (ed.), Organic matter in natural waters. Institute of Marine Science, Occasional Publications No. 1, College, Alaska.

Sigleo, A. C., Hare, P. E. & Helz, G. R. 1983. The amino acid composition of estuarine colloidal material. Est. Coast. Shelf Sci. 17: 87-96.

Sorokin, Y. I. & Pavaljeva, E. B. 1972. On the quantitative characteristics of the pelagic ecosystem of Dalnee Lake (Kamchatka). Hydrobiologia 40: 519-552.

Stabel, H.-H. 1977. Gebundene Kohlenhydrate als stabile Komponenten im Schöhsee und in Scenedesmus-Kulturen. Arch. Hydrobiol. Suppl. 53: 159-254.

Stabel, H.-H. 1978. Zur Molekulargewichtsverteilung gelter organischer Moleküle in verschiedenen Oberflächengewässern. Arch. Hydrobiol. 82: 88-97.

Strickland, J. H. D. & Parsons, T. R. 1977. A practical handbook of seawater analysis. J. Fish. Res. Bd. Can., Ottawa.

Søndergaard, M. 1984. Dissolved organic carbon in Danish lakes: Concentration, composition and lability. Int. Verein. Int. Limnol. 22: 780-784.

Søndergaard, M. & Schierup, H.-H. 1982. Dissolved organic carbon during a spring diatom bloom in Lake Mossø, Denmark. Wat. Res. 16: 815-821.

Søndergaard, M., Riemann, B. & Jørgensen, N. O. G. 1985. Extracellular organic carbon (EOC) released by phytoplankton and bacterial production. Oikos 45: 323-332.

Tait, R. V. 1972. Elements of marine Ecology. Butterworths, London.

Thomson, C. W. 1874. The depths of the sea. Macmillan, London.

Tuschall, J. R. & Brezonik, P. L. 1980. Characteristics of organic nitrogen in natural waters: Its molecular size, protein content, and interaction with heavy metals. Limnol. Oceanogr. 25: 495-504.

Vaccaro, R. F., Hicks, S. E., Jannasch, H. W. & Carey, F. G. 1968. The occurrence and role of glucose in sea water. Limnol. Oceanogr. 13: 356-360.

Weinmann, G. 1970. Gelöste Kohlenhydrate und andere organische Stoffe in natürlichen Gewässer und in Kulturen von *Scenedesmus quadricauda*. Arch. Hydrobiol. Suppl. 37: 164-242.

Wetzel, R. G. 1983. Limnology. Saunders college Publishing, Philadelphia.

Wetzel, R. G., Rich, P. H., Miller, M. C. & Allen, H. L. 1972. Metabolism of dissolved and particulate detrital carbon in a temperate hard-water lake. Mem. Ist. Ital. Idrobiol. Suppl. 29: 185-243.

Williams, P. M., Oeschger, H. & Kinney, P. 1969. Natural radiocarbon activity of the dissolved organic carbon in the North-east Pacific Ocean. Nature 224: 256-258.

Wright, R. T. 1970. Glycollic acid uptake by planktonic bacteria, p. 521-536. In: Hood, (ed.), Organic matter in natural waters. Institute of Marine Science, Occasional Publication No. 1, College, Alaska.

Wright, R. T. & Hobbie, J. E. 1966. Use of glucose and acetate by bacteria and algae in aquatic ecosystems. Ecology 47: 447-468.

Zygmuntowa, J. 1981. Free amino acids in cultures of various algae species. Acta Hydrobiol. 23: 283-296.

# Chapter 3. PHYTOPLANKTON

by Morten Søndergaard and Lars Møller Jensen
with contributions from Jørgen Kristiansen (section 3.2) and Wayne Bell (section 3.3.5).

## 3.1. General introduction

Lakes have two major sources of organic carbon: 1. An autochthonous input from photosynthetic organisms within the lake and 2. Allochthonous organic carbon transported into the lakes. In the majority of eutrophic lakes lacking large littoral zones and in hypereutrophic lakes, the planktonic microalgae totally dominate the input of organic carbon (Wetzel 1983).

Photosynthetic carbon reduction and the subsequent phytoplankton biomass build up are driving forces for all heterotrophic activity. Zooplankton graze the biomass of algae and bacteria and bacterial production is basically due to utilization of dissolved organic products of algal origin. To illustrate the pathways of organic carbon cycling in the pelagic zone a simplified scenario is presented in Fig. 3.1. Some biotic pathways such as the influence of flagellates and abiotic processes such as co-precipitation of $CaCO_3$ and dissolved organic carbon (DOC) have been omitted for reasons of clarity.

Two major carbon routes from phytoplankton to other biological component-scan be depicted from Fig. 3.1.
1. A direct utilization of particles by zooplankton grazing
2. The detrital pathway, which represents the organic carbon transformed to DOC and utilized by bacteria for growth and maintenance energy. Via the bacterial biomass part of the DOC becomes available to organisms grazing bacteria.

Using the broad definition of Wetzel (1983), detritus includes DOC and dead particulate carbon; the latter can be grazed directly by zooplankton (Saunders 1972a) and the former utilized by bacteria (Cole et al. 1984). Finally, particulate organic carbon (POC) can also sediment. The biological and chemical processes in the sediment have major influence on lake metabolism, but are not treated here.

In the following chapter we have addressed our interest in the phytoplankton community (section 3.2), primary production and methods to measure the flow of extracellular organic carbon (EOC) to bacteria (section 3.3) and the fate of algal carbon upon death (section 3.4). By such an approach we hope to demonstrate and discuss the most important features and processes directly involving the phytoplankton and to enlighten some of the methodological difficulties in studies of the pathways illustrated in Fig. 3.1. A quantitative presentation and analysis of the changing importance of each specific pathway and some pelagic carbon budgets are included in chapter 4-7.

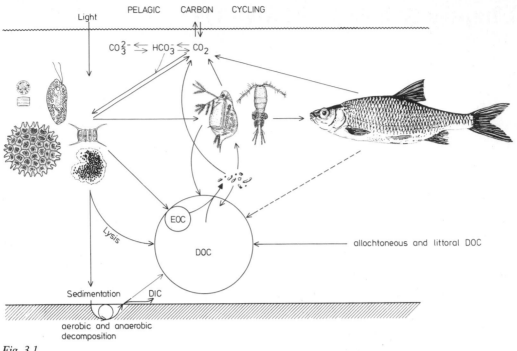

PELAGIC  CARBON  CYCLING

*Fig. 3.1.*
*Simplified scenario of pelagic carbon cycling. Abbreviations: EOC = extracellular carbon, DOC = dissolved organic carbon, DIC = dissolved inorganic carbon. Sizes of organisms and carbon pool indications are not drawn to scale.*

## 3.2. Phytoplankton in eutrophic lakes - community structure and succession

### 3.2.1. Introduction

In this chapter a survey of the phytoplankton in eutrophic, temperate lakes is given. The composition of the phytoplankton and its seasonal succession are discussed, and changes in composition caused by shifts in trophic status due to eutrophication processes or manipulation experiments, are demonstrated. Examples from Danish lakes will be used extensively for illustration of these topics. Finally, a survey will be given of the most important phytoplankton classes in eutrophic lakes, and their occurrence.

The phytoplankton studies of such lakes ("Baltic lakes") were initiated just before the turn of the century, mainly in Northern Germany at Plön, by the numerous investigations by Zacharias, Lemmermann, and others. Investigations in the English Lake District contributed to a more clear definition of the eutrophic phytoplankton (Pearsall 1932).

The above mentioned German investigations were the background for the first organized phytoplankton investigations in Denmark, resulting in the monumental work of Wesenberg-Lund "Undersøgelser over de danske søers plankton" (Plankton investigations in Danish lakes) from 1904. Here nine lakes from various parts of the country were regularly examined through one or two years, to obtain a general knowledge of the composition and seasonal succession of the phytoplankton.

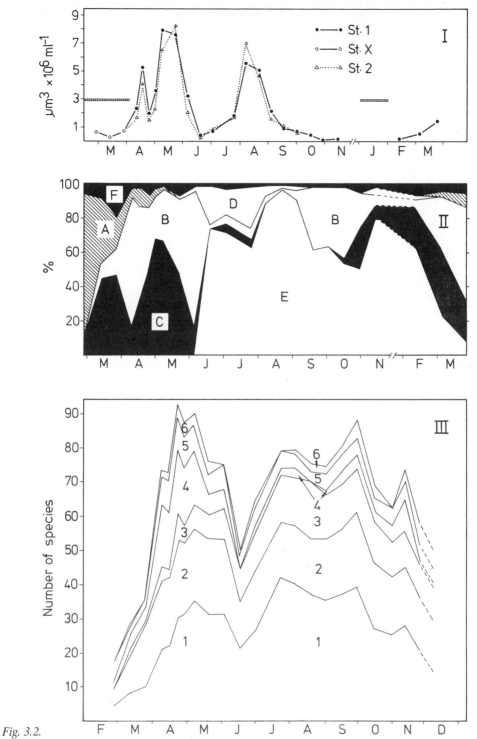

*Fig. 3.2.*
*Seasonal changes in (I) biomass, in half a Secchi- depth in Lake Tystrup 1980-81, calculated as volume, corresponding to freshweight (10⁶ μm³ ml⁻¹ = mg l⁻¹), (II) percentual composition of the phytoplankton biomass in Lake Tystrup. A: Chlorophytes (Volvocales), B: Diatoms, C: Cryptophytes, D: Chlorophytes (Chlorococcales), E: Cyanophytes, F: Other Groups. (From Kristiansen et al. 1982),(III) species number, total and of individual classes. 1: Chlorophytes, 2: Diatoms, 3: Cyanophytes, 4: Chrysophytes, 5: Cryptophytes, Dinophytes, and others, 6: Euglenophytes.*

Text and tables were accompanied by a unique collection of original photomicrographs showing representative phytoplankton samples. Wesenberg-Lund's paper is a valuable source of information on the condition of eutrophic lakes and the composition of their phytoplankton at the turn of the century, before the radical changes due to further eutrophication and pollution had taken place, and it demonstrates clearly the general trends in such phytoplankton changes.

### 3.2.2. Phytoplankton composition and succession

A generalized pattern for the phytoplankton composition and seasonal succession in the various lake types has been given by Reynolds (1984, his Fig. 43). The eutrophic lakes are characterized mainly by diatoms, chlorophytes, and cyanobacteria, sometimes also dinoflagellates. A vernal diatom maximum is replaced by summer populations of chlorophytes. They are succeeded by *Ceratium* and/or cyanophytes. In the autumn, the diatoms obtain their second maximum. An example is the eutrophic Lake Tystrup, situated in Denmark about 50 km W of Copenhagen. It will be considered here in some detail, and will be used as a base for comparisons with other lakes.

Lake Tystrup is one of the best investigated lakes in Denmark as regards the phytoplankton. It was included in Wesenberg-Lund's survey (1904). Composition and biomass of its phytoplankton was studied in relation to primary production by Kristiansen and Mathiesen (1964), and recently phytoplankton composition, biomass, and production have been studied in relation to several important environmental parameters (Kristiansen et al. 1982, 1983, Riemann 1983). In this section the seasonal variation in biomass, species number, and importance of the various main phytoplankton groups (Fig. 3.2) will be discussed.

In winter and early spring the biomass is very low, and it corresponds also to a low number of species, c. 20. Among these only cryptophytes and small chlorophytes (mainly Volvocales) are of any importance. The cryptophytes, *Cryptomonas* and *Rhodomonas,* reach their maximum in early spring, contributing about half of the total biomass.

The vernal maximum, in April-May, is the most important of the two annual maxima, and is equivalent to about 8 mg freshweight $l^{-1}$ (calculated from phytoplankton counts and volume estimates). It is almost exclusively dominated by small centric diatoms, e.g. *Stephanodiscus hantzschii* (Fig. 3.3) and species of *Cyclotella,* which may constitute 80% of the biomass. The double top of the maximum is mainly due to the occurrence pattern of S. *hantzschii.* These very small forms reach very high cell numbers, up to $40 \times 10^3$ cells $ml^{-1}$. Also the species number exhibits a vernal maximum, especially caused by chlorophytes, diatoms, and chrysophytes, altogether more than 90 species (Fig. 3.2).

Early summer (June) is characterized by a pronounced decline in biomass, to 0.5 mg $l^{-1}$, i.e. similar to winter levels. This decrease is accompanied by a similar drop in species number (Fig. 3.2), although least in the number of chlorophytes, which are at their greatest importance in the biomass in summer. The Chlorococcales have their main occurrence now, the genera *Sphaerocystis, Coelastrum, Ankyra,*

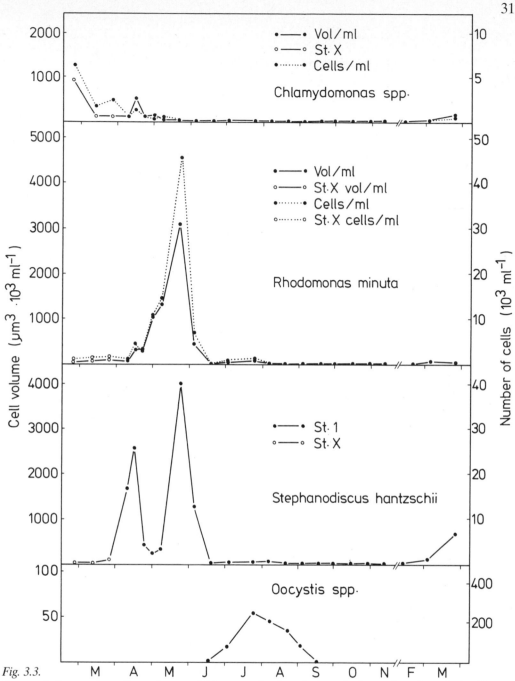

*Fig. 3.3.*
*Seasonal changes in cell numbers and biomass of selected phytoplankton algae from Lake Tystrup (from Kristiansen et al. 1982). (NB! The Lower most diagram shows number of cells ml⁻¹).*

*Oocystis,* and *Dictyosphaerium* becoming especially prominent (Fig. 3.3). The desmids, represented by *Closterium acutum,* have their maximum a little later (Fig. 3.4).

In late summer and early autumn, the second phytoplankton biomass maximum occurs, this being slightly smaller than the vernal maximum. As in the spring, the rise in biomass coincides with a maximum in species number, again more than 90

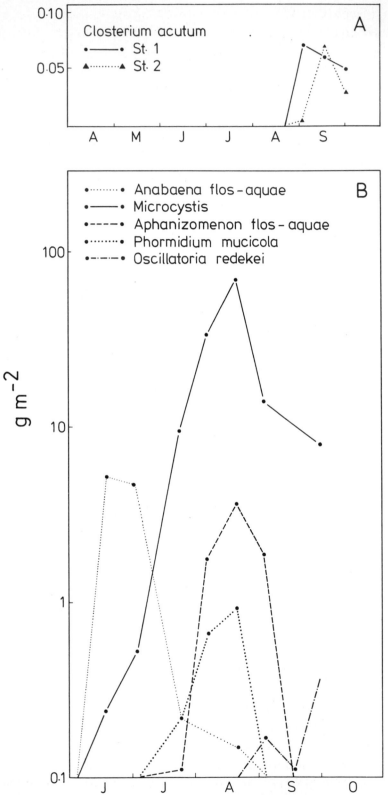

*Fig. 3.4.*
*Seasonal changes in the biomass of Closterium acutum in Lake Tystrup (A) and of different species of cyanophytes (B) (from Kristiansen et al. 1982).*

species are present (Fig. 3.2). Contrasting with the earlier maximum, however, the biomass is now markedly dominated by the cyanobacteria, up to 90% (Fig. 3.2). Already by June there is a peak of *Anabaena flos-aquae,* but this is succeeded in July and August by *Microcystis* species, mainly *aeruginosa* at first, and *M. wesenbergii* a little later. Meanwhile there are shorter secondary maxima of *Aphanizomenon* and *Oscillatoria* (Fig. 3.4). From this cyanobacteria-dominated late-summer maximum there is a gradual transition directly into the autumnal maximum. This consists largely of diatoms, especially *Melosira granulata,* which account for 40% of the biomass (Fig. 3.2). Then there is a gradual decline in biomass down to levels typical of winter. The decrease in biomass is accompanied by a decline in species number, with cryptophytes as one of the few groups to maintain some importance in the sparse phytoplankton. Most other phytoplankton groups survive as resting stages in the sediment.

Lake Esrom, examined 1961-62 (Jónasson and Kristiansen 1967), showed a similar seasonal cycle in phytoplankton biomass as Lake Tystrup, but with pronounced differences in the roles played by the most important algal groups (Fig. 3.5) corresponding to its lower nutritional load. Altogether 131 species were found. There were also two annual maxima, but lower peak values (3 mg l⁻¹) than in Lake Tystrup. The maxima were to a great extent determined by the diatoms. The vernal maximum consisted almost exclusively of diatoms, first *Stephanodiscus hantzschii* and then *Asterionella.* Prior to this, even before icebreak, there was a numerically high, but as biomass considered negligible, maximum of small chlorophytes, mainly *Monoraphidium.* Chrysophytes played a significant part in the vernal maximum. In summer, there were many chlorophytes and also dinophytes; the desmids (10 species) occurred a little later. The late-summer and autumnal maximum consisted mainly of diatoms, and although cyanobacteria were common, altogether about 1 mg l⁻¹, the cyanobacteria component comprised mainly *Anabaena* species, and later *Gomphosphaeria.* The diatoms were mainly *Asterionella* and *Stephanodiscus* at first, later succeeded by *Melosira.*

The lakes of the Mølleå River system Farum Sø, Furesø, Lyngby Sø, Bagsværd Sø near Copenhagen were examined during the years 1969-70 (Olrik 1973). Lake Farum had a total of 94 species. Its vernal maximum (8.5 mg l⁻¹) consisted almost exclusively of diatoms. Also during early summer the biomass was rather high, with cryptophytes or chloroccalean chlorophytes dominating. However, soon the cyanobacteria took over, constituting almost exclusively the autumnal maximum, at 25 mg l⁻¹ and continuing with very high values during the rest of the year.

Lake Furesø, with 110 species, exhibited a quite similar pattern, although with differences between the various areas and basins of the lake. A vernal maximum at 26.5 mg l⁻¹, caused by diatoms, volvocaleans and cryptophytes during summer, and large populations of cyanobacteria during the autumn maximum.

In Lake Lyngby with a total of 97 species, very large phytoplankton concentrations developed. The maximal biomass was found during the autumn maximum of 1969, 432 mg freshweight l⁻¹. The cyanobacteria were dominating most of the summer and autumn, only in spring there was a minor diatom maximum. Also the

34

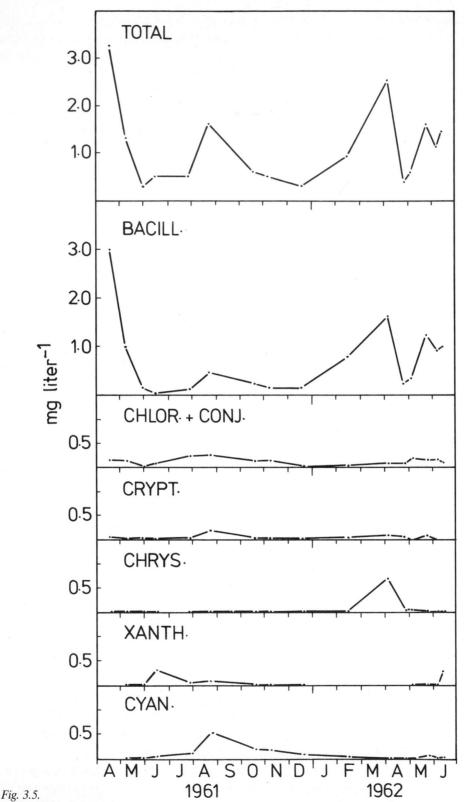

Fig. 3.5.
*Seasonal changes in total phytoplankton biomass and of the dominating groups in Lake Esrom 1961-62 (From Jónasson and Kristiansen 1967).*

chlorococcalean chlorophytes occurred in mass developement during autumn, but in comparison with the biomass of the cyanobacteria they were only of minor importance.

Lake Bagsværd, the most eutrophicated of the lakes, with 86 species, had a more specialized succession pattern. In early spring the lake was dominated by cryptophytes and volvocalean chlorophytes, whereas the cyanobacteria, mainly species of *Microcystis,* dominated the rest of the year and developed huge maxima, e.g. 99% of the July maximum 1970 at 517 mg $l^{-1}$.

While the lakes discussed above show evident similarities in their seasonal patterns, eutrophic Lake Arresø in the northernmost part of Zealand deviates considerably (Olrik 1981). It is by Danish standards a large lake, 41 km$^2$, but it is very shallow, with a mean depth of only 3 m. It is surrounded by agricultural country and receives biologically treated sewage. The phytoplankton is very dense, as a consequence the transparency is always less than 1 m. Biomass expressed as chlorophyll *a* reaches maximal values of about 400 mg $l^{-1}$ in September-October. In total, 79 species have been recorded, in winter as low as 18, but more than 30 in summer samples. During the whole year this lake is dominated by chlorococcalean chlorophytes, mainly species of *Scenedesmus.* Diatoms only occur in secondary maxima in spring and autumn, and cyanobacteria are of minor importance in autumn. The obvious question is now why this highly eutrophic lake is dominated by chlorococcaleans and not by cyanobacteria, even if N-concentrations are always low. It is proposed (Olrik 1981) that because in such a shallow wind-exposed lake the chlorococcaleans are always kept in suspension, the cyanobacteria in spite of their gas-vacuoles have no clear competitive advance to them. Further, pH never reaches the high values preferred by cyanobacteria, perhaps also due to the effective water circulation resulting in stable high $CO_2$ concentrations (see also Shapiro 1984). It is an open question what other factors might be responsible. It is noteworthy that in the summer of 1985, the lake shifted completely into a cyanobacteria lake.

### 3.2.3. General characteristics of the phytoplankton in eutrophic lakes.

As discussed above, there is a general, seasonal development of phytoplankton in eutrophic, temperate lakes. In early spring there is an occurrence of cryptophytes and small chlorococcalean chlorophytes. Diatoms constitute the vernal maximum. The chlorococcaleans have their main occurrence during summer. In late summer the cyanobacteria maximum takes place, succeeded by the autumnal diatom maximum. Among the examples only Arresø differed significantly from this pattern. In general, the spring phytoplankton is dominated by colonizing species with intrinsic high growth rates, taking advantage of the combination of high available nutrients and increasing light. During summer, the phytoplankton is dominated by slower growing species which most efficiently utilize available resources. The former species are r-strategists as opposed to the k-strategists dominating during summer (Sommer 1981).

Higher nutrient loads normally result in increased importance of chlorophytes and especially of cyanobacteria. The biomass ranges over two orders of magnitude from 5 to more than 500 mg freshweight $l^{-1}$. The lakes also differ considerably, especially in the magnitude and composition of the summer-autumn maximum. The characteristic pattern in moderately eutrophic lakes includes well defined spring and autumn maxima, with a low-production summer period in between, as in Lake Erken (Rodhe et al. 1958). In highly productive lakes there tend to be successive peaks during the whole summer (Reynolds 1973).

In most eutrophic lakes the species number remains around 100. Only in Lake Tystrup many more species were recorded, 196. This may be due to the fact that this investigation had focused more on species composition than most other investigations (Kristiansen et al. 1982). It must also be noted that electron microscopy was used in identification work, and has resulted in the detection of 33 species of silica-scaled chrysophytes.

A method for comparing different lakes by their phytoplankton composition is given in the Quotient Hypothesis proposed by Nygaard (1949). The so-called Compound Quotient is based on the relation between the number of species of eutrophic and oligotrophic groups of planktonic algae. For Lake Furesø: values of 10.5 - 30 during July were obtained, according to Nygaard they indicate a very high degree of eutrophy. For Lake Esrom, corresponding values from July-August are 6-7, indicating moderate eutrophy. For Lake Tystrup and for Arresø, values about 13 for the year as a total have been recorded.

### 3.2.4. Changes in phytoplankton composition during increasing or decreasing eutrophy.

Lake Furesø is one of the best known Danish lakes, and during the present century it has suffered severely from pollution and eutrophication.

At the beginning of the century, as demonstrated by Wesenberg-Lund (1904) and by Nygaard's (1950) renewed examinations of Wesenberg-Lund's original, preserved samples, Lake Furesø was a clear-water alkaline lake, dominated by diatoms, and with a rich, submerged macrophyte vegetation (see also Wesenberg-Lund 1912). It would thus be characterized as mesotrophic or weakly eutrophic.

Investigations in 1950-52 showed considerable changes in phytoplankton composition (Nygaard 1958). Several species have appeared since 1904, especially cyanobacteria, such as species of *Anabaena* and *Microcystis,* and some diatoms. Other species have disappeared, such as the previously dominating *Tabellaria fenestrata ,* *Melosira islandica var. helvetica,* and several desmids. This indicates a change from mesotrophy to eutrophy. Moreover, the depth distributions of submerged macrophytes had changed considerably, indicating a deteriorated light climate, which must have been caused by increased phytoplankton and epiphyte biomasses and shading (Sand-Jensen and Søndergaard 1981).

Olrik's (1973) investigations in Furesø during 1969-70 revealed further drastic changes. Cyanobacteria were by then dominating the phytoplankton for long periods, as already mentioned. Chlorococcalean green algae had further increased in species number, from 6 in the beginning of the century to 21 in 1950-51 and now to 32. Also some euglenophytes had appeared.

Many species of desmids and chrysophytes recorded during 1952 had disappeared by 1969, species of *Cosmarium* and *Staurastrum*, of *Dinobryon* and *Mallomonas*. *Ceratium hirundinella*, which formed summer maximum in 1951, had almost disappeared in 1969-70. The Compound Quotient had increased considerably, values between 10 and 30 were calculated from 1970 summer samples, indicating pronounced eutrophy.

Similar long-term investigations have made it possible to demonstrate quite comparable changes in Grosser Plönersee in northernmost Germany (Hickel 1975). Just like Furesø, this lake is a naturally eutrophic lake which during this century suffered increasingly from sewage and nutrient input from the surrounding agricultural land. The phytoplankton composition has changed considerably during a 75-year period. The number of species of chlorophytes has increased, those of diatoms and especially of chrysophytes and desmids have decreased. Several species typical of clean water have disappeared, such as *Attheya*, *Rhizosolenia* and species of *Dinobryon*. Also the composition of the cyanobacteria flora has changed, *Gloeotrichia* has almost disappeared, whereas species of *Microcystis* have become common.

The long-term development of the phytoplankton in Lake Tystrup does not show similar changes as those described for Furesø (Wesenberg-Lund 1904, Kristiansen and Mathiesen 1964, Kristiansen et al. 1982). During the earliest investigation the lake had two diatom maxima, the vernal maximum consisting of *Asterionella*, and the autumnal one of *Melosira*. In summer there was a well developed peak of *Ceratium hirundinella*. The cyanobacteria were of very minor importance. In 1961-62, however, the cyanobacteria had increased in occurrence with especially dense blooms of *Aphanizomenon* during August-September.

The most recent investigation 1979-81 showed pronounced changes and deviations from earlier records. Most importantly, the blooms of the cyanobacteria now occurred during a longer period, from June to October, and now dominated by *Microcystis* spp. The *Ceratium* maxima had disappeared, whereas the cryptophytes had increased their numbers and biomass.

Lake Hampen in Central Jutland is situated on diluvial sand, and it was thus previously a weakly buffered *Lobelia* lake, with a phytoplankton flora of desmids and some species of diatoms and chrysophytes. Because of its role as a tourist resort, and due to minor sewage inlets from surrounding houses and farms, it has suffered some eutrophication, and it now develops water blooms of cyanobacteria, mainly *Gomphosphaeria*, *Anabaena* and *Microcystis* (Kristiansen 1980), but in summer also of the chrysophyte *Uroglena*.

The effects of decreasing eutrophy have often been monitored by changes in phytoplankton biomass and primary production. However, there are also a few cases, where the effects on phytoplankton composition have been studied in detail. One of the best investigated cases is the restoration (1970-71) of Lake Trummen in southern Sweden (Cronberg 1982). The lake was so heavily polluted that even when the sewage affluent was diverted, this was of no effect. Before the restoration, the lake was characterized by chlorophytes (especially chlorococcaleans) and cyanobacteria, the latter forming dense blooms during most of the summer and autumn. The restoration included removal of 0.5 m bottom sediment and of the emergent littoral vegetation, the only vascular plants present. In this way the internal nutrient load was minimized.

As a result, a number of pronounced effects were observed: The maximal annual phytoplankton biomass was reduced from 80 to 15 mg freshweight l⁻¹. The dense blooms of cyanobacteria were much reduced, several of the cyanobacteria-species disappeared completely, many large-sized species were replaced by smaller ones. The species number of Chlorococcales diminished, whereas the number of chrysophytes increased considerably, and several more oligotrophic species appeared. As for the diatoms: their biomass decreased, especially as regards the *Melosira* species, whereas the less eutrophic *Rhizosolenia* increased in number. On the whole, the changes were reverse to the effects of eutrophication, and there were now considerable similarities in phytoplankton composition with the adjacent more oligotrophic lakes.

Oligotrophication may also be effected by increasing the zooplankton grazing pressure on phytoplankton by elimination of the fish populations in highly eutrophic lakes. In a Norwegian experimental lake (Reinertsen and Olsen 1984), the phytoplankton biomass during summer and autumn was lowered considerably, and the phytoplankton composition was totally altered. Most apparent was the reduction in the cyanobacteria (especially *Anabaena flos-aquae)*, which before the treatment had been absolutely dominant the whole summer, but after the elimination of fish were reduced to smaller quantities in August and September. The desmids, and small forms such as cryptophytes increased considerably instead.

Manipulations in large experimental enclosures have contributed much to our understanding of phytoplankton dynamics, as in the studies in Blelham Tarn in the English Lake District (Lund and Reynolds 1982). In continuously isolated enclosures, the nutrients are depleted and the phytoplankton composition becomes more oligotrophic than in the surrounding lake, e.g. in spring and early summer chrysophytes instead of chlorophytes, and during summer colonial chlorophytes instead of cyanobacteria.

Enclosures have proven useful in many other types of experiments, e.g. the effects of addition of controlled amounts of nutrients. Enrichment with phosphate caused increase of cyanobacteria, and during spring an increase of the diatom maximum was effected by the addition of Si. (Lund and Reynolds 1982).

### 3.2.5. Remarks on the occurrence and ecology of important phytoplankton groups in eutrophic lakes.

In the following survey major features of the occurrence and ecology of the various phytoplankton classes in eutrophic lakes will be discussed.

The *cyanobacteria* are among the most characteristic organisms of eutrophic lakes, and they may form extremely dense blooms during late summer and early autumn. The magnitude and duration of these blooms are dependent on the nutrient concentrations. In highly eutrophic lakes they may start in the middle of summer and continue until the lake becomes covered with ice. *Gomphosphaeria* occurs in mildly rich waters, whilst with increasing eutrophy, *Aphanizomenon* and species of *Anabaena* and *Microcystis* will dominate. There are also differences between the species in their seasonal occurrence (Fig. 3.5). Some red-coloured *Oscillatoria* species (e.g. *O. rubescens)* prefer low temperature; they are restricted to hypolimnion during summer, but may attain surface maxima in winter (Findenegg 1943).

Olrik (1978) found that the largest maxima occurred in lakes with highest mean concentrations of P and N, and a maximal depth of 3 m. However, the N-fixing cyanophytes will be favoured in comparison with the chlorophytes, when N becomes limiting, with P still available (see also Reynolds 1984).

*Cryptophytes* occur mainly in winter and spring. They may attain high numbers (Fig. 3.3), but as they are mostly small forms, their biomass seldom reaches noteworthy values. On a percentual basis, however, they may be important in the sparse plankton during winter (see also Fig. 3.2).

Contrastingly, *dinophytes* are mainly summer forms, although certain colourless phagotrophic species (e.g. *Gymnodinium helveticum*) can be found throughout the year. *Ceratium hirundinella* may develop large maxima in summer, often showing obvious patchiness (see Kristiansen and Mathiesen 1964). Its capability of active vertical movements makes it especially adapted for the summer stratification period (Heaney 1976). There is apparently an antagonism between dinophytes and cyanophytes, so the former tend to disappear when blooms of cyanobacteria become denser and more prolonged (Olrik 1978).

*Chrysophytes* have previously been thought to be characteristic of oligotrophic lakes. However, the genera *Dinobryon, Synura,* and *Mallomonas* often contribute a

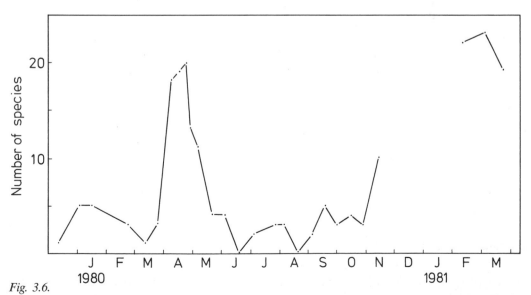

Fig. 3.6.
*Seasonal changes in number of scale-bearing chrysophytes in Lake Tystrup (from Kristiansen 1985).*

large part in spring and summer plankton of mesotrophic and slightly eutrophic diatom lakes. Moreover, a chrysophycean flora rich in species can also be found in highly eutrophic lakes. Most of these species are more or less ubiquitous, but some do prefer nutrient rich waters (Kristiansen et al. 1982, Kristiansen 1985, Kristiansen 1986). There is some increase in the number of species with increasing eutrophy, although a smaller increase than that shown by diatoms.

A peak in species number during spring is evident, after which most of the species encyst and survive in bottom sediments, and a smaller number is found the rest of the year (Fig. 3.6). Colourless forms, such as *Paraphysomonas* spp. occur during the whole year (Kristiansen 1985).

The *diatoms* occur in almost the same number of species during most of the year; the maxima are due to an increase in cell number of rather few species. Some diatoms are restricted to mesotrophic to slightly eutrophic lakes, where the genera *Attheya* and *Rhizosolenia* are characteristic. Other centric diatoms *(Stephanodiscus, Cyclotella, Melosira)* are more important in nutrient rich lakes. The ratio between species number of centric and pennate diatoms increases with increasing degree of eutrophy (Nygaard 1949). The diatoms typically reach their biomass peaks during the vernal and autumnal maxima, and so they may be largely responsible for the seasonal pattern of variation in the total biomass (Jónasson and Kristiansen 1967). There is often a difference in species composition between the maxima: *Asterionella, Cyclotella,* and *Stephanodiscus* being most prominent in spring, whereas the *Melosira* species dominate in autumn. The rapid decline of these maxima is mainly due to silica limitation and sedimentation, but other factors, as grazing and attack by parasitic fungi are involved (Lund l950, Canter and Lund 1948).

The *chlorophytes* also have two characteristic occurrence periods. Very early in spring, often before ice-break, a large maximum of very small species may appear, either *Chlamydomonas* spp. or *Monoraphidium.* During summer, however, other chlorophytes (often colonial forms) are important, mainly Chlorococcales *(Oocystis, Coelastrum, Pediastrum, Scenedesmus* etc.), but sometimes also Volvocales, e.g. *Eudorina, Pandorina.* The size and duration of the chlorophyte summer maximum seems to depend on the nutrient status of the lake; thus prolonged and dense growth occurs in highly eutrophic lakes (Berg and Nygaard 1929).

The *euglenophytes* are generally more important in ponds than in lakes. Nevertheless, several species of *Euglena, Phacus,* and *Trachelomonas* may occur in lakes, but seldom in large quantities.

Nygaard (1949) considered *desmids* as a group to be characteristic of oligotrophic conditions. However, a few species, mainly of the genus *Closterium,* but also of *Staurastrum* and of *Cosmarium,* are more restricted to eutrophic conditions, and they may attain high cell numbers in late summer plankton.

## Acknowledgements

The author is indebted to Dr. A. Bailey-Watts, Edinburgh, for linguistic help. The typing of the manuscript was done by Anne-Grete Kreiborg and Kirsten Pedersen.

## 3.3. Primary production, extracellular release and algal-bacteria interactions

### 3.3.1. Introduction

Phytoplankton assimilation of dissolved inorganic carbon (DIC), its reduction to organic compounds and the build up of algal biomass are the origin of the carbon cycle. Due to the crucial ecological importance of phytoplankton activity, considerable attention has been directed at the quantificaiton of photosynthetic rates and primary production ( = the biomass production).

Algal growth occurs simultaneously with procesess removing biomass, e.g. zooplankton grazing, cell lysis and sedimentation. Thus, it is usually impossible to measure net algal production as the accumulation of biomass over time. An important ecological question is: What is the current level of primary production? Two physiological methods are most widely used to measure instantaneous photosynthetic rates. 1) The $^{14}$C-method, introduced by Steemann Nielsen (1952), which traces the algal uptake of DIC and 2) The photolytic production of oxygen, which was published by Gaarder and Gran (1927). Many technical improvements have been applied to both methods (see e.g. Harris 1978). However, the basic approaches have remained unchanged over the years.

Due to its ease in use, the $^{14}$C-method has today become the most widely used way to measure primary production, although data interpretation is by no means easy. In addition to possible technical error sources (filtration, use of fixatives, counting procedures, bottle size etc.), the transport of $^{14}$C among different carbon compartments in natural environments being fast, can result in significant respiration losses of organic $^{14}$C at several trophic levels. The $^{14}$C-method is not the ultimate answer to primary production measurements (Lean and Burnison 1979, Carpenter and Lively 1980, Peterson 1980, Dring and Jewson 1982, Smith and Platt 1984). As no direct estimate of respiration is available from the normal procedure, uncertainty still exists whether net or gross photosynthesis is measured. This is in contrast to the oxygen method (Harris 1978, 1980).

The basic principle of the $^{14}$C-method is to add inorganic $^{14}$C as $NaH^{14}CO_3$ to an enclosed water sample, trace the $^{14}$C and estimate the movement of $^{12}$C into algae and other photoautotrophic organisms. A detailed description can be found in most current manuals on aquatic ecology (Strickland and Parsons 1972, Vollenweider 1972, Gargas 1975, Wetzel and Likens 1979). Uptake of $DI^{12}C$ is calculated by solving the following equation with respect to $DI^{12}C$ uptake:

$$\frac{DI^{14}C\text{-uptake}}{DI^{14}C\text{-added}} = \frac{DI^{12}C\text{-uptake}}{DI^{12}C\text{-present}}$$

An isotopic discrimination of 5 or 6% is used to correct for a slower $^{14}$C than $^{12}$C uptake (Steemann Nielsen, 1952, 1955, Goldmann et al. 1972). The soundness of this correction has been questioned (Harris 1980).

$^{14}$C in organisms is normally measured in particles retained on 0.2 or 0.45 μm pore size filters. It has hitherto been assumed that this radioactivity represents $^{14}$C fixed by and into algae. Alternatively, Schindler et al. (1972) introduced a method to measure $^{14}$C-fixation without filtration. Inorganic $^{14}$C in the sample is removed quantitatively by acidification and bubbling, leaving the remaining activity representing $^{14}$C-labeled organic products less volatile components. The technique has been discussed by Theodòrsson and Bjarnason (1975), Gächter and Mares (1979), Gieskes and Kraay (1980), Gächter et al. (1984) and Søndergaard (1985).

The carbon flow diagram in Fig. 3.1. depicts DIC and $^{14}$C assimilated in algal photosynthesis moving into other carbon compartments than the algae. Thus, the most unambigeous interpreation of a $^{14}$C experiment is: The $^{14}$C, which at some time during the incubation was fixed by an alga or fixed anaplerotically (e.g. Wood-Werkman reaction) by bacteria or other heterotrophs is present in an organic form, when the experiment is terminated. The organic form can either be particulate or dissolved. This "net community $^{14}$C-fixation" can now be converted to $^{12}$C units. An excellent review on the history of the $^{14}$C-method was given by Peterson (1980). No standard procedure for primary production has yet been agreed upon in the international scientific community.

The reliability of published production values, especially those based on $^{14}$C measurements, have been discussed and questioned since their introduction. The debate has not been subdued lately (Gieskes and Kraay 1984). It is our impression, however, that no one has proved primary production measurements are of a wrong order of magnitude, but many questions are still unanswered. One main reason to trust results based on $^{14}$C-techniques is the general agreement between $^{14}$C and $O_2$ measurements, $^{14}$C giving carbon equivalents and $O_2$ providing a measurement of energy turnover. We view primary production measurements as neither being very accurate, nor totally unreliable.

It has long been recognized that microalgae release dissolved organic matter into the environment, both in cultures and in nature (Harder 1917, Krogh et al. 1930, Fogg, 1952, Fogg and Westlake 1955, Lewin 1956, Antia et al. 1963, among others). Since Fogg (1958) initiated his investigations into phytoplankton release of recently fixed $^{14}$C and discussed the implications for measurements of primary production, there has been an increase in interest in this area. Not only as a new theme of discussion concerning the $^{14}$C-method, but also because of the ecologically important concept that algal photosynthetic products are not necessarily accumulating in the algal cell, but are also available to bacteria. The list of published investigations is long. A few - mostly *in situ* studies from freshwater - shall be listed here (Watt and Fogg 1966, Thomas 1971, Saunders 1972b, Berman 1976, Nalewajko and Schindler 1976, Smith et al. 1977, Lancelot 1979, Tilzer and Horne 1979, Mague et al. 1980, Blaauboer et al. 1982, Cole et al. 1982, Coveney 1982, Larsson and Hagström 1982, Riemann et al. 1982, Søndergaard and Schierup 1982a, Wolter 1982, Chrost 1983). General reviews on extracellular release have been published by Fogg (1966, 1971, 1983), Nalewajko (1977), and Sharp (1977).

Over the years doubt has been raised whether the published results showing high relative values of extracellular release (release as percent of total primary production) are artefacts and that actual release rates always are low. Sharp (1977) summarized the critique and pointed to shock effects, filtration errors, $^{14}$C-impurities, incomplete $^{14}CO_2$ removal and low radioactivity in samples as some of the more serious errors undermining the reliability and credibility of most, if not all previous results on extracellular release. Sharp was promptly answered (Fogg 1977). Recent investigators being aware of the potential errors, have shown the release to be a real phenomenon. We belive that algae release extracellular organic carbon as an inevitable process linked with their photosynthesis. The release rates differ with respect to species, locality, season, and physiological state of the algae. However, it should also be emphasized that errors are easily intro- duced and no sample should be handled as a routine sample.

Nalewajko (1977) in her review on extracellular products defined the term "extracellular release" synonymous with "excretion" as "the loss of soluble organic compounds by healthy cells of microorganisms". This definition is theoretical in the sense that it is not possible to know exactly whether release by moribund or lys- ing cells is involved, and to what extent algae can be given the prefix "healthy".

In practical work the $^{14}$C-method is used to measure organic release. Extracel- lular release is then defined as the $^{14}$C-labeled organic products present in a fil- trate (= dissolved), usually a 0.2 µm filtrate. The rationale for this practical defi- nition is that the labeled compounds, at some time during the incubation, were photosynthetically fixed by the algae. Besides the "true" release, the $^{14}$C-labeled dissolved organic carbon can originate from algal lysis, bacterial release of assim- ilated labeled products, or zooplankton activity. These items are treated in section 4.6. All products released do not stay in solution, but are partly taken up and meta- bolized by bacteria. Further, it has been shown by Jensen and Søndergaard (1982) that released dissolved products can reenter a particulate fraction by abiotic means.

Many different names have been applied to *Extracellular Organic Carbon*. In this chapter and throughout the book the abbreviation EOC will be used.

From the previous description of the $^{14}$C-method it has become clear that $^{14}$C does not necessarily move in a two compartment system; that is from DI$^{14}$C into an algal cell, where the tracer is immobilized. $^{14}$C can take several routes and end up in several compartments. Many studies have shown that release of recently fixed carbon as EOC is one major pathway. To investigate the algal release and bacterial utilization of EOC several different methods have been used. In the fol- lowing sections current methods are presented, with emphasis on methodology, theoretical and practical assumptions, limitations and recent improvements.

### 3.3.2. Differential filtration (particle size fractionation)

In most EOC studies from the sixties and seventies, but also in recent pub- lications, a single filter has been used to separate $^{14}$C-labeled particles and EOC (Fogg et al. 1965, Watt 1966, Nalewajko and Schindler 1976, Smith et al. 1979, Lancelot 1979, Tilzer and Horne 1979, Faust and Chrost 1981, Blaauboer et al. 1982). In the previous section we suggested that such an approach could not be used to estimate the release of EOC. Simultaneous algal release and bacterial uptake means that most filter-retained radioactivity represents $^{14}C$ in algae and in bacteria utilizing labeled EOC. Thus, traditionally measured EOC is the amount not assimilated by bacteria at the time of filtration. In measurements of total primary production using the $^{14}C$-method, algal respiration and bacterial respiration of EOC cannot be evaluated and so represent an inherent error. The release of $^{14}C$-labeled EOC is, therefore, underestimated. In order to minimize the effect of bacterial respiration, Nalewajko et al. (1976) have suggested very short-time ($<30$ minutes) incubations to give an estimate of gross release. The respiration of EOC can, however, be estimated indirectly with several procedures (Wiebe and Smith 1977, Cole et al. 1982, Iturriaga and Zsolnay 1983). A discussion of this subject is represented in section 3.3.6.

Several methods have been suggested to solve the basic problem: How much EOC is actually released, and how much is metabolized by bacteria during incubation? One way to answer the question is to uncouple release and uptake by antibiotics (section 3.3.3). Another way is to follow the labeling sequence in a $^{14}C$-experiment; that is to try to separate $^{14}C$ incorporated into algae and bacteria. Derenbach and Williams (1974) made the first attempt to differentiate between heterotrophic and photosynthetic populations with filters of different pore-size. An increasing number of investigations have used this technique (Berman 1975, Larsson and Hagström 1979, 1982, Berman and Gerber 1980, Cole et al. 1982, Coveney 1982, Riemann et al. 1982, Wolter 1982, Chrost and Faust 1983, Jørgensen et al. 1983, Bell and Kuparinen 1984, Kato and Stabel 1984, Rai 1984, Jensen and Søndergaard 1985, Søndergaard et al. 1985).

Differential functional groups (bacteria, algae, heterotrophic flagellates, zooplankton) overlap in size (Sieburth et al. 1978), so the size fractionation is no straightforward procedure. Many aquatic bacteria occur in aggregates and/or on larger particles in a similar size fraction as many phytoplankters ($>1.0$-$2.0$ μm, Riemann 1978, Cole et al. 1982, Kogure et al. 1982, Søndergaard et al. 1985). In most aquatic systems, including the marine environment, a quantitative separation cannot be carried out with respect to bacteria. The fractionation has to take the size of the smallest photoautotrophic organisms as the starting point. The first requirement of a differential filtration using two filter pore-sizes is to retain *all* algae on the first filter = the "algal filter" and to retain the rest of the bacteria on the second filter = the "bacterial filter". It is now possible to calculate the total activity, if the bacterial fraction activity on the bacterial filter to the total is known. The outcome of a differential filtration experiment is more sensitive to algae on the "bacteria filter" than to some variations in the bacterial size distribution. The choice of "algal filter" is thus of utmost importance.

The procedure for a $^{14}C$-experiment followed by differential filtration is in principle simple to perform, but does involve a series of assumptions and critical technical manipulations. These items are now discussed with reference to the schematic drawing of the experimental procedure given in Fig. 3.7.

The first step is to perform a "normal" $^{14}C$-incubation in light and dark bottles. Addition of fixatives to stop $^{14}C$-uptake can perhaps give too high EOC values

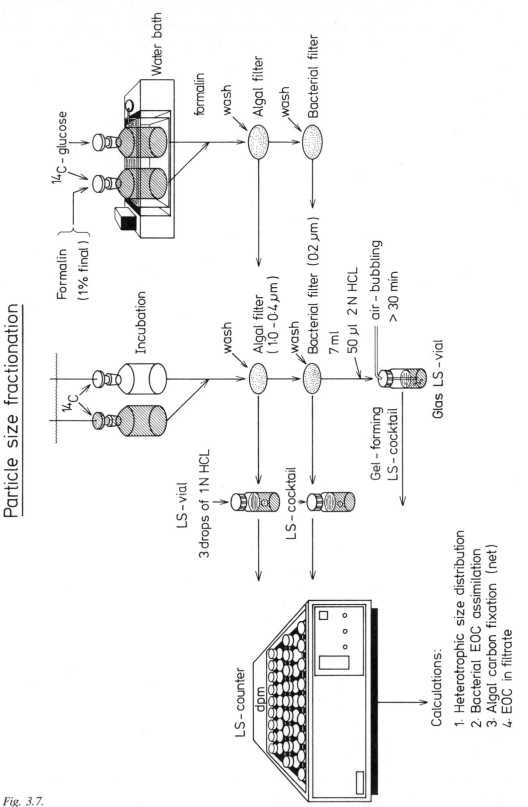

Fig. 3.7.
*Outline of the procedure for an experiment measuring algal carbon fixation, extracellular release and
bacterial uptake of extracellular organic carbon. Assumptions and calculations are explained in the text.*

(Silver and Davoll 1978, Watanabe 1980), although Bell and Kuparinen (1984) found no loss of particulate radioactivity using formalin (0.4% final concentration). The use of fixatives has not been systematically investigated and should probably be avoided unless investigated in each case. Samples should be kept dark and refrigerated until filtration, which should be performed as soon as possible. The incubation time is considered in section 3.3.2.2.

Subsamples are serially filtered through the chosen "algal filters" and 0.2 μm pore-size filters. The use of a 0.2 μm filter to retain bacteria is common practice in studies of aquatic bacteria (e.g. Hobbie et al. 1977). The first filtration in particular must be performed at a very low vacuum (<5 cm Hg), use of a minor volume set by particle density and filtering area and minimize shear-forces due to filtration rates. Cell breakage with outflux of labeled organics have been suggested, if the algae are exposed to physical stress due to high vacuum, filter loading, or shear-stress created by a fast flow rate (Arthur and Rigler 1967, Nalewajko and Lean 1972, Williams et al. 1972, Sharp 1977, Wolter 1980, Fuhrman and Bell 1985). This potential error can, however, be reduced by careful sample treatment (Berman 1973, Smith 1975, Mague et al. 1980). Both vacuum and pressure have been used to effect filtration, however, no systematic investigation on preference has been published.

The filtrations should be carried out stepwise. Serially connected filtration devices make it difficult to wash funnels for adhering particles and exchange water with $DI^{14}C$ in and on filters. Rinsing is merely a practical step in the procedure and should only be carried out with particle free water from the locality in question. We have never found any effects due to rinsing. A general recommendation cannot be given, and effects should be tested in each specific case. Decreasing total activity as a function of number of filters can occur (Larsson and Hagström 1982), but not always (Cole et al. 1982, Søndergaard 1985). Before counting, the filters must be acid-fumed or added a small volume of dilute acid (e.g. 0.5 N HCl) in the scintillation vial to kill the organisms and remove any residual $DI^{14}C$ (Lean and Burnison 1979, Mague et al. 1980).

Besides "physiological shock" effects on phytoplankton, Sharp (1977) considered lack of effective removal of $DI^{14}C$ from the filtrates and organic $^{14}C$ contamination of added $DI^{14}C$ to be the most important factors resulting in erroneous and often too high EOC values. Several studies have shown that quantitative $DI^{14}C$ removal can be achieved. Using a strongly dissociated acid, the sample is lowered to pH around 2.5-2.8 (always <3) and bubbled vigorously with air, $CO_2$ or $N_2$ for >20 minutes (Smith 1975, McKinley et al. 1977, Søndergaard 1980, Mague et al. 1980). $DI^{14}C$ removal is most conveniently carried out in a glass scintillation vial, which can be used afterwards for counting. Polyethylene vials must not be used as $DI^{14}C$ irreversible sticks to the walls (Søndergaard 1980). Organic $^{14}C$ impurities in used $DI^{14}C$-stocks solutions can be a problem. The effectiveness of removal of $DI^{14}C$ should always be checked carefully, diluting $^{14}C$ in particle free water from the study locality. As a result of some abiotic mechanism(s) the residual counts are always higher in "natural" water than in distilled water or in the ampule itself

(Søndergaard 1985, unpublished results). In a study of commercially available ampules, Bresta et al. (submitted) found only one, which could be recommended. Redistillation of $^{14}$C stocks can be recommended (J. J. Cole, personal communication).

Subtraction of dark samples is one way to correct for residual activity in the filtrate and non-photosynthetic activity on the filters. Dark values of $^{14}$C in particles represent anaplerotic fixation (Overbeck 1979). As there is a carry over of reducing capacity when algae are transferred from light to dark, which can also be traced to "EOC"-activity, DI$^{14}$C should be added to dark bottles after keeping the samples about one hour in darkness (Cole and Søndergaard, unpubl.). In prolonged incubations (8-12 hours) the carry over problem is negligible for both filters and filtrates.

Volatile and acid-unstable compounds may escape the EOC fraction as a result of acidification and bubbling. Examinations of this error using radioactive organic isotopes have shown that glucose, glycolic acid, acetate and a protein hydrolysate are not lost by the procedure (Anderson and Zeutschel 1970, Mague et al. 1980).

At this point it should be mentioned that one has to make a choice on filter-types. The most specific pore sizes are found in polycarbonate membranes (e.g. Nuclepore, Sheldon 1972), although the holes seemingly overlap and create larger pores than stated. As the pores are not made at right angles to the surface of the filter, the manufacturer claims that overlap on the surface does not affect the stated pore sizes. Membrane filters (cellulose-ester types) and glass fiber filters effectively retain particles smaller than their stated pore-sizes (Sheldon 1972) and some retain EOC, probably by adsorption (Wolter 1980). In a comparative study Salonen (1974) recommended polycarbonate membranes for the separation of algae and bacteria. There is no doubt that polycarbonate filters should be used in differential filtration experiments.

The choice of pore-size for the "algal filter" is more difficult. The optimal achieves the retainment of all photoautotrophs and allows passage of all bacteria. Algae, including photosynthetic cyanobacteria, on the 0.2 µm filter would lead to an overestimation of EOC uptake by bacteria. Several sizes have been used including 3.0 µm (Larsson and Hagström 1979, 1982, Berman and Gerber 1980, Wolter 1982, Rai 1984), GF/C filters (Coveney 1982), 1.2 µm (Chrost and Faust 1983), 1.0 to 0.6 µm (Kato and Stabel 1984, Søndergaard et al. 1985) and 0.4 µm (Cole et al. 1982). Cole and co-workers argued that no one filter pore size would generally work.

Both eukaryotic species and chroococcoid cyanobacteria belonging to the picoplankton size fraction (0.2-2.0 µm Sieburth et al. 1978) are important components of oceanic plankton communities (Johnson and Sieburth 1979, 1982, Waterbury et al. 1979, Krempin and Sullivan 1981, Li et al. 1983, Platt et al. 1983, Douglas 1984). Differential filtration experiments in the marine environment excluding algae from the "bacteria-filter" should be performed with 0.4 or 0.6 µm filters, unless the effect of photosynthetic organisms can be assayed and corrected for. Larsson and Hagström (1982) and Wolter (1982) used light mediated $^{14}$C-uptake in a prefiltered <3 µm fraction to correct for picophytoplanktonic carbon fixation. In the Baltic, Larsson and Hagström (1982) found that about 60 to 70% of the radioactivity in the fraction <3 >0.2 µm from their normal separation was due to photoautotrophic activity.

The evidence for presence of picophytoplankton in lakes is more scarce than in marine environments. However, there is a growing list of publications concerning chroococcoid cyanobacteria (Chang 1980, Cronberg and Weibull 1981, Craig 1983, 1984). Caron et al. (1985) in Lake Ontario showed cyanobacterial size to be 0.7-1.3 µm. In Lake Constance, Kato and Stabel (1984) depicted from pre-filtration experiments that a mean of 2.6% of the total photoassimilated $^{14}C$ was in the size fraction <1.0 >0.2 µm. This amounted to a mean of 30% of the radioactivity found in this fraction. It must, however, be emphasized that the seasonal variations were large. Further, pre-filtration might influence the activity of the organisms (Venrick 1977, Søndergaard et al. 1985). Presence of these small photoautotrophs should be confirmed from epifluorescence- and electron microscopy. Lake Ontario and Lake Constance are rather oligotrophic and have that in common with the sea. In two eutrophic Danish lakes the size fraction <1.0 >0.2 µm was examined for chlorophyll autofluorescence by epifluorescence microscopy (Søndergaard et al. 1985). No chlorophyll or phycoerythrin containing organisms were observed. This area certainly needs more research.

The present knowledge forces every investigator using differential filtration for EOC flux studies to examine the "bacteria-filter" for photoautotrophs. Use of 0.6 or 0.4 µm filters as algal screens (Cole et al. 1982) might prove to be the most appropriate. The real question in these studies is not the specific size of any picoalga, but whether or not the cells pass the filters. It is the filtrate that must be examined.

### 3.3.2.1. Heterotrophic size distribution

In a differential filtration experiment a variable part of the bacterial population is retained on the "algal-filter", so particulate radioactivity on this filter represents $^{14}C$ in both algae and bacteria. In order to estimate the total bacterial EOC assimilation and calculate $^{14}C$ in algal cells it is necessary to determine the size distribution of bacterial activity. The heterotrophic size distribution can be measured with a simple experiment using e.g. $^{14}C$-glucose (see Fig. 3.7). A mixture of substrates can also be used. The procedural assumption is that the assimilated substrate(s) behaves like the EOC products and that assimilation per unit bacterial biomass is constant.

The optimal procedure is to use $^{14}C$-labeled EOC released by the investigated algal populations. Such an approach is at best tedious and difficult, but a more serious drawback is that the qualitative composition of the EOC products change during exposure to bacteria (e.g. Jensen 1983, Kato and Stabel 1984), and this can change the apparent size distribution of uptake. Obviously, it is difficult to get a reference against which size distribution obtained from other substrates can be evaluated.

Most often radioactive glucose is chosen as an alternative to natural EOC (Derenbach and Williams 1974, Riemann 1978, Cole et al. 1982, Riemann et al. 1982, Jørgensen et al. 1983, Kato and Stabel 1984, Rai 1984, Søndergaard et al. 1985).

The heterotrophic size distribution of glucose uptake was compared with a soluble cell extract by Cole et al. (1982) and several different compounds (including an amino acid mixture, fructose, $^{14}$C-EOC from a *Microcystic aeruginosa* culture and $^3$H-thymidine incorporation) by Søndergaard et al. (1985). Both investigators found reasonable agreement between glucose and the other substrates. Large differences, however, have been recorded, comparing especially amino acids (e.g. glutamic acid) with glucose and thymidine (Berman and Gerber 1980, Novitsky 1983, Palumbo et al. 1984). There seems to be a tendency for amino acids to underestimate the heterotrophic activity of free bacteria, possibly due to a higher amino acid turnover per bacterial cell for particle-associated bacteria (Palumbo et al. 1984, see also Table 3.1). Although as yet no conclusive statements can be made, either glucose or thymidine may be regarded as reasonable choices.

In the oceans and coastal waters without major influence of sediment resuspension, the heterotrophic activity is mainly confined to free living bacteria in a $< 1$ µm fraction (Azam and Hodson 1977, Palumbo et al. 1984). In lakes there is a

Table 3.1. Examples of heterotrophic size distribution in marine and freshwater environments expressed as the percentage activity or number of the total. The table is arranged with increasing productivity from above to below.

| Location | Time | Substrate/ number | Particle fraction (µm) | | % | Reference |
|---|---|---|---|---|---|---|
| Atlantic | Feb-Mar | glucose, serine acetate | < 1 | > 0.2 | 90 | Azam & Hodson (1977) |
| Baltic | March | number | < 3 | > 0.2 | 95 | Larsson & Hagström (1979) |
| Mirror Lake | Aug Sep | | < 1 | > 0.2 | 72 | Cole et al. (1982) |
| Lake Almind | May | glucose | < 1 | > 0.2 | 80 | Søndergaard et al. (1985) |
| Lake Constance | May-Nov | glucose | < 1 | > 0.2 | 87 | Kato & Stabel (1984) |
| Lake Esrom | May | glucose | < 0.6 | > 0.2 | 30 | Søndergaard et al. (1985) |
| Lake Mossø | April | glucose | < 1 | > 0.2 | 57 | do. |
| | Oct | glucose | < 1 | > 0.2 | 35 | do. |
| Lake Hylke | May | glucose | < 1 | > 0.2 | 67 | do. |
| Lake Hylke | Sep | glucose | < 1 | > 0.2 | 18 | Søndergaard, unpubl. |
| Lake Ørn | Jun | glucose | < 0.8 | > 0.2 | 40 | do. |
| Frederiksborg Slotssø | Jun | glucose amino acid | < 1 < 1 | > 0.2 > 0.2 | 60 20 | do. do. |
| Lake Bysjön | | number | GF/C-filter | | 65 | Coveney (1982) |

trend towards a higher fraction of free bacteria in oligotrophic than in eutrophic lakes (Table 3.1).

The size distribution of heterotrophic activity is a function of locality and season (Table 3.1), whereas diel changes are minor in eutrophic lakes (Riemann 1980). One example of seasonal changes was observed in eutrophic Lake Mossø, Denmark. During a diatom bloom the contribution of glucose uptake in the size fraction $<1.0 >0.2$ μm declined from about 60% during exponential algal growth to about 40% during algal senescence (Riemann et al. 1982). Very low values, $<20\%$, in the size fraction $<1.0 >0.2$ μm can be found during blooms of cyanobacteria like the September situation in Lake Hylke (Table 3.1). Many of the bacteria are present in the matrix of the cyanobacterial colonies (Bern 1985).

Knowing the hetetrotrophic size distribution, the $^{14}$C-distribution between algae and baceria can be calculated:

$$\text{Net bacterial EOC uptake (assimilation) } (B_n) = \frac{DPM_{\text{bacterial-filter}}}{H}$$

Where H is the fraction of the total organic uptake (e.g. $^{14}$C-glucose) on the "bacteria-filter" and DPM is disintegrations per minute. Now, "net" algal fixation (A) is:

$$DPM_{\text{algal-filter}} - B_n (1\text{-}H) = A$$

The quotation mark on "net" is to avoid confusion with true net photosynthesis. Inserting the concentration of $DI^{12}C$ and added $DI^{14}C$, it is now possible to recalculate the data to carbon values. An example of original and recalculated data from Lake Esrom with an H value of 0.3 (Table 3.1) is shown in Fig. 3.8. It should be emphasized that with a low H, the correction alone accounts for the larger part of calculated bacterial EOC uptake. Although the choice of filter pore size must be made with respect to the smallest algae, an uncritical lowering of "algal-filter" pore size will lower the heterotrophic size distribution factor, thereby lowering the part of bacterial EOC uptake actually measured.

The outcome of a differential filtration experiment is the amount of $^{14}$C accumulated in three carbon pools during the incubation: Algae (A), bacteria ($B_n$) and a dissolved fraction ($EOC_n$). The bacterial respiration of EOC products ($B_r$) cannot be estimated by this method. An estimate of "gross" release imply use of a growth yield factor. Bacterial respiration of EOC products is very variable and is treated more fully in section 3.3.5. Assuming respiration is 50% (see Cole et al. 1982, Jensen 1985a), the calculated "gross" release is:

$$\text{"Gross" EOC release} = \frac{B_n}{0.5} + EOC_n$$

The lability of the EOC products and the heterotrophic activity can be characterized by the transport of EOC to bacteria (Cole et al. 1982, Coveney 1982, Riemann et al. 1982):

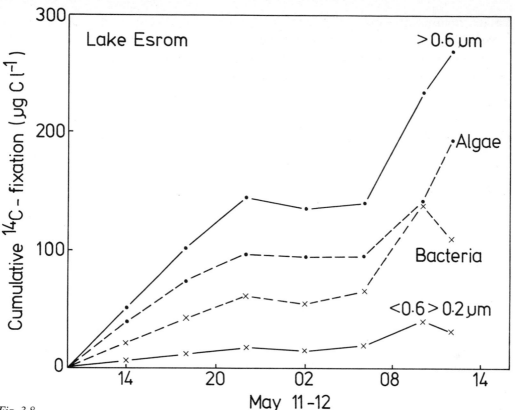

Fig. 3.8.
*Basic data showing the time dependent size distribution of $^{14}C$-fixation into particles >0.6 μm and <0.6 >0.2 μm and data recalculated to net fixation into algae (A) and bacterial EOC uptake ($B_n$). The heterotrophic size distribution factor H was 0.3. Experiment carried out in Lake Esrom, May 11-12, 1982. Data from Søndergaard et al. 1985.*

$$\frac{\dfrac{B_n}{0.5} \times 100}{\dfrac{B_n}{0.5} + EOC_n} = \% \text{ transport (T)}$$

As the respiration is not known, it might be better to present data as measured net values, ($B_n + EOC_n$). Since $B_r$ is about equal to $B_n$, it would, however, also be useful to include some estimate of $B_r$, as it doubles the carbon flux to bacteria.

Percent extracellular release is an often used expression which aids comparison among localities. It was introduced by Fogg (1952, 1958), and is usually defined as:

$$\text{p.e.r.} = \frac{EOC_n \times 100}{^{14}C\text{-particulate} + EOC_n} \%$$

This term simply describes the amount of EOC not yet utilized by bacteria in relation to total fixation. At the very least, p.e.r. calculations should include net bacterial uptake of EOC in the numerator. This subject is also treated in section 3.3.6.

### 3.3.2.2. Incubation time

Since Rodhe (1958) observed that the summation of several short-term $^{14}$C-incubations during a day yield higher primary production than one whole-day incubation, the choice of incubation time has been a matter of debate. Photoinhibition, photorespiration, effects of turbulence, gross versus net production and "bottle" effects have all been considered in this discussion (Harris 1978, Gieskes and Kraay 1984). The influence of labeling time on the rate of EOC release can now be added.

Time-course studies of extracellular release in $^{14}$C-experiments have been used to analyze different processes related to release. The experience has been that release rates can change as a function of incubation time. Nalewajko et al. (1976) suggested the use of initial release rates as an estimate of gross release. Short-term kinetics have shown that release rates can increase or decrease with time, probably as a result of a lag-phase in EOC production and heterotrophic uptake, respectively (Watt 1966, Saunders 1972b, Lancelot 1979, Larsson and Hagström 1979, Riemann et al. 1982). Such features urged Lancelot (1979) and with her, Fogg (1983), to suggest always to use time-course studies. Lancelot (1979) showed that gross release can be underestimated by up to 60%, using a single 4 h incubation

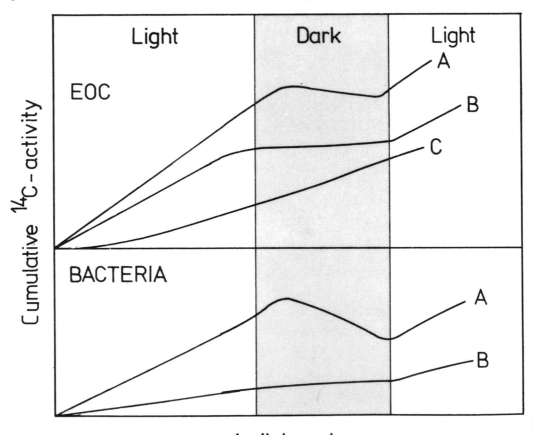

*Fig. 3.9.*
*Generalized time dependent patterns of $^{14}$C accumulation into a dissolved extracellular pool (above) and into bacteria (below).*

compared with time-course based calculations. Other investigators have followed the cumulative development of EOC in subsequent light/dark cycles while investigating EOC lability and the quantitative importance to bacteria (Smith et al. 1977, Cole et al. 1982, Wolter 1982, Kato and Stabel 1984, Søndergaard et al. 1985).

In Fig. 3.9 we have tried to summarize the different release patterns found by several authors *in situ* and in axenic algal cultures (Watt 1966, Saunders 1972b, Nalewajko et al. 1976, 1980, Lancelot 1979, Mague et al. 1980, Riemann et al. 1982, Kato and Stabel 1984, Søndergaard et al. 1985). Two types of EOC curves are most often found in the light: a "lag-phase" (Fig. 3.9, EOC curve C) and a linear form (A and B). The occurrence of lag-phase kinetics is obtained, if the specific activity ($^{14}C/^{12}C$) of the released products does not instantaneously reach the level of specific activity in the DIC pool. The more frequently experienced constant release rate in the light (Coveney 1982, Riemann et al. 1982, Kato and Stabel 1984, Søndergaard et al. 1985) is due to the dominance of EOC products labeled very early in the photosynthetic sequence, thus isotopic equilibrium is achieved rapidly.

The patterns become more complex entering the dark situation. In 14 diel experiments in Danish lakes and a coastal area, Søndergaard et al. (1985) found three different curve progresses, all of which have been experienced by others (e.g. Saunders 1972b, Kato and Stabel 1984). A slight decrease in EOC release rate during darkness indicates faster EOC disappearance than production (Fig. 3.9A). An almost constant level (B), often reached just before sunset, might reflect balanced uptake and release. The continuous increase (C) can be interpreted as a constant difference between release and uptake.

Using natural water samples, Nalewajko et al. (1976) and Larsson and Hagström (1979) have observed that EOC can reach a constant level after 1-2 hours' incubation. Larsson and Hagström (1979) suggested that after an initial phase of isotopic dis-equilibrium, "steady state" isotopic equilibrium with balanced release and bacterial uptake is achieved. Such rapidly reached constant EOC levels have, however, not been found in more recent time-course studies, e.g. Fig. 3.9.

The assumptions of specific activity of the released products are treated in detail in section 3.3.4.

Few experiments have involved *in situ* conditions during diel time-course studies of bacterial labeling with $EO^{14}C$ (Kato and Stabel 1984, Søndergaard et al. 1985). Thus, the generalization we offer here must be viewed as a preliminary statement. Two time-course curves seem to be the rule. Curve A and B (Fig. 3.9, below) both depict the bacterial labeling following a linear development parallel with the EOC labeling in the light. During subsequent darkness the labeling either decreases (A) due to higher respiratory loss than assimilation gain or the curve increases slightly (B) due to a continued uptake exceeding loss. To what extent these patterns are related to continuous EOC release in darkness, after $^{14}C$-labeling of the algae in light is not known. In axenic cultures of *Skeletonema costatum* Mague et al. (1980) found constant release during successive 2 hours light/dark periods. Conversely, Nalewajko et al. (1980) observed release by axenic *Chlorella pyrenoidosa* ceases in darkness.

It is evident that time-course patterns of release of EOC and bacterial uptake are rather complex. Some of the differences, e.g. a lag-phase or instantaneous release are most probably due to true differences between algal species and their physiological state. Along with Lancelot (1979) and Fogg (1983), Søndergaard et al. (1985) argued that diel or at least a time-course approach should be used in these studies. It should then be possible to avoid an underestimation of release due to a short incubation time. The use of and theoretical interpretation of time-course patterns of labeling are explored in details in section 3.3.4.

Differential filtration is seemingly an easy technique to use in studies of EOC flow, however, basic shortcomings are tied to the universal overlapping sizes of microalgae and bacteria and that bacterial respiration of EOC cannot be measured directly. The antibiotic approach presented in the next section does not suffer these inherent problems.

### 3.3.3. The antibiotic approach to measurements of extra-cellular release.

The trophic transfer of energy and organic matter is fundamental for the existence of nature - the members of a higher trophic level relying upon the former as a food source. In this transfer, a fraction of the organic matter is used for growth and reproduction, while another part is dissipated through respiratory processes.

It is now axiomatic that mineral consumption and regeneration are tightly coupled and that phytoplankton and heterotrophic bacteria are of prime importance in these processes. Furthermore, recent studies have shown that algal EOC (extra-cellular organic carbon) products may be made up of as much as 50% of total autotrophic carbon fixation (Crost and Faust 1983, Jensen 1985a, Lancelot 1983, Larson and Hagström 1982). Therefore, in an ecological context, two questions are of importance with respect to the fate of released EOC products: i. to what extent is EOC used for bacterial nourishment? and ii. what is the bacterial efficiency of converting EOC to biomass which can be exploited by higher trophic levels?

Unfortunately, the commonly used $^{14}$C-procedure of measuring algal release of EOC and its subsequent uptake by bacteria (see paragraphs 3.3.1 and 3.3.2) can only give us an estimate of net assimilation of EOC by bacteria (B). The fraction of EOC respired ($B_r$) cannot be measured. An alternative method of measuring bacterial mineralization of algal EO$^{14}$C has been to separate production and utilization of EOC products with respect to time and space. This means that algal utilization of an added NaH$^{14}$CO$_3$ solution is "separated" from bacterial release of $^{14}$CO$_2$ through respiratory processes (Fig. 3.10). Although such an approach has been widely used (e.g. Herbland 1975, Iturriaga and Hoppe 1977, Wiebe and Smith 1977, Crost and Faust 1983) to measure bacterial mineralization, this method and the differential filtration technique as outlined in paragraph 3.3.2 have in common the serious drawback that what is measured as EOC release only represents *net* apparent release, i.e. the fraction of EO$^{14}$C not taken up by the bacteria during the incubation. Consequently, the addition of the isolated and prelabeled EOC solution to a new sample containing no $^{14}$CO$_2$ will yield underestimates of both bacterial assimilation and mineralization rates.

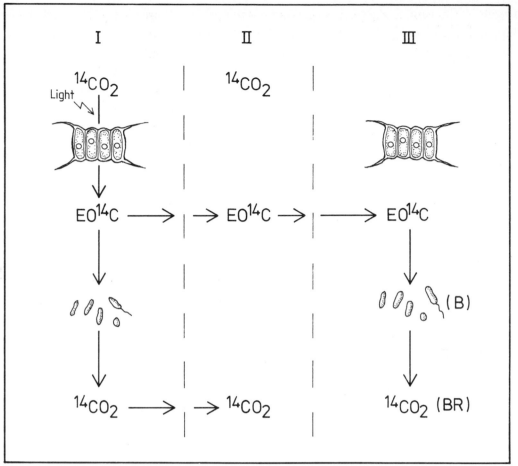

*Fig. 3.10.*
*Steps involved in preparation of $^{14}C$-labeled EOC solution for determination of bacterial assimilation and mineralization rates. I. A natural assemblage of planktonic algae and bacteria is incubated in light (in situ or in the lab.) with a NaH$^{14}CO_3$ solution. The phytoplankton photoassimilated $^{14}CO_2$ is partly released as EO$^{14}$C. Only particulate primary production and net EOC release are determined by II. filtration through a 0.2 μm filter to remove particles greater than the nominal pore diameter of the filter. The filtrate containing EO$^{14}$C and $^{14}CO_2$ is partly acidified to a pH less than 2.8 and bubbled with atmospheric air or N$_2$ for 30 min. to remove $^{14}CO_2$. The filtrate, now containing EO$^{14}$C only, is adjusted to the original pH of the sample and the filtrate is re-filtered through a 0.2 μm filter before III. addition of the EO$^{14}$C solution to a new sample containing a natural assemblage of plankton organisms. The incubation is performed in darkness in order to avoid reassimilation of respired EO$^{14}$C by the phytoplankton. The incubation is terminated by acidification to pH 2.8. Bacterial respiration (B$_r$) is determined by capturing $^{14}CO_2$ in a carbon dioxide absorber (e.g. ethanolamine/methanol, Carbosorb) and bacerial net assimilation (B) by filtering the sample through a 0.2 μm filter.*

A more promising and direct approach of circumventing this problem is to use batericidal agents as, for example, antibiotics. Antibiotics have been used for decades in the isolation and maintainance of axenic algal cultures (e.g. Spencer 1952, Berland and Maestrini 1969) but have, also, been applied with success in terrestrial (e.g. Anderson and Domsch 1973, 1975) and aquatic sediments (e.g. Hargrave 1969) to examine and measure the contribution of bacterial microheterotrophs to the total metabolism of a complex microbial system. In studies of phytoplankton release of EOC and the subsequent assimilation of these compounds by the bacterioplankton, antibiotics have, however, yielded ambiguous results (e.g. Anderson and Zeutschel 1970, Derenbach and Williams 1974, Berman 1975). In addition, primary production in most instances has been affected by the addition of antibiotics. Crost (1978), however, showed that the antibiotic gentamycin at concentrations of 10-40 $\mu$g ml$^{-1}$ could be used to selectively inhibit bacterial uptake of EOC without dramatically altering phytoplankton carbon metabolism for several hours, thus making it possible to measure "gross" release.

Jensen (1983) took the next step and introduced the antibiotic approach as a method of obtaining information about the quantitative interrelationship between phytoplankton and heterotrophic bacteria. The general principles of the antibiotic approach are outlined in Fig. 3.11. Instead of one sample, as is the case in a "traditional" primary production experiment, two samples are incubated in parallel - one with and one without antibiotics. In the sample with no added antibiotics, $^{14}CO_2$ will be assimilated by the phytoplankters and, through the release of $EO^{14}C$, tracer will move to the bacterial compartment. In contrast to the fractionation technique (section 3.3.2.), no attempt is made to separate phyto- and bacterioplankton after termination of the experiment. Instead, the whole sample is filtered directly through a 0.2 $\mu$m membrane filter in order to distinguish between $^{14}C$ incorporated into biomass (phytoplankton and bacteria) and $^{14}C$ released as EOC and not taken up by the bacteria during the incubation. Samples incubated with antibiotics are filtered directly through a 0.2 $\mu$m membrane filter. Since, under optimal conditions, bacterial activity is inhibited by the addition of antibiotics, no transport of EOC to bacteria occurs. The EOC measured in the filtrate represents, therefore, total release of EOC while the $^{14}C$ retained by the filter is attributable to phytoplankton $^{14}C$-fixation only (Fig. 3.11).

According to Jensen (1983), bacterial assimilation and respiration of EOC then can be calculated by the following equations:

1.  $A = A' + B + B_r$
2.  $P = P' - B$

where A represents the total release of EOC as measured in the sample incubated with antibiotics, and A' represents residual EOC, i.e. the fraction of EOC not taken up by bacteria during the incubation ($= EOC_n$ given in 3.3.2.1). A' is measured in the sample with no antibiotics added, B is bacterial net uptake of EOC, and $B_r$ is bacterial respiration. P and P' are particulate $^{14}C$ from the sample with and without antibiotics added, respectively. Note that $B_r$ and B are unknowns, the latter being part of the total activity retained by the filter from the sample with no antibiotics (P'). By rearranging the equations 1 and 2, we are able to calculate the bacterial parameters of EOC utilization as follows:

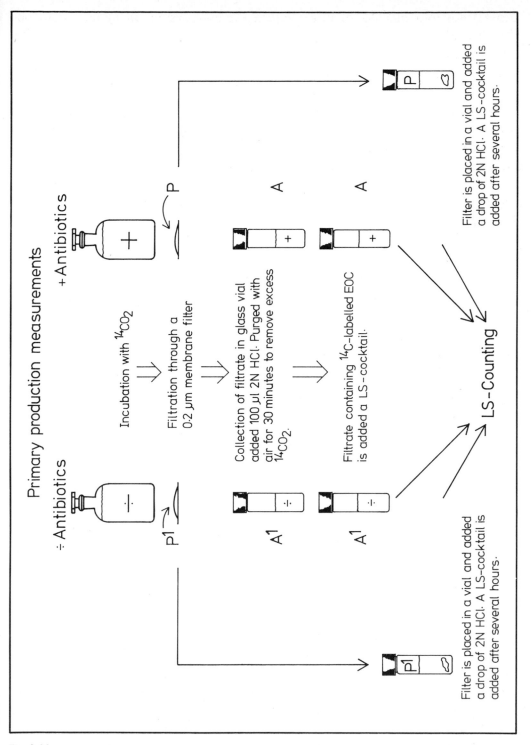

*Fig. 3.11.*
*Flow chart showing the steps involved in determination of particulate primary production, EOC release, and bacterial uptake of EOC by the antibiotic approach. Inserting the results obtained by LS-counting of the different components in eqs. 1-5 yields estimates of net (A') and total (A) release of EOC, particulate primary production (P), bacterial net (B) uptake of EOC, bacterial respiration ($B_r$) and gross uptake (C) of EOC.*

3.  $B = P' - P$
4.  $Br = C - B$
5.  $C = A - A' = B + Br$

where C is gross bacterial uptake of algal EOC.

Two major assumptions underlie the above calculations. Firstly, that no uptake of EOC occurs in samples with antibiotics added, and secondly that the phytoplankton community is not susceptible to the action of antibiotics. Both assumptions are, however, difficult to prove. Concerning bacterial inhibition, two approaches have been used: 1. tests involving culturing of the bacteria on an appropriate agar medium containing the relevant antibiotics followed by a counting of colony forming units (CFU). 2. following the fate of different radioactive labelled organic products. In both cases, the results are expressed relative to a sample containing no antibiotics.

Since culturing of bacteria on a nutrient rich medium may be selective for one or a few species, usually those with high $k_m$, the second approach should be preferred. Ideally, $^{14}C$-labelled EOC products from the algal species on which the bacteria derive nourishment should be used. However, as previously mentioned, the EOC products isolated from a natural sample containing bacteria will have a different composition than the products released initially, thus giving rise to a different uptake pattern by the bacteria to which the EOC solution is offered after isolation. As a consequence, properly chosen labelled organic solutes such as, for example amino acids and monosaccharides, are to be preferred. Alternatively, either the extract from cell homogenates prepared from $^{14}C$-labelled algal cultures (Schleyer 1981) or $EO^{14}C$ isolated from cultures of one or several of the dominating algal species can be used (see section 3.3.5).

Unfortunately, all antibiotics tested so far have proved to be insufficient in totally preventing bacterial growth. Even highly specific procaryotic inhibitors, for which the primary target is the blocking of bacterial cell wall synthesis, have been shown not to inhibit bacterial activity by more than up to 75% - even at concentrations as high as 100 µg ml$^{-1}$ (Jensen 1984). At lower concentrations, some antibiotics may even enhance bacterial activity (Yeatka and Wiebe 1974, Jensen 1985c). Moreover, bacterial assimilation and respiration is often affected to a different degree (Jensen 1984) by a given antibiotic. Further complications arise from the fact that bacterial susceptibility to antibiotics seems to be a function of growth phase - early log-phase cells exhibiting the greatest susceptibility (Yetka and Wiebe 1974).

Thus, if antibiotics are used to estimate bacterial uptake and respiration of algal EOC products, it is necessary to perform tests of susceptibility in order to establish a correct estimation of the parameters given in eqs. 1-5. If the tests show that complete inhibition is not achieved, this must be corrected for. The easiest and most convenient way to do this is to incubate parallel samples, to half of which have been added antibiotics, with a $^{14}C$-labelled organic solution of, for example, amino acids, glucose or EOC (Fig. 3.12).

Assuming that it is possible to extrapolate from the observed degree of inhibi-

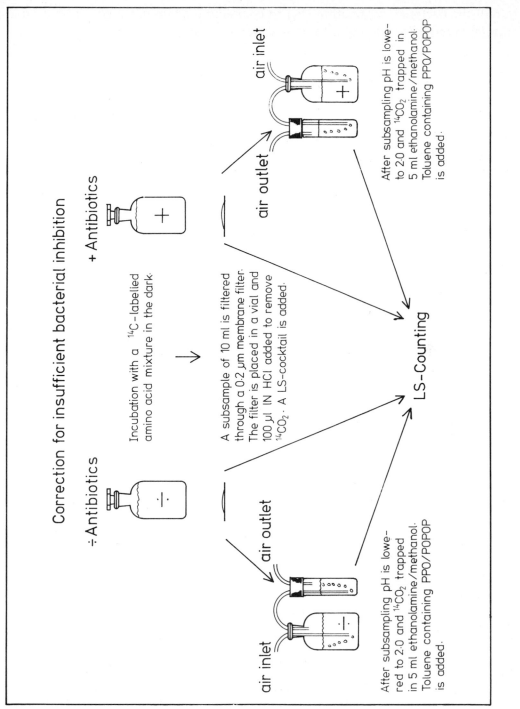

*Fig. 3.12.*

*Flow chart for correction of incomplete bacterial inhibition. The magnitude of bacterial inhibition ($F_1$) is calculated as the depression of bacterial uptake of a $^{14}C$-amino acid mixture in the sample added antibiotics relative to the uptake in a parallel sample with no antibiotics added. Bacterial respiration in the sample containing antibiotics is measured also, and expressed as a percentage ($F_2$) of total uptake. The parameters $F_1$ and $F_2$ are used to calculate corrected particulate primary production ($P_c$), total release ($A_c$), bacterial gross uptake of EOC ($C_c$), and bacterial net uptake of EOC ($B_c$). The corrected bacterial respiration ($B_{rc}$) is determined as the difference between $C_c$ and $B_c$ (see eqs. 6-9).*

tion of the bacterial assimilation ($F_1$) to complete inhibition, we get the following correction terms for particulate and dissolved primary production:

6.  $A_c (A - A')/F_1 + A'$
7.  $P_c = P - (A_c - A)(1 - F_2)$

where $A_c$ is the corrected "gross" release and $P_c$ is the corrected primary production in samples incubated with antibiotics. Since a given bacterial activity in the sample containing antibiotics will result in an uptake of released EOC, the particulate primary production (P) will be overestimated by that magnitude minus the fraction respired ($F_2$) during the incubation. Hence, we have to subtract this. By inserting equations 6 and 7 into equations 3 and 4 the following equations emerge:

8.  $C_c = A - A'/F_1$
9.  $B_c = P' - P + (A_c - A)(1 - F_2)$

where $C_c$ and $B_c$ are bacterial gross and net uptake of EOC corrected for insufficient inhibition of the bacteria (Jensen and Søndergaard 1985).

As discussed by Jensen (1983, 1985c) it is even more difficult to assay algal susceptibility in a sample containing a mixed microbial population, especially when only small inhibitions of primary production occur. In an attempt to judge whether or not a phytoplankton population is inhibited, Jensen (1983) suggested the use of the production ratio P + A/P' + A'. Depending upon bacterial growth efficiency, the production ratio will attain values greater than or equal to unity. This means that a decreasing ratio during the course of an experiment is indicative of an algal inhibition. However, it must be stressed that the ratio can only be used as a rough indication and only as long as the algal inhibition (expressed in carbon equivalents) does not exceed bacterial inhibition.

The question of interest is then: Is it possible to find antibiotics which inhibit bacterial uptake of EOC without imposing secondary effects on algal carbon metabolism? Although field experiments with gentamycin (Crost 1978) and streptomycin (Jensen 1983) showed that these antibiotics occasionally could be used to selectively inhibit bacterial activity, algal response was found variable and probably dependent on the species present and the physiological state of the algae. Furthermore, laboratory experiments with algal cultures revealed that algal incorporation of $^{14}C$ into biomass (P) and EOC (A) responded in different ways and as a function of the antibiotics used, the concentration applied, and contact time with the antibiotics (Jensen 1984). However, two antibiotics, vancomycin and mefoxithin, were found not to impose secondary effects on algal carbon metabolism at concentrations of less than 5 µg/ml - even after 24 hours of contact with the antibiotics (Jensen 1984). As complete bacterial inhibition is not possible at such low concentrations of antibiotics, it is necessary to correct for this by measuring $F_1$ and $F_2$ as given above.

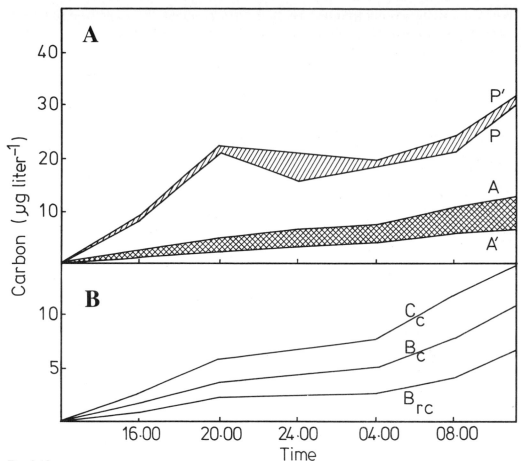

*Fig. 3.13.*
*Primary production and carbon transport to bacteria during a diel study in mesotrophic Lake Almind, Denmark, as determined by the antibiotic approach.*
*A: P' = particulate production in samples without antibiotics. P = particulate production in samples with antibiotics added. A = release of EOC as measured in samples with antibiotics, and A' = EOC release in samples without antibiotics. The difference between P' and P represents bacterial net uptake of EOC (B), while the difference between A and A' represents gross uptake (C).*
*B: Corrected net (B$_c$) and gross (C$_c$) bacterial uptake of EOC. Corrected bacterial respiration (B$_{rc}$) is calculated as the difference between C$_c$ and B$_c$.*

An example of the time course of $^{14}$C incorporation into algal biomass, EOC products, and bacteria as determined by the antibiotic approach, is shown in Fig. 3.13. Since the $^{14}$C incorporated into particles $>0.2$ µm in samples incubated without antibiotics represents the sum of algal $^{14}$C-fixation into cell biomass and bacerial assimilation of $^{14}$C-labeled EOC products, particulate activity (P') is greater than in samples (P), where bacterial activity is inhibited. Because of this, the opposite is true for the dissolved activity, i.e. the amount of EO$^{14}$C released, but not taken up by the bacteria during the course of the experiment is greater in samples with antibiotics (A) than in those without (A') (Fig. 3.13A). Thus, according to eqs. 1 and 2, the differences between P' and P, A and A' represent net (B) and gross (C) bacterial uptake of EOC, respectively (Fig. 3.13B). However, in the example shown, bacterial activity was not totally depressed, thereby giving rise to an overall under-estimation of the bacterial parameters B, Br and C (Fig. 3.13B).

Jensen and Søndergaard (1985) compared the antibiotic approach and the size-fractionation technique in several lakes ranging from mesotrophic to highly eutrophic. Generally, the two methods were found to agree well as long as no algal inhibition occurred. Moreover, the antibiotic approach yielded bacterial growth efficiencies of ca. 30-50%, well inside the range commonly found in the literature (e.g. Chrost 1981, Williams 1981, Bell 1983, Iturriaga and Zsolnay 1983), thus rais-ing confidence in the results obtained by the antibiotic approach.

However, the use of antibiotics to selectively inhibit bacterial activity has two serious shortcomings. It is a time consuming procedure, and it may be difficult to assay toxic effects, if not lethal, on algal carbon metabolism. Nevertheless, if knowledge of bacterial respiration of *in situ* produced EOC products is desired, the antibiotic approach would seem to be the preferred method.

### 3.3.4. Specific activity, isotopic equilibrium and compartmentalization in tracer experiments for measurements of EOC

In the preceding paragraphs (3.3.2. and 3.3.3.) different methods to measure algal release of newly photosynthesized organic carbon as EOC and the subse-quent assimilation of these products by the indigeneous bacterial population have been presented and discussed. It was indicated that the primary production may be severely underestimated, if the release of EOC is not taken into account, and that a substantial part of the released EOC can be converted to particulate organic car-bon by the osmotrophic activity of the bacterioplankton. Later on (4.6.) it will be shown that EOC is of importance for bacterial secondary production and thus for the carbon flow in aquatic environments.

The correct quantification of the release is, therefore, not only important for the assessment of total primary production, but is also essential for a better under-standing and description of major pathways involved in primary production utilization.

We will now consider the use of the $^{14}$C-method to measure dissolved primary production (EOC) in terms of its quantitative reliability to calculate true release rates, i.e. how far can the measured tracer ($^{14}$C) flux be expected to be definitely related to the overall properties of the tracee ($^{12}$C)?

Although such a question is almost a self-evident attribute in experiments using radioactive tracers, most investigators working with phytoplankton release of EOC have implicitly assumed that the intracellular precursor pool(s) is very small com-pared to the total algal cell carbon pool. A consequence of this view or theory is that no timelag between tracer accumulation in the intracellular precursor pool from which the exchange occurs and the dissolved pool of EOC is predicted.

There is some experimental support for this "classical" view, since intermediates of photosynthesis have been shown to attain steady state conditions in a few sec-onds (Bassham and Calvin 1957). Further, several examples of a close superposition of photosynthesis and EOC release have been published (Nalewajko 1977, Lancelot 1979, Larsson and Hagström 1979, Coveney 1982).

The "classical" theory operates with a catenary reaction sequence, where the intracellular exchange pool and the EOC products fated for release equilibrate almost instantaneously with the inorganic carbon pool (DIC) (Fig. 3.14). Hence, the conversion of tracer flux to the actual flux of tracee can be done according to the following equation:

$$^{14}CO_2/^{12}CO_2 = EO^{14}C/EO^{12}C$$
$$EO^{12}C = EO^{14}C \times {}^{12}CO_2/^{14}CO_2$$

This is the equation employed in all calculations of EOC release (Peterson 1980), whether or not the assumptions of the "classical" theory have been mentioned explicitly in the analyses of the $^{14}C$-EOC estimates.

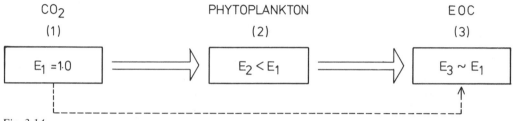

*Fig. 3.14.*
*Model compartments of algal carbon exchange as postulated by the "classical" model. The model assumes a small exchange pool (2), consequently no loss of $^{14}C$-activity from the phytoplankton pool occurs. The specific activity of the EOC pool ($E_3$), therefore, equals that of the $CO_2$ pool ($E_1$). Diagramatically, this means that $^{14}C$ will move directly from the $CO_2$ to the EOC pool. This is indicated by the stippled line.*

Historically, the "classical" theory may be regarded as having its background in the early work of Fogg and co-workers (e.g. Watt 1966), who found that glycollate, which is an intermediate product labeled rapidly and early in the photosynthetic sequence, is a dominating product of the extracellular EOC pool. If glycollate or other low molecular weight products exhibiting the same labeling pattern constitute a significant part of the EOC products, a minimal timelag between $^{14}C$ incorporation into algal POC and EOC would be expected. On the other hand, if the released substances consist exclusively of high molecular weight products labeled late in the metabolic sequence, a lag between tracer incorporation into algal POC and EOC will be introduced (Fogg 1966). In that case, the labeling kinetics will not trace the real flux of tracee because of isotopic dilution in the intracellular precursor pool. Release with timelags is no uncommon phenomenon.

Wiebe and Smith (1977) rejected the possibility that the specific activity of the internal exchange pool did not reach isotopic equilibrium within several hours. In time course experiments with phytoplankton populations from an Australian estuary, they found non-linear kinetic curves of the EOC release after a rapid linear increase. By performing a separate uptake experiment with $^{14}C$-labeled EOC products, Wiebe and Smith (1977) concluded that the plateau region of the curve occurred as a consequence of an EOC compartment of constant size (bacterial consumption equaled algal release rate) for the duration of the experiment (approximately 11 hours), and because the specific activity of the EOC pool became equal to that of the DIC pool. Wiebe and Smith (1977) reported that more than 95% of the released material consisted of substances with a molecular weight less than 3,500 daltons. This may indicate that the release occurred from a low molecular weight intracellular pool at instantaneous isotopic equilibrium. Smith and Higgins (1978) and Larsson and Hagström (1979) arrived at the same conclusions. From tracer kinetic incorporation experiments it thus seems possible to show whether or not isotopic equilibrium in the internal EOC precursor pool is achieved. However, it must be emphasized that most kinetic studies have failed to show plateau-like curves (e.g. Saunders 1972, Nalewajko et al. 1976, Lancelot 1979, Mague et al. 1980, Coveney 1981, Jensen and Søndergaard 1985) and as discussed by Coveney (1982) steady state conditions for algal EOC product pools are not universal in natural plankton communities. Further, Coveney(1982) argued that if non-plateau curves were obtained (i.e. linear kinetics), this could euqally be explained by non-equilibrium conditions in the precursor pool, even though the extracellular pool is at steady state. Linear kinetics could be achieved due to 1. release at a rate faster than bacterial uptake or 2. release of EOC to a large pool whose specific activity is still increasing linearly during the experiment.

Several examples of linear kinetics with an initial lag-phase followed by a rise in the [14]C-accumulation rate in the EOC pool are known (Saunders 1972, Nalewajko 1977, Lancelot 1979). Such lag-phase kinetics may be obtained if the specific activity of the intracellular exhange pool does not instantaneously reach isotopic equilibrium with the DIC (Lancelot 1979) or if high molecular weight products labeled late in the metabolic sequence are released along with more simple substances (Nalewajko 1977).

The chemical nature of the released products is after more than 20 years of research still a matter of controversy and is probably dependent on the physiological state of the phytoplankton (Nalewajko and Lean 1972, Hammer and Brockmann 1983) and the algal species (Hellebust 1965, Chrost and Faust 1980). Considering most of the published work on the qualitative composition of the EOC products released, the possibility for generalizations to be made is meagre, since both high molecular weight (Nalewajko and Schindler 1976, Watanabe 1980, Chrost 1981, Cole et al. 1982) and low molecular weight products (Wiebe and Smith 1977, Mague et al. 1980, Saks 1982, Søndergaard and Schierup 1982, Jørgensen et al. 1983, Carlucci et al. 1984) may be found. Moreover, most data show that both types of substances may be released simultaneously.

Storch and Saunders (1975) recognized the limitation of the "classical" model and adopted, instead, a tracer flow model (see Fogg 1966) which allowed them to take into account all gradations of release kinetics from simple organic molecules to more complex proteins labeled late in the metabolic sequence. Storch and Saunders (1975) calculated two limiting cases for the spectrum of EOC release kinetics by assuming: 1. release from a small pool at instantaneous isotopic equilibrium and 2. release from a large pool (i.e. intracellular protein pool) slow to achieve isotopic equilibrium. The "non-equilibrium" model resulted in EOC release rates 2 to 7 times higher than those obtained when the assumptions of the "classical" model were applied (Fig. 3.15). Similar results were obtained by Jensen (1985a).

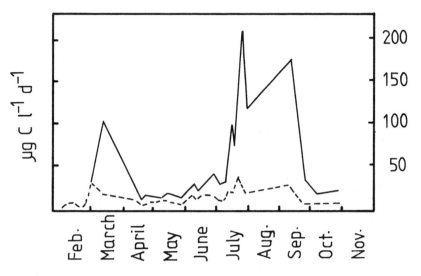

Fig. 3.15.
*Daily net release of EOC by phytoplankton in the surface of Frains Lake assuming either: 1. the release to occur from a small intracellular pool at instantaneous equilibrium with the inorganic carbon pool (i.e. the predictions of the "classical" model (------) or 2. the release occurs from the intracellular protein pool in which isotopic equilibrium is delayed for several hours (_____). Redrawn from Storch and Saunders (1975).*

The work of Storch and Saunders (1975) illustrates the size of the potential error which may be introduced, if it is assumed that the released EOC products mainly consist of simple molecules donated from a small intermediary pool in isotopic equilibrium with the inorganic carbon pool. As already pointed out, the released EOC consist rather of a continuous spectrum of organic molecules than either low or high molecular weight products. Consequently, the kinetics of $^{14}C$ -EOC release could become quite complex, and the true amount of released tracee ($^{12}C$-EOC) lies, probably, somewhere between that calculated by applying the "classical" model and the "non-equilibrium" model of Storch and Saunders.

It is obvious from the literature cited that the kinetic pattern obtained is determined by several factors, of which the more important are:

1. the nature of the intracellular exchange pool
2. the molecular complexity of the released material
3. the ambient size of the EOC pool
4. bacterial affinity for the released products, i.e. how fast the released material is removed from solution through the osmotrophic activity of the bacteria

While there is no disagreement in the literature about the fact that kinetic experiments with carefully chosen subsamplings combined with separate measurements of $^{14}$C-EOC uptake by the bacteria is a way to avoid artefacts in characterization of the release pattern (e.g. Nalewajko 1977, Lancelot 1979, Coveney 1982) there has been considerable discussion about how to interpret non-linear kinetics in quantitative as well as qualitative terms (e.g. Smith 1974, Wiebe and Smith 1977, Coveney 1982).

In the forthcoming discussion we will restrict the discussion to the first two subjects. Not in order to ignore subjects 3 and 4, which are related to the size and activity of the bacterial population, but rather because the importance of these subjects generally has been ignored and is unknown. This is so because the "classical" model, which most practitioners of the $^{14}$C-method implicitly have used in their interpretation and calculation of the $^{14}$C-data, does not recognize such an exchange pool or, if it does exist, is assumed not to constitute a significant part of total cell carbon (e.g. Wiebe and Smith 1977). Opposed to this model stands the view presented by Hobson et al. (1976), who treated the exchange pool as a part of a single well-mixed carbon pool. Consequently, an appreciable lag will be introduced before saturation with $^{14}$C occurs. Therefore, the model predicts that carbon fated for release and respiration will not be in isotopic equilibrium with the added tracer before several hours (more than 30 hours of continuous light) of incubation. Accordingly, such a model implies that conventional estimates (i.e. "classical" model) of EOC released would be strongly biased towards too low estimates. Formally, the Hobson et al. model corresponds to the limiting case of Storch and Saunders (1975). The released substances may, and perhaps always, consist of a mixture of low and high molecular weight products. The question of interest is then: To what extent are newly synthesized substrates preferentially released? Obviously, the magnitude of such a preference depends on the size of the pool from which the exchange occurs and the exchange rate between this pool and total cell carbon. Unfortunately, no unequivocal answer to this important and quite complex question exists. However, some pieces of indirect evidence do indicate that the magnitude of the exchange pool may vary considerably and may in some cases reach a very small value of total cell carbon. Wiebe and Smith (1977) concluded from kinetic tracer incorporation experiments in which $^{14}$C-accumulation in the extracellular EOC pool followed a plateau-curve, that the endogenous precursor pool was small and equilibrated rapidly, since no inflection point was found. Mague and co-workers (1980), using a more physiological approach, arrived at the same conclusion. They found the intracellular soluble precursor pool to saturate within 2 hours. However, isolation of free amino acids from the intra- and extracellular pools of EOC by ligand exchange revealed a more complex release pattern than the kinetics of $^{14}$C incorporation into the intracellular soluble pool indicated (Table 2 in Mague et al. 1980). Although the possible maximum specific activity for the released amino acids is not mentioned in their paper, the results presented by Mague et al. (1980) clearly emphasize that the extracellular pool of amino acids did not equilibrate within 5 hours despite saturation of the intracellular soluble pool after 2 hours (Table 3.2). Actually, a 16 fold difference between the maximum (serine) and minimum (tyrosine) specific activity was observed. The work of Wiebe and Smith (1977) and Mague et al. (1980) emphasizes the complexity of release kinetics, and that a small exhange pool does not necessarily imply that the released products are in isotopic equilibrium with the added tracer within several hours. However, the results of Mague et al. (1980) may indicate a relatively rapid exchange rate between the exchange pool and total cell carbon although the data are too scarce to make any definite conclusion.

Table 3.2. Release of EOC from **Ulothrix zonata** (3 incubations and 0.117 ly.min⁻¹light level) and **Melosira varians** (2 h incubation and 0.113 ly.min$^{-1}$ light level). Calc = release calculated from $^{14}C$-activity in filtrate by applying the assumptions of the "classical" model. Meas = release measured directly as DOC. Molecular weight distribution of the released EOC is also shown on basis of the $^{12}C$ and $^{14}C$ distribution. After Kaplan and Bott (1982).

| Alga | Temp. range (°C) | EOC released mg$^{12}C$ Chl-a$^{-1}$h$^{-1}$ calc. | meas. | EOC released mol. wt comp. ($\% < 10,000$) $12_C$ | $14_C$ |
|------|------------------|------|------|------|------|
| **Ulothrix** | 10.0-14.5 | 18" 7 | 132"44 | 53"2 | 91"3 |
| **Ulothrix** | 10.5-19.0 | 18"11 | 98"54 | 50"4 | 88"2 |
| **Melosira** | 6.0-11.8 | 11" 7 | 40"13 | 56"6 | 93"2 |
| Melosira | 11.2-20.2 | 5" 3 | 41"26 | 40"8 | 92"2 |

From work on the molecular size spectrum of phytoplankton extracellular products there exist several examples of a shift towards increased dominance of large molecular weight material with increasing incubation time (Steinberg 1978, Iturriaga and Zolnay 1983, Jensen 1983). Iturriaga and Zsolnay (1983) found a gradual increase in high molecular weight compounds from 13% to 18% and 41% after 1, 4 and 6 hours of incubation, respectively. Such a shift in the $^{14}C$ molecular size spectrum indicates the existence of a large exchange pool. However, no inflection point in the time course curve for $^{14}C$ incorporation into EOC was observed (Fig. 3.16). The work of Iturriaga and Zsolnay (1983) shows that linear kinetics do not necessarily imply release from a small intracellular exchange pool at instantaneous equilibrium as proposed by the predictions of the "classical" model (Wiebe and Smith 1977). Additional evidence for the inadequacy of the "classical" model to describe and hence estimate algal release of EOC comes from the convincing work of Kaplan and Bott (1982) who measured periphyton release of EOC in a piedmont stream. By measuring diel fluctuations in DOC they estimated algal release of $^{14}C$- (2-3 hours' incubation) as well as $^{12}C$-EOC, and provided a molecular weight characterization of the released material. Kaplan and Bott (1982) were able to demonstrate a 3 to 8-fold difference between estimates of release based on $^{12}C$ and $^{14}C$ measurements. Furthermore, the molecular size spectrum revealed that the $^{14}C$-labeled EOC products were dominated (more than 88%) by low molecular weight substances (i.e. less than 10,000 daltons) as opposed to approximately 50% if the patterns of $^{12}C$ distribution were used (Table 3.2).

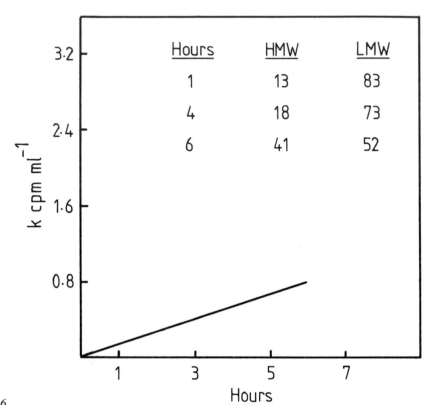

*Fig. 3.16.*
*Release and relative molecular weight distribution of EOC on high (HMW) and low (LMW) molecular weight products. Redrawn from Iturriaga and Zsolnay (1983).*

It appears now that the intracellular precursor pool either may constitute a rather insignificant part of total cell carbon as postulated by the "classical" model (Fig. 3.14), or a greater part, perhaps the whole cell carbon pool, as indicated by the work of Hobson et al. (1976) (Fig. 3.17). In the former case, released EOC is almost instantaneously in equilibrium with the added tracer (e.g. Wiebe and Smith 1977, Larsson and Hagström 1979), while in the latter case saturation is delayed for several hours (e.g. Hobson et al. 1976, Kaplan and Bott 1982, Iturriaga and Zsolnay 1983) with a severe underestimation of the $^{12}C$-EOC release (tracee) as a result (Marra et al. 1981, Smith 1982).

Though Mague et al. (1980) speculated about the functional role of the intracellular soluble pool as an exchange pool for carbon fated for release, relatively little effort has been made to interrelate the labeling kinetics at the cell level to those of coarsely defined biochemical fractions such as lipids, low molecular weight products, polysaccharides, and proteins (e.g. by the method of Morris 1980, 1981). Recently, Jensen et al. (1985) followed the $^{14}C$-labeling kinetics of two chlorophyceans, *(Chlorella vulgaris* and *Selenastrum capricornutum)* and one cyanophycean *(Microcystis aeruginosa)* species. By a chemical fractionation technqiue (Roberts et al. 1955, Morris et al. 1974) combined with simultaneous measurements of $^{14}C$-activity and standing stock of $^{12}C$, the specific activity ($^{14}C/^{12}C$) of four intracellular functional pools (low molecular weight product pool containing e.g. amino acids

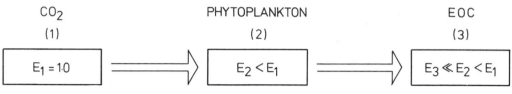

Fig. 3.17.
*Model compartments of algal carbon exchange as postulated by the "non-equilibrium" model. Since the exchange pool (2) equals total phytoplankton cell carbon, the specific activity of the EOC products ($E_3$) will be much less than that of the $CO_2$ pool ($E_1$) and equals the specific activity of the phytoplankton cell carbon pool ($E_1$).*

and sugars, lipids, polysaccharides, and proteins) and the extracellular pool of EOC was calculated. Evidence for non-equilibrium conditions in all pools was presented. The specific activity of the intracellular precursor pools first equilibrated after 24 hours of incubation under continuous light. Further, this study revealed that the low molecular weight fraction first saturated, followed by the polysaccharide, lipid, and protein fraction (Fig. 3.18). In a similar work on natural populations of phytoplankton, Jensen (1985a) concluded that the release may be seriously underestimated (up to 3 to 20 times) in short term incubations (less than 6 hours) as usually employed in primary production measurements.

One interesting feature emerged from the work of Jensen et al. (1985) on phytoplankton cultures. The time course of specific activities in the different intracellular pools and the extracellular pool of EOC suggested that the specific activity of EOC released by *S. capricornutum* and *M. aeruginosa* was in some way related to the carbon metabolism of the intracellular low molecular weight fraction. In the case of *C. vulgaris* a weak relationship between the specific activity of the EOC pool and protein fraction was suggested, although the released material mainly had a molecular weight of less than 900 daltons. The apparent relationship of the EOC and the protein pool in the *C. vulgaris* experiment was reflected in the amino acid composition of the intra- (free amino acids in the low molecular weight fraction and the protein fration after acid hydrolysis) and the extracellular fractions. However, the relationship appeared not to be causal, since the specific activity of the amino acids in the EOC pool was more like those of the intracellular low molecular weight fraction than those of the protein fraction (Fig. 3.19).

The work of Jensen et al. (1985) raises questions for the "classical" and the "non-equilibrium" model (e.g. Hobson et al. 1976) representing two opposite views on algal carbon exchange and the kinetics involved. As such the "classical" model

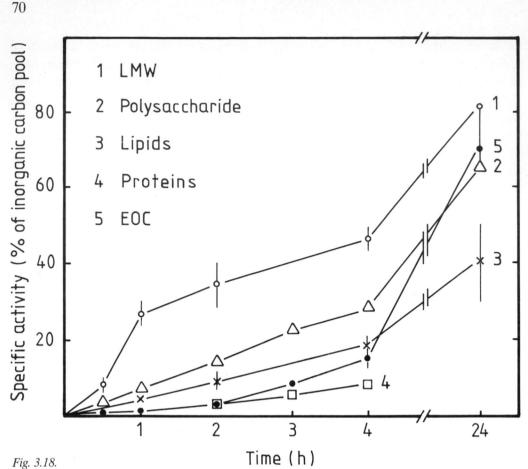

Fig. 3.18.
*Specific activity of EOC and different intracellular fractions by Chlorella vulgaris grown in batch culture at 22°C and a light intensity of 80 μE m⁻² s⁻ˢ. Redrawn from Jensen et al. (1985).*

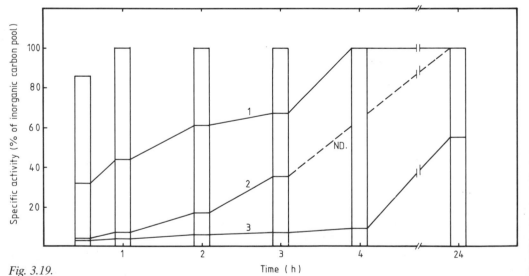

Fig. 3.19.
*Specific activity of amino acids in 1. intracellular low molecular weight fraction, 2. the EOC pool, and 3. the intracellular protein fraction. Total height of columns represent the specific activity of the metabolic active amino acid glutamic acid in the intracellular low molecular weight fraction. All values are given relative to the specific activity of the $CO_2$ pool. ND. = no data. Redrawn from Jensen et al. (1985).*

postulates the exchange pool to approach 0% of total cell carbon as opposed to 100% as postulated by the "non-equilibrium" model. It appears that neither of the models can be applied to the kinetics observed in the investigation of Jensen et al. (1985). The labeling patterns of amino acids rather suggested these to be classified according to three different kinetic classes: 1) a group of amino acids which equilibrated almost instantaneously (e.g. glutamic acid) and following the predictions of the "classical" model, 2) a group of amino acids which reached isotopic equilibrium after several hours (e.g. aspartic acid) following the predictions of the "non-equilibrium" model, and 3) a group of amino acids which followed a kinetic labeling pattern intermediate between the other two (see Fig. 3.19).

Since incorporation of $^{14}C$ into photosynthetic products is related to growth, the specific activity of the cell carbon pool can be described in terms of the specific growth rate ($\mu$) by assuming exponential growth.

At time zero of a $^{14}C$-primary production experiment the specific activity of the cell carbon pool is 0, after one doubling it will be 50% of the specific activity of the inorganic carbon pool, after two doublings it will be 75%, after three doublings it will be 87.5%, and after four doublings it will be 94%. This change in specific activity can be described by the equation (Jensen 1985a):

$$E = 1 - (2^{-n})$$

where n is the number of doublings and E is the specific activity as a decimal fraction of that in the inorganic carbon pool. This means that $E_{lim} \longrightarrow 1$ in which case the labeling is uniform. Because n can be described by the specific growth rate $\mu (= \ln 2 \times n)$ the equation given above can be written as (Welschmeyer and Lorenzen 1984):

$$\mu = \frac{-\ln (1-E)}{t}$$

where t is the duration of the experiment. From the equation follows that the specific activity of the cell carbon pool is defined by the growth rate and hence related to environmental conditions (e.g. light and nutrients) and the physiological state of the cell. Consequently, the time before equilibration occurs ("non-equilibrium" model) may be in the order of 2-3 days under conditions favouring rapid algal growth (Sommer 1981). It is hard to imagine such long equilibrium times for the EOC precursor pool, and there exist indeed no data supporting this to be the case. Nevertheless, this excersize shows that whatever the nature of the precursor pool the time for equilibration to occur will be a function of incubation time and turnover rate of the precursor pool. The smaller the standing stock of the carbon in the precursor pool is - the shorter incubation time is necessary to achieve isotopic equilibrium.

Since neither the "non-equilibrium" nor the "classical" model is sufficiently dynamic to describe and explain the time course nature of specific activity in terms of the physiological state of the algal cell carbon pool(s) a more detailed model is needed.

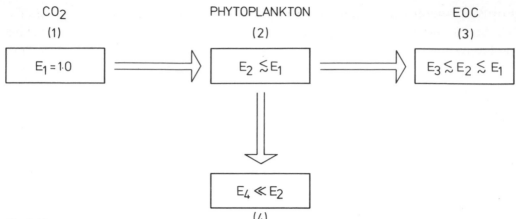

Fig. 3.20.
*Model compartments of algal carbon exchange as proposed by Smith and Platt (1984). The exchange pool (2) may comprise from 0 to 100% of total phytoplankton cell carbon. The exchange pool also donates carbon for biosynthetic reactions, i.e. growth. (4) Depending upon the size of the exchange pool and the physiological state of the cell the specific activity of the EOC ($E_3$) will vary between the limiting cases of the "classical" model (Fig. 3.14) and the "non-equilibrium" model (Fig. 3.17).*

Recently, Smith and Platt (1984) based on work with nitrogen-limited chemostat cultures of the diatom *Thalassiosira pseodonana* proposed a third model, the "4-compartment" model. Opposite to the static models which have been prevalent in theory and application of $^{14}$C-methodology in primary production studies, the "4-compartment" model is more dynamic, since it recognizes that the exchange pool may vary between 0 and 100% of total cell carbon. Moreover, the "4-compartment" model assumes that the phytoplankton carbon pool is comprised of 2 functionally different carbon pools: an exchange pool and a synthetic pool. The former exchanges carbon with the inorganic carbon pool and donates carbon both to the EOC and to the synthetic pool. The latter pool represents net production (Fig. 3.20). The "4-compartment" model of Smith and Platt (1984) operates with a unidirectional flow of carbon to the synthetic pool, i.e. carbon incorporated into e.g. proteins is immobilized and do not participate in overall cell metabolism. Continued incorporation of $^{14}$C into protein in the night period at the expense of continued polysaccharide synthesis supports this veiw (e.g. Mague et al. 1980, Jensen and Søndergaard 1985). However, direct measurements of the specific activity of the intracellular pools, including the protein fraction and the extracellular pool of EOC rather suggest some exchange of carbon from the protein pool (Jensen et al. 1985). Therefore, we propose that incorporation of carbon into the synthetic pool does not necessarily imply permanent immobilization, i.e. some exchange of carbon between the protein pool and e.g. the low molecular intracellular soluble fraction may occur. Additional evidence for this hypothesis arrives from a diel study in a coastal area in the Limfjord, Denmark. In this study both the kinetics of tracer incorporation into the intracellular pools and the kinetics of tracee incorporation into the intracellular low molecular weight and protein fraction were followed. The incorporation of $^{14}$C into proteins followed the predicted pattern for an actively growing phytoplankton population (Mague et al. 1980) with a continued accumulation of $^{14}C$ into the proteins during night at the expense of polysaccharides. Taking the kinetics of tracee into consideration revealed a more complex kinetic pattern (Fig. 3.21). However, overall specific activity of algal POC declined from sunset to just before sunrise, while the specific activity of the intracellular ethanol soluble (low molecular weight) fraction continued to increase several hours after sunset, followed by a slight decrease in the next hours before sunrise. The protein pool showed a continued increase in specific activity throughout the diel period (Fig. 3.21). Although the interpretation of the kinetics of specific activity in natural populations may be complicated by the presence of detrital organic matter, the results of Fig. 3.21 do, nevertheless, indicate that the flux of tracer and tracee was partially uncoupled and/or that "new" substrates were respired to a greater extent than "old" substrates.

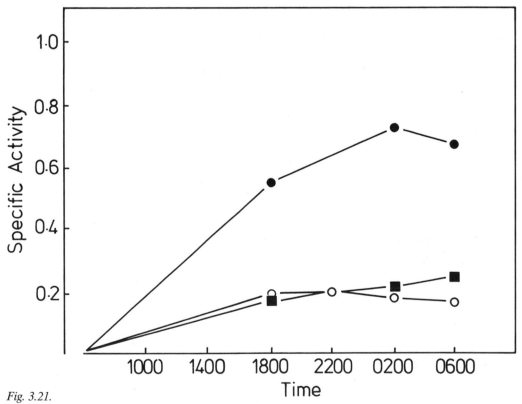

Fig. 3.21.
*Specific activity of particulate organic carbon (○), low molecular weight intracellular soluble fraction (■), and the protein fraction (●) of phytoplankton in the Limfjord, Denmark, during a diel study 24-25 August 1984. The specific activity is expressed as a fraction of the specific activity in the inorganic carbon pool.*

It appears with the present state of knowledge that estimates of algal release of EOC are at best underestimates. Further, the kinetics of release seems to be much more complicated than assumed hitherto. Release of low molecular weight products ($^{14}$C) does not necessarily mean that the release occurs from a small intermediary pool at instantaneous isotopic equilibrium, since exchange between the intracellular low molecular weight fraction and other intracellular pools as the protein pool may occur. In view of the important ecological implications of a correct quantification of microalgal release of EOC, not only for the determination of primary productivity (3.3.1.), but also as a carbon source for pelagic bacteria, it is important in future investigations to look more carefully at the release kinetics. Combination of the theoretical approach of Smith and Platt (1984) with direct measurements of specific activity of coarsely defined biochemical fractions (Jensen et al. 1985) may prove to be successful in getting more precise answers to the question: What is the exact magnitude of EOC release from phytoplankton?

### 3.3.5. Alga-bacterial interactions: Applications of multi-substrate kinetic theory in the analysis of heterotrophic activity.

Multi-substrate kinetic analysis was originally developed to analyze heterotrophic uptake of substrate mixtures having unknown chemical composition, particularly algal extracellular organic carbon (Bell 1980). The technique has proven useful in evaluating the "phycosphere" concept, i.e. the potential stimulation of bacterial activity by algal blooms in natural waters (Bell and Mitchell 1972, Cole 1982). Bell (1980) presented a method of preparing $^{14}$C-labeled EOC ($^{14}$C-EOC) for multi-substrate kinetic studies of native and laboratory populations of heterotrophic bacteria. The specific activity of the products can be increased (Jensen 1985b) to enhance the sensitivity of the assays or decrease experimental incubation times.

### 3.3.5.1. Multi-substrate kinetic theory

"Pure-" and mixed-substrate kinetic experiments are misleadingly similar in experimental design and data analysis. Both approaches measure heterotrophic activity as a function of the concentration of labeled substrate added to a water sample or bacterial suspension. However, with a mixture such as $^{14}$C-EOC the investigator does now know n, the number of different substrates utilized, nor $p_i$, the relative proportion of each such substrate in the mixture.

One can easily calculate the fraction (f') of total added substrate (A') removed as

$$f' = \frac{Q}{A'},$$

where Q is the measured radioactivity and refers to the sum of *all* compounds taken up. The actual fraction ($f_i$) of each *individual* substrate removed is in fact

$$f_i = \frac{Q_i}{p_i \cdot A'}$$

where $Q_i$ is the radioactivity removed for each substrate. The two fundamental "pure-substrate" kinetic equations must be expanded to reflect the additive effect of different compounds on $v_{A'}$, the total rate of substrate uptake at each substrate A'. Water samples are assumed to contain naturally occurring (unlabeled) counterparts to the labeled substrates at concentrations $S_{n_i}$. The first equation makes no assumption as to the mechanism of substrate uptake, thus:

$$V_{A'} = \sum_{i=1}^{n} \frac{f_i \cdot (S_{n_i} + p_i \cdot A')}{t}$$

where t is the experimental incubation time. The second assumes that uptake proceeds by Michaelis-Menten kinetics, with the usual parameters ($K_{t_i}$ and $V_{max_i}$) for each of the substrates:

$$V_{A'} = \sum_{i=1}^{n} V_{max_i} \cdot \frac{p_i \cdot (A' + S_{n_i})}{K_{t_i} + p_i \cdot (A' + S_{n_i})}$$

Although the rates $v_A$, of substrate uptake predicted by equations (3) and (4) are equal, the equations cannot easily be combined and simplified to produce the equivalent of the modified Lineweaver-Burke "pure-substrate" equation (Wright and Hobbie 1965). Bell (1980) derived limiting cases in which such simplification is possible in order to explore the general predictions of multi-substrate kinetic theory. As an example, assume all substrate-specific variables are equal for those compounds used. There will be n terms in each equation, but the substrate specific subscripts i can be dropped; further, f' = n·p·f. The resulting modified Lineweaver-Burke equation is:

$$\frac{t}{f'} = \left( \frac{1}{n \cdot V_{max}} \right) \cdot A' + \frac{(K_t + S_n)}{n \cdot p \cdot V_{max}}$$

This limiting example predicts a linear relationship between t/f' and A'; these are the multi-substrate analogs of t/f and A in single-substrate experiments (Wright and Hobbie 1965). Computer analysis (Bell, unpublished) indicates that linearity itself (but not the precision of the kinetic parameter estimates) is quite robust even if the constraints of various limiting cases are not closely approximated, provided the assumption of Michaelis-Menten kinetics remains valid.

Although linear Lineweaver-Burke plots are predicted in both "pure-" and multi-substrate analyses, the resulting estimates of kinetic parameters are quantitatively different. Fig. 3.22 compares these estimates. In short, the inverse of the slope of the multi-substrate line will overestimate $V_{max_i}$ for any given compound in the mixture, while the absolute value of the multi-substrate X-intercept also overestimates the value that would be obtained in an experiment with a single, "pure" substrate.

### 3.3.5.2. Experimental design and analysis

Fig. 3.23 provides a flow chart for multi-substrate kinetic experiments with [14]C-EOC. As opposed to tracer experiments (Hobbie and Rublee 1977), the range of 20-200 µg C liter[-1] added as [14]C-EOC represents a significant elevation in substrate concentrations relative to those of naturally available (unlabeled) compounds (Bell 1984). Because different amounts of [14]C-EOC are added, separate blanks must be employed at each A'.

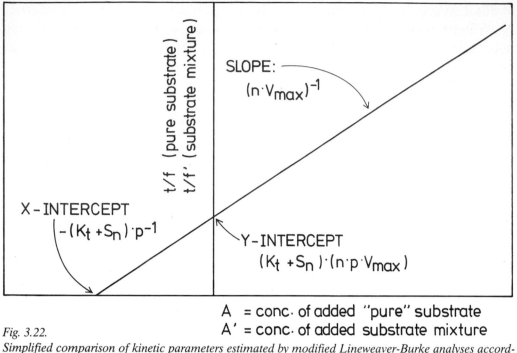

*Fig. 3.22.*

A = conc. of added "pure" substrate
A' = conc. of added substrate mixture

*Simplified comparison of kinetic parameters estimated by modified Lineweaver-Burke analyses accord-
ing to "pure-" and mixed-substrate theories. The limiting mixed-substrate case depicted assumes the relative proportions of each substrate in the mixture of n utilized compounds are equal, i.e. $p_1 = p_2 = ...$
$= p$. This mixed-substrate case reduces to the "pure-substrate" condition when $p = 1.0$.*

Smith et al.(1984) have critisized "pure-" substrate kinetic studies in part for shortcomings in experimental design. They correctly point out that modification of the
method to minimize increases in natural substrate concentrations - i.e. a tracer
approach with substrates of high specific activity (Azam and Holm-Hansen 1973) -
no longer requires the assumption of Michaelis-Menten kinetics to explain linear
Lineweaver-Burke plots. However, a wealth of non-tracer data support the
Michaelis-Menten model for substrate uptake (Hobbie and Rublee 1977, Bell
1980). Preparation of $^{14}C$-EOC suitable for kinetic studies (Jensen 1985b) reduces
the possibility of influencing bacterial activity through increased A' and prolonged
incubation times (Wright 1973). These purported advantages remain to be evau-
lated for multi-substrate studies.

It is no longer acceptable to attempt accurate estimates of kinetic parameters
directly from the Lineweaver-Burke transformation of the Michaelis-Menten
equation. Parameters can be estimated with greater precision by computerized
non-linear curve fit to untransformed data (Li 1983, Bell 1984). However, visual
inspection of experimental results is facilitated by comparing the straight lines (or
deviations therefrom) of modified Lineweaver-Burke plots (Smith et al. 1984).

### 3.3.5.3. Interpretation of multi-substrate kinetic data

Despite its quantitative basis, the multi-substrate kinetic theory yields information which is largely qualitative in nature. Multi-substrate estimates of $V_{max}$,

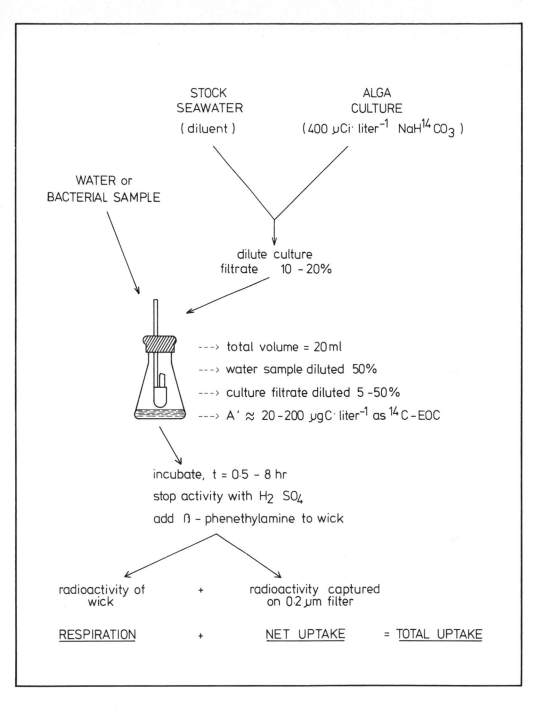

Fig. 3.23.

*Flow chart for mixed-substrate experiment with algal* [14]*C-EOC. Samples are normally run in triplicate, with a zero-time blank, at each A'. Dilution of water sample by 50% can reduce kinetic parameters $V_{max}$ and $S_n$ (Bell, 1980), but has minimal effect on positions of resulting net uptake and respiration lines relative to one another.*

$(K_t + S_n)$, and $T_t$ are at best only crude approximations of the parameters for *individual* substrates (Fig. 3.22), and provide only minimal insight into the true flux of algal EOC into bacterial food chains. While a repetitive series of experiments with the same labeled substrate mixture may permit estimation of fluxes from diurnal or daily changes in kinetic parameters (Bell and Sakshaug 1980), such laborious studies have been superseded by improvements in differential filtration (Larsson and Hagström 1982) and other methods for estimating *in situ* heterotrophic activity (Lancelot 1984; Ducklow and Hill 1985). Perhaps the most useful comparison involves changes in $V_{max}$ which, if other potential sources of variation are controlled, is sensitive to changes in heterotrophic population size (Bell 1980, Bell and Sakshaug 1980).

Multi-substrate kinetic studies can provide valuable descriptive insights into the nature of bacterial activity under natural conditions. Straight lines of positive slope on modified Lineweaver-Burke plots are evidence that substrate utilization is described by Michealis-Menten kinetics and is therefore enzyme-mediated (Hobbie and Rublee, 1977). A common deviation is a line approaching zero slope, evidence of low substrate affinity (high $K_t$); in the limiting case the rate of substrate utilization is directly proportional to A', i.e., a diffusion-limited process.

Lineweaver-Burke lines of positive slope occasionally break toward the horizontal at higher A'. Wright and Hobbie (1965) originally attributed this behavior for uptake of glucose in lake water samples to assimilation by algae. This almost certainly is not a general explanation. Experiments which employ a very wide A' range can produce lines of progressively decreasing slope as uptake systems having different substrate affinities are recruited (Azam and Hodson 1981). Line breaks toward the horizontal are common when monitoring bacterial utilization of $^{14}$C-EOC in alga-free systems (Bell 1983, 1984), presumably for the same reason. This necessitates careful statistical testing for data agreement with kinetic models. It is usually sufficient to test Lineweaver-Burke transformed data for linearity and to drop delinquent data at higher A' if they are encountered (Bell 1980, 1983). Improper use of 0-time blanks can also produce this line-break artifact (Smith et al. 1984).

Simultaneous measurement of released $CO_2$ (Crawford et al. 1973) permits generation of 3 sets of kinetic data from a single experiment. *Respiration* and *net uptake* are potentially independent kinetic processes; *total* uptake is their sum. When the method is applied in single-substrate studies, the resulting Lineweaver-Burke lines (if such are obtained) invariably intersect at the same X-intercept (Hobbie and Rublee 1977, Bell 1984), a physiological phenomenon worth exploring. Since the X-intercept, $(K_t + S_n)$, represents the sum of factors outside the cell cytoplasm ($K_t$ is the Michaelis constant for the substrate transport system; $S_n$ is the natural concentration of substrate in the water sample), it follows that net uptake of label is governed by the transport system - i.e. net uptake is transport-limited. When the respiration shares the same X-intercept, respiration of the substrate in question must also be transport-limited. This leads to the interesting generalization that the rate of an external substrate's metabolism is limited by its rate of uptake. Adaptations which increase a bacterial cell's ability to assimilate compounds would facilitate growth and survival at low substrate concentrations; these include increased substrate affinity (lower $K_t$) and increased surface to volume ratio (increased $V_{max}$ (unit vol.)$^{-1}$) (Morita 1982).

When multiple substrates are employed the potential independence of net uptake and respiration is reintroduced, for each process could be limited by the different kinetic parameters of different compounds. The resulting Lineweaver-Burke lines can thus exhibit significantly different X-intercepts; such data have been reported (Bell 1983, 1984). However, it is not at all uncommon to obtain multi-substrate lines which share the same X-intercept (Bell and Sakshaug 1980), implying once again that net uptake and respiration are transport-limited - in this case, *by* the same spectrum of compounds in the same relative proportions in both the external (transport) and internal (respiratory) pools (Bell 1983). Coincidental X-intercepts for net uptake and respiration are specifically related to the choice of substrates employed in an experiment and the recent history of the bacterial population under study. These considerations will be discussed shortly.

One final observation, not dependent upon kinetic theory, is that % substrate respiration obtained in mixed-substrate studies is typically 20-40% of total uptake (Bell and Sakshaug 1980, Bell 1984, Jensen 1985b). This approaches the maximum growth efficiency for bacteria maintained on complex laboratory media (Payne and Wiebe 1978) and suggests that the spectrum of compounds present in most EOC pools will support balanced cellular metabolism. Further, values of 20-40% can be obtained in experiments lasting as short as 30 min (Bell 1984), implying rapid approach of the respiratory pool to isotopic equilibrium.

### 3.3.5.4. Applications of multi-substrate kinetics

Table 3.3 summarizes field and laboratory kinetic studies which have employed natural or artificial mixtures of labeled substrates. Not all involved correct interpretation of the multi-substrate kinetic parameters, but even these are useful in demonstrating how similar the approach is to the widely-employed "pure-substrate" method and the reliability with which linearity can be expected.

An extensive multi-substrate kinetic analysis of natural heterotrophic activity was performed during a bloom of the marine alga, *Skeletonema costatum,* in Norway's Trondheimsfjord (Bell and Sakshaug 1980). Linear Lineweaver-Burke plots for uptake and respiration of *S. costatum* $^{14}$C-EOC were obtained in each of 10 experiments over the 6-week bloom period. Further, the X-intercepts for net uptake and respiration were statistically coincidental (Fig. 3.24). In contrast, heterotrophic utilization of $^{14}$C-EOC prepared from cultures of *Chaetoceros affinis* - an alga which also blooms in the fjord - produced net uptake and respiration lines inconsistent with kinetic theory and apparently reflecting independent physiological processes.

The estimated total flux of EOC during the Trondheimsfjord bloom, 0.08 µg C liter$^{-1}$ h$^{-1}$, was less than 10% of the maximum observed multi-substrate $V_{max}$ for total uptake. This is consistent with findings from single-substrate studies indicat-

Table 3.3. Survey of published studies which include mixed-substrate experiments.

| Reference & Location | Substrate (s) | Results |
|---|---|---|
| Seki et al. (1972)<br>  Open Pacific Ocean | **Chlorella** protein<br>hydrolysate<br>(commercially extracted) | Analyzed incorrectly as single-substrate; net uptake only. Linear kinetics; avg. $V_{max}$(92 mg $m^{-2}$ $h^{-1}$) 4-10 times greater than values obtained with individual amino acids. |
| Seki et al. (1974)<br>  Philippine Sea | **Chlorella** protein<br>hydrolysate<br>(commercially extracted) | Analyzed incorrectly as single-substrate; net uptake only. Linear kinetics; $V_{max}$(90 mg $m^{-2}$ $h^{-1}$) 10-20 times greater values obtained with individual amino acids. |
| Maita et al. (1974) | **Chlorella** protein<br>hydrolysate | Kinetic plots presented as if single-substrate; parameters not reported. |
| Bell (1980)<br>  Laboratory study | [14]C-EOC from:<br>  **S. costatum**<br>  **C. affinis** | Test of multi-substrate kinetic theory; respiration included. Test bacterium maintained with **S. costatum** produced horizontal kinetic lines for utilization of **C. affinis** [14]C-EOC. |
| Bell and Sakshaug (1980)<br>  Trondheimsfjord | [14]C-EOC from:<br>  **S. costatum** | Linear kinetics and common X-intercepts for net uptake and respiration only obtained using **S. S. costatum** [14]C-EOC. |
| Schleyer (1980)<br>  Tropical sub-tidal reef | [14]C-cell extract<br>from **Chlorella** | Analyzed incorrectly as single-substrate. Linear kinetics for net uptake; $V_{max}$ (48 mg C liter$^{-1}$ $h^{-1}$) up to 100 times greater than corresponding estimates for glucose uptake. |
| Bell (1983)<br>  Laboratory study | [14]C-EOC from:<br>  **S. costatum**<br>  **D. tertiolecta**<br>  **T. pseudonana** | Bacteria physiologically adapted to host alga through selection in continuous culture developed common X-intercepts for net uptake and respiration lines. |
| Bell (1984)<br>  Eel Pond<br>  Vineyard Sound | [14]C-EOC from:<br>  **S. costatum**<br>  **D. tertiolecta**<br>  **T. pseudonana** | Linear kinetics for natural bacterial populations with no recent previous exposure to these algal species; net uptake and respiration lines with different X-intercepts. |
| Jensen (1985b)<br>  2 Danish lakes | [14]C-EOC from:<br>  **M. aeruginosa** | Bacteria adaptation to **M. aeroginosa** EOC developed over course of summer bloom. |

ing that the substrate transport systems of heterotrophic populations are undersaturated at substrate concentrations typically encountered in natural waters (Hobbie and Rublee 1977). Therefore, the bacteria were able to track *in situ* increases in EOC simply by increased rates of subtrate utilization mediated by enhanced uptake via unsaturated transport systems. Less than a 4-fold incrase in $V_{max}$ was sufficient to prevent significant accumulation of *S. costatum* EOC during the bloom (Bell and Sakshaug 1980).

The mathematical underpinnings of multi-substrate theory do not fully explain the kinetic patterns observed during the course of the Trondheimsfjord bloom. In particular, why should the net uptake and respiration lines for *S. costatum*

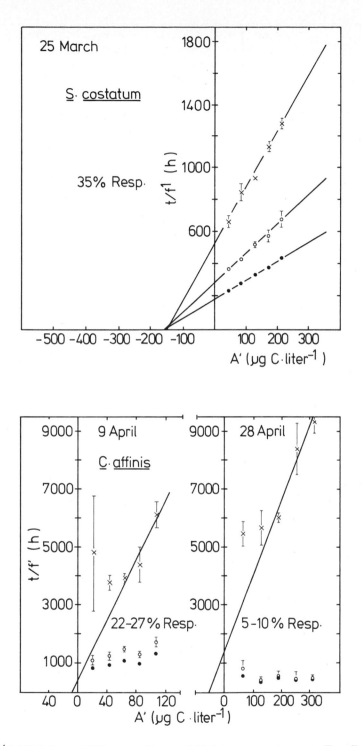

*Fig. 3.24.*
*Utilization of [14]C-EOC during 1977 spring bloom of Skeletonema costatum in Trondheimsfjord, Nor-way (data from Bell and Sakshaug, 1980). Upper: Representative kinetic plot of data obtained using S. costatum [14]C-EOC as added substrate; experiment was performed approximately 1 week after beginning of bloom. Lower: Results of two experiments with Chaetoceros affinis [14]C-EOC as added substrate, performed during mid- and late-bloom. Data for net uptake (o) and respiration (×) plotted as mean of triplicate determinations "standard error; data for total uptake (●) are mean only. % respiration is range across all A'.*

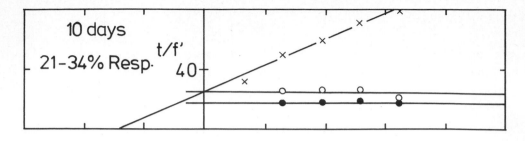

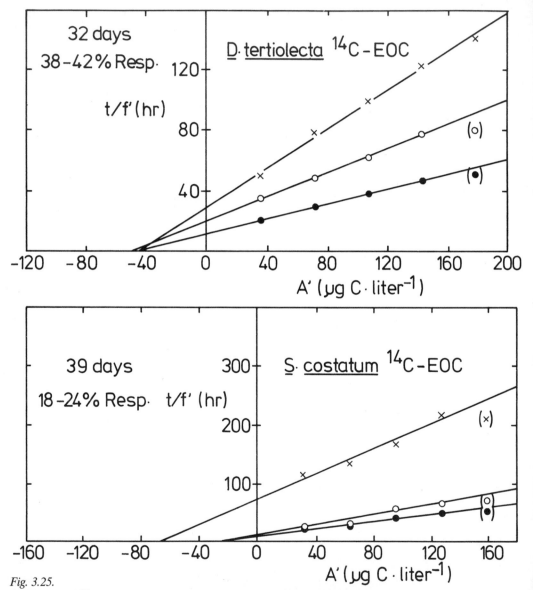

Fig. 3.25.

*Utilization of $^{14}C$-EOC by bacterial population which developed from a 1-ml inoculum of natural seawater into continuous culture of the alga, Dunaliella tertiolecta (data from Bell 1980). Upper: Kinetic patterns obtained with D. tertiolecta $^{14}C$-EOC, showing how X-intercepts for net uptake and respiration lines converge on common lower value as bacteria respond to the alga-mediated environment. Lower: Utilization of S. costatum $^{14}C$-EOC after prolonged bacteria exposure to D. tertiolecta in continuous culture. Age of association in days since seawater inoculation is indicated; symbols as in Fig.3.24.*

[14]C-EOC utilization consistently exhibit statistically identical X-intercepts while those for utilization labeled EOC from other algal species do not? This phenomenon was subsequently investigated in detail using continuous culture to select for development of algal-bacterial associations under the low substrate concentration regimes characteristic of natural waters.

Bell (1983) followed the heterotrophic activity which developed following inoculation of a small volume of natural water into a formerly axenic algal continuous culture, using [14]C-EOC prepared from the algal host species as substrate in kinetic analysis (Fig. 3.25). The heterotrophic activity which developed within 2 weeks (and in control cultures without an algal host) was characterized by straight lines on modified Lineweaver-Burke plots, but the X-intercepts for net uptake and respiration were statistically distinct. Over the next 2-3 weeks, however, the kinetic pattern gradually developed until the net uptake and respiration lines decreased to intersect at the same X-intercept. Kinetic experiments performed after this time using [14]C-EOC from an algal species other than the culture host typically resulted in straight net uptake and respiration lines with statistically distinct X-intercepts.

Bell (1983) concluded that prolonged exposure of a natural population of heterotrophic bacteria to a given pool of algal EOC selects for a microbial population physiologically adapted for growth on compounds available in that pool. Data dramatically consistent with this hypothesis were obtained in a study of heterotrophic activity in Lake Hylke, Denmark, by Jensen (1985b). Primary production in this eutrophic lake is dominated by a prolonged bloom of *Microcystis aeruginosa* which begins in early April and persists through late summer. Fig. 3.26 shows that linear kinetic patterns for bacterial utilization of *M. aeruginosa* [14]C-EOC were obtained throughout the course of the study. Prior to the bloom the lines for net uptake and respiration exhibit relatively high and statistically distinct X-intercepts. As the bloom progresses, the X-intercepts decrease and become coincidental. In early fall, when the lake is dominated by another cyanobacteria species *(M. wesenbergii)*, the X-intercepts for uptake and respiration of *M. aeruginosa* [14]C-EOC again become statistically distinct and appear to be increasing.

Physiological adaptation to a given algal EOC pool suggests that both net uptake and respiration become transport-limited processes, or, stated more boldly, bacterial growth becomes limited by the rate of uptake of available substrates. When coupled with the fact that substrate transport systems are operating well below $V_{max}$, physiological adaptation leads to two important ecological consequences which have been observed in nature. First, changes in the rate of algal EOC release will be closely tracked by metabolic changes in associated heterotrophic activity (Keller et al. 1982). Second, the concentration of EOC will not significantly increase during the course of an algal bloom (Riemann et al. 1982, Søndergaard and Schierup 1982) because such increases would be accompanied by concomitant enhancement of heterotrophic uptake. Estimates suggest bacteria utilize up to 80% of the EOC released during algal blooms in the North Sea (Ittekot 1982); 25% or more may be removed per hour during periods of enhanced algal activity (Lancelot 1984).

84

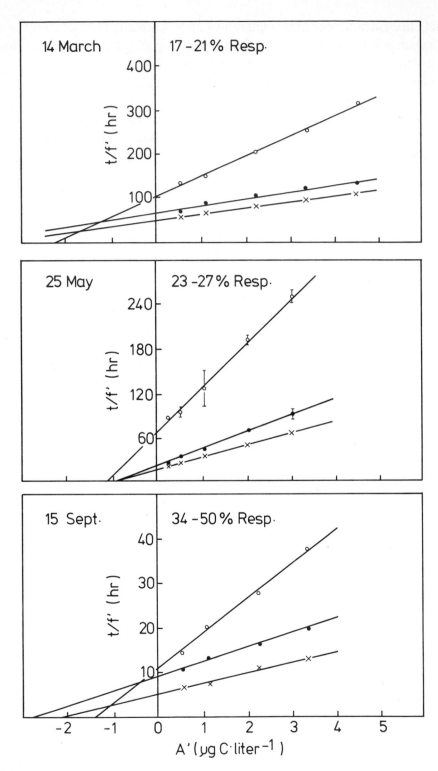

Fig. 3.26.
*Utilization of Microcystis aeruginosa $^{14}$C-EOC by native bacterial populations present in eutrophic Lake Hylke, Denmark, during 1983 (data courtesy of L. Møller Jensen). M. aeruginosa dominated the algal population from approximately 5 April to September, after which time M. wesenbergii became the most abundant primary producer. Symbols as in Fig. 3.24.*

The fact that natural bacterial populations exhibit linear multi-substrate kinetics when tested with $^{14}$C-EOC prepared from a variety of algal sources strongly suggests that possession of enzyme-mediated transport systems with high substrate affinity is an adaptation for existence under the low-nutrient conditions characteristic of most natural waters (Bell 1984). Equality of the X-intercepts for net uptake and respiration becomes superimposed upon this fundamental adaptation as metabolism becomes adjusted - "balanced", if you will - to the rate of assimilation of available substrates.

### 3.3.5.5. Suggestions for future research

The reality of the phycosphere as a zone of algal influence on the activity of surrounding bacteria is now generally accepted (Cole 1982). The concept is helpful in suggesting the appropriate choices for substrate mixtures in kinetic studies of natural populations. Multi-substrate analysis would be an excellent qualitative supplement to quantitative studies of EOC fluxes between algae and bacteria in nature. In estuarine and freshwater environments with a high background of dissolved organic C derived from terrestrial sources, multi-substrate kinetic analyses may be useful in evaluating the relative importance of exogenous and endogenous substrates in supporting heterotrophic activity. It would also be interesting to use multi-substrate kinetics to investigate the effect of pollutants and putative toxins on bacterial utilization of algal EOC using multi-substrate kinetics. One can hypothesize that toxins could disturb both the processes of substrate transport and physiological adaptation, the latter leading to chronic disruption of *in situ* nutrient regeneration.

Of more fundamental interest is the nature of bacterial adaptation to low-nutrient environments. Although the existence of "oligotrophic" bacteria has been known for some time (Poindexter 1981), these organisms are difficult to study under the relatively high-nutrient conditions characteristic of laboratory batch cultures. Consequently, there have been relatively few studies of transport systems on native bacteria isolated from nature and maintained under environmentally realistic substrate concentrations (Morita 1982). Continuous cultures permit selection of bacteria adapted to a specific algal phycosphere, a process which can be followed using multi-substrate kinetic analysis (Bell 1984). Substrate transport systems and the physiological adjustments made to produce balanced growth on the available algal EOC are amenable to study using bacteria sampled or isolated from such systems. Particularly intriguing is the apparent coupling of substrate uptake to respiration as suggested by the rapid production of $^{14}$CO$_2$ in kinetic experiments lasting less than 0.5 hr (Bell 1984). It is not yet known whether physiological adaptation results from metabolic adjustment or from selective stimulation of pre-adapted bacteria in an alga-mediated environment (Bell et al. 1974). The processes are not mutually exclusive.

It is unlikely that the multi-substrate kinetic approach will ever see the wide field application formerly enjoyed by "pure-substrate" experiments. Nevertheless, data obtained using this method will continue to serve as a constant reminder to aquatic scientists and microbial ecologists alike that there is much yet to be learned from experiments which recognize and attempt to evaluate the selective factors that influence bacterial activity under natural conditions.

Contribution No. 1673 from Center for Environmental and Estuarine studies, Univ. of Maryland.

### 3.3.6. Evaluation of methods and data

Estimates of true EOC release are biased for several reasons. The most fundamental error is related to the normally used assumption of instantaneous isotopic equilibrium in released products; an assumption which is probably erroneous in most studies (section 3.3.4). Carefully performed time-course studies can give some but not all answers to the dimension of the error. Until this problem is solved, release measured with the $^{14}C$-method should be named *apparent* release as oppoosed to true release. Apparent release then includes EOC in solution $(EOC_n)$, EOC incorporated into bacteria $(B_n)$, and EOC respired by bactería $(B_r)$. To emphasize that it consisted of substances not removed by heterotrophs during incubation, Nalewajko et al. (1976) used the term net release $(EOC_n)$ for the dissolved fraction. It can be argued that the term apparent release only should include $EOC_n$ and $B_n$, as $B_r$ only can be estimated using an antibiotic method (section 3.3.3) or by measuring the growth yield on substrates like EOC.

In most traditional *in situ* studies only $EOC_n$ is measured due to retainment of all particles on a single filter. Most published values of percent release are calculated from $EOC_n$ values. The underestimation of apparent release depends on the amount of bacterial EOC uptake and the respiration. The differential filtration technique has the potential to give $B_n$ and $EOC_n$, but since $B_r = B_n$, for most products resembling EOC, $B_r$ can be estimated.

To illustrate the marked difference between release in studies measuring $EOC_n$ and studies using size fractionation, the ratio

$$\frac{B_n}{B_n + EOC_n}$$

can be used. The ratio is an expression of EOC transport to bacteria (see also section 3.3.2.1.) without assuming any specific respiration value. The ratios calculated for several fractionation studies are presented in Table 3.4.

Transport can be dependent on incubation time. In eutrophic Lake Bysjön Coveney (1982) found one case out of three where the transport increased from 25 to 60% after ½ and 5 h incubation, respectively. Small changes during 24 h incubation were found by Cole et al. (1982), but the transport increased when bottles were incubated for 48, 72 and 84 hours. Similarly, Søndergaard et al. (1985) observed only minor variations in diel time-course experiments.

Vertical changes of EOC transport were not found by Coveney (1982) and Søndergaard et al. (unpubl.), although they might be expected to occur whenever stratification within the photic zone exists. Seasonal changes have been found and can also be seen in the data from Frederiksborg Slotssø presented in Table 3.4. An

Table 3.4. Bacterial net assimilation of EOC as a fraction of $EOC_n + B_n$ (transport percentage). N = number of samples.

| Locality | Date | $\dfrac{B_n \times 100}{B_n + EOC_n}$ (%) | N | Trophic status[6] | Reference |
|---|---|---|---|---|---|
| Mirror Lake | Seasonal | 40[1] | - | oligo | Cole et al. (1982) |
| Lake Almind | May | 36[3] | 37 | oligo/meso | Søndergaard (1986) |
| Lake Constance | Jul-Oct | 9-36[2] | 3 | meso | Kato and Stabel (1984) |
| Lake Erken | Apr-May | 4[3] | 13 | meso | Bell and Kuparinen (1984) |
| Lake Esrom | May | 74[3] | 7 | eu | Søndergaard et al. (1985) |
| Lake Mossø | Mar-Apr | 25-70[5] | 12 | eu | Riemann et al. (1982) |
| Lake Hylke | May | 83[4] | 56 | eu | Søndergaard (1986) |
| Lake Bysjön | Apr-Sep | 40-70[5] | 16 | eu | Coveney (1982) |
| Frederiksborg Slotssø | Sep | 79[3] | 9 | eu/hyper | Søndergaard et al. (1985) |
| Frederiksborg Slotssø | Jun | 47[4] | 112 | eu/hyper | Søndergaard (unpubl.) |

1. From Table 3, 24 h incubation in Cole et al. (1982)
2. From Fig. 2 in Kato and Stabel (1984)
3. Diel experiments
4. Day-time experiments
5. 4-5 hour incubation
6. Based on the general production classification by Wetzel (1983)

indication of the time-scale for expected changes can be depicted from the EOC transport during a diatom bloom in Lake Mossø (Fig. 3.27). During the initial exponential growth of the phytoplankton, the transport varied between 40 and 70%. In late April during the diatom senescence and the appearance of a mixed phytoplankton community (Riemann et al. 1982), the transport decreased to about 20%.

There seems to be a general trend towards higher transport values in eutrophic than in oligotrophic lakes (Table 3.4). This may be related to the generally higher level of bacterial activity in eutrophic lakes (Chapter 4), but could also be related to limitation of bacterial growth due to low concentrations of inorganic nutrients. The extremely low transport of about 4% in Lake Erken (Table 3.4) was found during a spring bloom of *Stephanodiscus hanzchii* and was explained by low temperature and a suggested dominance of refractory, high molecular weight EOC products (Bell and Kuparinen 1984). The latter aspect, however, was not investigated. As dominance of low molecular weight EOC products by *S. hantzchii* has been reported (Watt 1969, Søndergaard and Schierup 1982a, Jensen 1985b), this seems not to be a valid suggestion.

The conclusion is: In most eutrophic lakes the EOC transport is higher than 50%, so $EOC_n$ values should at least be multiplied by 2 to give $EOC_n + B_n$. The underestimation of apparent release in traditional studies is even more severe, taking $B_r$ into account. Setting (conservative) bacterial respiration at 30-40% of the total carbon uptake would imply a transport rate of EOC from 65 to 90% in eutrophic lakes. The magnitude of $B_r$ is shortly discussed in section 3.3.6.2..

The use of differential filtration allows calculation of $EOC_n + B_n$. An antibiotic method should - in theory - make it possible to calculate $B_r$ (see section 3.3.3). Thus, values of apparent release should be of comparable magnitude with these two independent methods. So far, the only direct comparison between the methods has been published by Jensen and Søndergaard (1985). Diel time-course studies in four Danish lakes and one coastal area revealed reasonable agreement as long as phytoplankton photosyntheis was not affected by the antibiotics. Apparent release and transport values of similar magnitude give credibility to both methods. However, differential filtration probably should be preferred, since the basic approach is simple and does not involve estimation of bacterial inhibition and judgment of algal susceptibility for applied antibiotics.

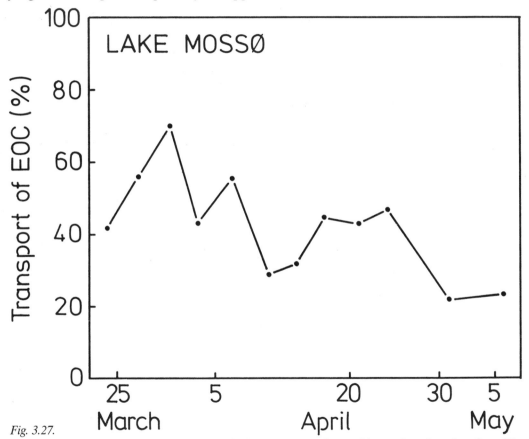

Fig. 3.27.
*Transport of EOC in Lake Mossø, Denmark, during a spring diatom bloom. Based on data from Riemann et al. 1982.*

An evaluation of ecological methods should not only be made from a comparison among different methods, but should also include a carbon mass balance. Measured EOC production must over time be balanced by loss or accumulation. As any EOC accumulation has to be measured against a very high background level of dissolved organic carbon, an ecological evaluation in practical work can only be carried out by measuring the loss of EOC from the dissolved pool of organics. Since this loss is equivalent to bacterial uptake, a first step in an evaluation sequence can be made by a comparison of bacterial EOC assimilation ($B_n$) with independent measurements of bacterial net production, $B_p$. $B_n$ should never exceed $B_p$. $B_n$ is a measure of instant bacterial activity, so $B_p$ should also be an instant productivity measurement. The thymidine method seems to represent such a requirement and is today the most used method (Chapter 4).

In order to avoid an overestimation of EOC as a bacterial carbon source a direct comparison of $B_n$ with $B_p$ should only be done on an area basis using the mixing depth of the waterbody as the lower vertical border. If the whole water body is overturning freely, the mean depth of the lake should be used. Generally, EOC is released as a function of the photosynthetic rate (see section 3.3.7), which generally decreases with increasing depth. In a mixed water column an algal cell has relatively high release at one time and low at another time. The bacteria are thus exposed to a theoretical mean release rate, depending on turbulence, mixing depth and the vertical extension of the photic zone. Enclosing a water sample in a bottle, fixed at the surface for hours to measure EOC release and uptake with the $^{14}$C-method, might yield high $B_n$ values compared with a $B_p$ estimate in a water-sample from the freely mixing watermasses and enclosed for say 30 minutes.

An example can illustrate the point. Carbon fixation, extracellular release, and

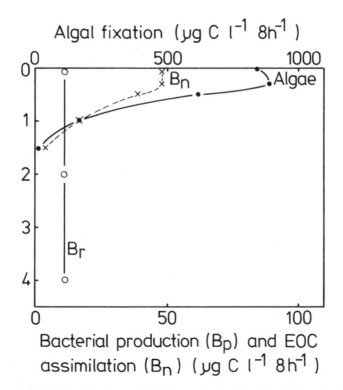

Fig. 3.28.
*Depth distribution of primary production, bacterial EOC assimilation ($B_n$) and bacterial production ($B_p$) measured with $^3$H-thymidine incorporation. Lake Ørn, Denmark. Based on data from Søndergaard et al. (1985).*

bacterial EOC uptake in eutrophic Lake Ørn took place in the upper 2 m (Fig. 3.28). At this time Lake Ørn was freely mixing. Total bacterial production for the whole lake can be described to take place in a water column representing the mean depth of the lake, 4.5 m. Due to the fixed position of the primary production bottles, the bottle in the surface had a $B_n$ of 6 µg C $l^{-1}$ $h^{-1}$ (Fig. 3.28). $B_p$ was 1.4 µg C $l^{-1}$ $h^{-1}$. However, integrating the values with depth showed $B_p$ to be 6 and $B_n$ to be 4 µg C $m^{-2}$ $h^{-1}$. Just using the photic zone to calculate a mean $B_n$ gives 2 µg C $l^{-1}$ $h^{-1}$, which is higher than $B_p$ (1.4 µg C $l^{-1}$ $h^{-1}$). The apparent discrepancy between $B_n$ and $B_p$ in the surface layers is artificial and a function of an erroneous data treatment neglecting the physical behaviour of the waterbody and the biological activity exposed to this behaviour.

Studies including simultaneous measurements of $B_n$ and $B_p$ are relatively few. Using differential filtration to measure $B_n$ and thymidine incorporation to measure $B_p$, $B_n$ is most often lower than $B_p$ in lakes (Bell and Kuparinen 1984, Søndergaard et al. 1985, Riemann and Søndergaard 1986, and marine systems (Larsson and Hagström 1982, Lancelot and Billen 1984). Comparing $B_n$ with $B_p$ estimates from $^{35}SO_4$ uptake, Cole et al. (1982) also found ecologically reasonable $B_n$ values in oligotrophic Mirror Lake. Although $B_p$ values were based on summation of single substrate measurements, Bölter et al. (1982) never found $B_n$ to be higher than their $B_p$ values.

One experiment with $B_n$ and $B_p$ of a similar magnitude was presented by Bell and Kuparinen (1984). For a 20-days period during early spring overturn in Lake Erken, the means of $B_n$ and $B_p$ was 1.7 and 1.2-1.7 µg C $l^{-1}$ $day^{-1}$, respectively. Differential filtration and thymidine incorporation were used for the measurements. From Bell and Kuparinen's description it seems as if $B_n$ was calculated for the photic zone only, and not for the mixed zone. This might give too high $B_n$ values compared with $B_n$. However, it must here be emphasized that estimates of bacterial production are only good within the same order of magnitude (Chapter 4).

The only published case of a major discrepancy between $B_n$ and $B_p$ was presented by Søndergaard et al. (1986). In nutrient enriched enclosures in oligotrophic Lake Almind, $B_n$ (differential filtration) exceeded $B_p$ (thymidine incorporation) by a factor two viewed over a 19 days' period. Several suggestions were forwarded to explain the discrepancy. The most important suggestion was that any such direct comparison is theoretically erroneous. In Chapter 4 we outline the reasons why an evalaution of the importance of substrates to bacterial production must be based on the gross uptake of the substrates in question and bacterial gross production.

The conclusion is now that the $^{14}C$-method, followed by a differential filtration procedure, produces ecologically reasonable results. How a $B_n + EOC_n$ underestimation due to isotopic dis-equilibrium will affect this conclusion is an interesting subject in future research.

This evaluation and suggested data treatment could be very long. Only three additional subjects will be shortly treated: 1. The presentation of EOC data as percent release, 2. Bacterial respiration of EOC products, and 3. Dark release.

### 3.3.6.1. Percent extracellular release (p.e.r.)

In a number of studies there has been found a tendency to high p.e.r. at low photosynthetic rates (Watt 1969, Mague et al. 1980, Watanabe 1980). Two reasons have been forwarded to explain high p.e.r. The first reason is purely technical in that the overall uncertainty at very low acitivities increase and especially would increase p.e.r. as the background level of counting systems are approached and procedural errors and contamination increase in significance. The second reason could be interference from dark release. High p.e.r. figures have been reported for dark-samples (e.g. Watt 1969, Watanabe 1980, Søndergaard and Schierup 1982e). At low light intensities the relative importance of dark processes increase and can be seen as high p.e.r. It must, however, be emphasized that in most cases with elevated p.e.r. due to high or very low light intensities, the quantitative importance for estimating EOC release is trivial, as low amounts of carbon are involved. It should also be pointed out that high p.e.r. in samples incubated at dim light is not a general phenomenon (Blaauboer et al. 1982). In his review, Sharp (1977) was very critical to any use of p.e.r., but his objections seem to be based more on some cases of misuse and rather than theoretical reasons. We will also warn against using p.e.r. too rigorously, but it must also be realized that p.e.r. has comparative value as demonstrated in section 3.3.7.

### 3.3.6.2. Bacterial EOC mineralization

Because respiration cannot be measured in a system with $^{14}CO_2$ as a tracer, it is not possible to measure the bacterial respiration, $B_r$, in differential filtration experiments. If an estimate of apparent release is wanted, a reasonable bacterial growth yield has to be assumed. Respiration of EOC products from algal cultures has been measured in kinetic tracer experiments (Bell and Sakshaug 1980, Jensen 1985a). Respiration values between 30 and 50% of the total uptake were most common. Respiration of a complex algal cell extract is about 55%, and Cole et al. (1982) used this value as their $B_r$. In a more detailed study from a marine locality, Iturriaga (1981) fractionated natural EOC products by ultra-filtration into different molecular weight groups and measured the bacterial mineralization (Iturriaga and Hoppe 1977). After 12 hours of incubation, the respiration of small molecules $<500$ Daltons was 95% of the total uptake (Fig. 3.29). Larger molecules were respired to a lesser extent (~60%). From Fig. 3.29 it is also seen, that uptake of small molecules proceeded with a higher rate (10% $h^{-1}$), compared to 4 to 8% $h^{-1}$ for larger molecules.

Bacterial respiration of EOC should be compared with the overall bacterial growth yield. Recent measurements in the pelagic zone of lakes (Bell and Kuparinen 1984 and in bacteria chemostate cultures (Bjørnsen 1986) have shown rather low bacterial growth yields of 0.1 to 0.3 ($= 70$-90% respiration). The range of $B_r$ can thus be from 30 to 95% depending on substrate composition and incubation time. The use of organic radiotracers to evaluate actual growth yield is erroneous. Incubation time approxing zero will give respiration values close to zero - an infinitely high growth yield, and long incubation times would provide high respiration values. Most published respiration values are from radiotracer incubations lasting less than a few hours. Compared with a true long-term respiration, most of these values are probably too low.

Measurements of phytoplankton release of EOC and bacterial EOC metabolism are most often done in incubations of 4 to 12 h duration. Taking account of these variable incubation times and the often complex composition of EOC, $B_r$ values from EOC uptake of 40-50% could be suggested without a serious overestimation of apparent release. A comparison of EOC uptake and the overall bacterial activity should thus be based on $B_n + B_r$ and gross bacterial production in order to minimize the error from different respiration values.

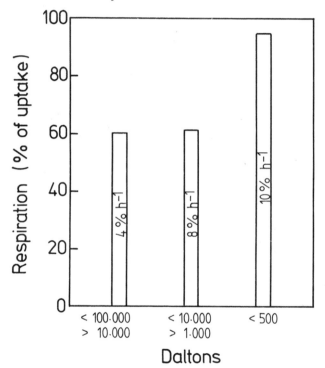

*Fig. 3.29.*
*Bacterial respiration (% of total uptake) and uptake rates (inside columns) of different molecular weight products released by a marine phytoplankton community. Redrawn after Iturriaga (1981).*

### 3.3.6.3. Release in darkness

The final comments are addressed to the question if dark values should be subtracted light values. One argument to make such subtractions is a wish to present values of release directly related to photosynthesis. The soundness of this argument is dependent on the continuation of dark processes in light. Watanabe (1980) found EOC released in darkness to deviate qualitatively from light products, a feature also observed by Watt (1969). This made Watanabe suggest not to subtract dark values. [14]C-fixation in darkness is a function of both non-biological and biological processes. It has been suggested to use zero-time filtrations to represent non-biological "fixation" (Morris et al. 1971, Mague et al. 1980). Use of fixatives in incubated samples have also been used (Bell and Kuparinen 1984).

In most cases the quantitative importance of dark fixation and dark EOC release is negligible, so it seems reasonable not to subtract any dark values. This argument is further supported by the fact that we do now know whether dark processes continue in light. Moreover, if they do, they are still part of a biological fixation of carbon and should then not be subtracted.

Carbon flux from algae to bacteria via EOC can be investigated by the $^{14}$C-method followed by differential filtration. Most recent investigations have shown the results of bacterial EOC uptake to fit into an ecological frame set by the overall bacterial production. The evaluation and lengthy discussion on data treatment involving both basic questions and more trivial technical subjects are due to the fact that no standardization is present and probably will not be within several years. The $^{14}$C-method became widespread in the late fifties; however, there is still no international standard performance. The reason is obviously to be found in the complex series of events that occur and are reflected in the outcome of $^{14}$C-incubations.

### 3.3.7. Extracellular release: Examples and controlling factors
### 3.3.7.1. Examples of extracellular release in lakes

A positive relationship between production of particulate and dissolved carbon was discussed early in the short history of EOC measurements. The only example of a constant release rate, independent of photosynthetic rates, was reported by Smith and Wiebe (1976). In a series of experiments they changed the photosynthetic rates of a natural phytoplankton population and three algal cultures by altering DIC concentrations. The release of EOC was found to be constant. Such results have not been observed by other investigators (e.g. Mague et al. 1980, Cole et al. 1982).

The positive relationship between particulate and dissolved production means higher absolute values of release in eutrophic compared with oligotrophic environments. It must here once more be put into mind that most previous studies only included $EOC_n$ measurements. Concomitant with the accumulation of quantitative results, it was also concluded that relative release expressed as a percentage of the total fixation (p.e.r.) increased along a gradient from productive to less productive areas. This conclusion is strengthened by the fact that both marine and freshwater areas could be fitted into the general scheme (Watt 1966, Anderson and Zeutschel 1970, Thomas 1971, Berman and Holm-Hansen 1974, Nalewajko and Schindler 1976, Chrost 1983). In a recent review on the ecological importance of EOC, Fogg (1983) referred to relative $EOC_n$ values between 1 and 10% in eutrophic and 30-40% in oligotrophic areas. Most earlier $EOC_n$ values were summarized by Fogg (1975).

The level of relative release is not merely a function of a general trophic level, but is also a function of algal species and environmental factors as light and nutrients, which control photosynthesis and growth. Before treating some of these biological and environmental variables in more detail, the general theory of high versus low p.e.r. in oligotrophic vs. eutrophic areas, respectively, shall be re-examined using recent data from lakes. Since the provoking, but justified question by

Table 3.5. Examples of extracellular release in lakes of different trophic levels. Values are relative for comparative purposes and meaned. Abbreviations: $EOC_n$ = net release of EOC, $B_n$ = bacterial assimilation of EOC, $B_r$ = bacterial respiration of EOC, A = fixation into algal cells.

| Lake | Trophic level | $\dfrac{EOC_n \times 100}{\text{Particles} + EOC_n}$ (%) | $\dfrac{(B_r + B_n + EOC_n) \times 100}{A + B_r + B_n + EOC_n}$ (%) | Remarks[3] | Reference |
|---|---|---|---|---|---|
| Ørn | eu | 4 | 11 | Diel exp. June surface, $B_r = 30\%$ | Søndergaard et al. (1985) |
| Esrom | eu | 10 | 30 | Diel exp. May surface, $B_r = 30\%$ | do |
| Mossø | eu | 7 | 18 | Diel exp. April surface, $B_r = 30\%$ | do. |
| Mossø | eu | 4 | 11 | Diel exp. Oct. surface, $B_r = 30\%$ | Jørgensen et al. (1983) |
| Mossø | eu | 3 (2-6)[4] | 5 (3-17) | Mar. 24-Apr. 24 surface, $B_r = 30\%$ | Riemann et al. (1982) |
| Frederiksborg Slotssø | eu | 0.5 | 4 | Diel exp. Sep. surface, $B_r = 30\%$ | Jørgensen et al. (1983) |
| Frederiksborg Slotssø | eu | 2 | 5 | Diel exp. Nov. surface, $B_r = 30\%$ | Søndergaard et al. (1985) |
| Frederiksborg | eu | 4 (2-6) | 9 (5-23) | June 2-21, photic zone, $B_r = 40\%$ | Riemann & Søndergaard (1986) |
| Hylke | eu | 5 (4-6) | 37 (24-48) | May 20-Jun. 2, photic zone, $B_r = 40\%$ | Søndergaard (1986) |
| Hylke | eu | 6 (2-12) | 11 (6-29) | Sep. 3-22, photic zone, $B_r = 40\%$ | Søndergaard (unpubl.) |
| Mendota | eu | 22 (0-75) | 41 (24-75) | Season, Lab. incubation, $B_r$ not included | Brock & Clyne (1984) |

| Lake | Trophic | | | Notes | Reference |
|------|---------|---|---|-------|-----------|
| Bysjön | eu | 1-4 | 2-7 | Scattered, photic zone, $B_r$ not incl. | Coveney (1982) |
| Jorzec | eu | 6 (2-14) | - | Season, surface | Chrost (1983) |
| Kinneret | eu | 4 (0.5-8) | - | Aug.-Dec., photic zone | Berman (1976) |
| Nakanuma | eu | 15 (8-54) | - | Apr.-Sep., photic zone | Watanabe (1980) |
| Glebokie | meso/eu | 17 (2-29) | - | Season, surface | Chrost (1983) |
| Vechten | meso/eu | 26 (0-55) | - | Season, photic zone | Blaauboer et al. (1982) |
| Erken | meso | 38 (13-49) | 40 | Apr. 18-May 7, calculated for the photic zone | Bell & Kuparinen (1984) |
| Constance | meso | 23 (10-60) | 30 | May-Nov., surface, $B_r$ not included | Kato & Stabel (1984) |
| Majcz | oligo | 34 (7-49) | - | Season, surface | Chrost (1983) |
| Almind | oligo | 33 (18-37) | 56 (43-61) | May 20-June 2, photic zone, $B_r = 40\%$ | Søndergaard et al. (1986) |
| Almind | oligo | 25 (19-30) | 48 (41-58) | Sep. 3-22, photic zone, $B_r = 50\%$ | Søndergaard et al. (1986) |
| Tahoe | ultra- | 13 | - | Scattered, photic zone | Tilzer & Horne (1979) |

1. Mean values are, where possible, based on integrated absolute values.
2. According to definition in Wetzel (1983), eu = eutrophic, meso = mesotrophic, oligo = oligotrophic.
3. Time of investigation, way of incubation, assumed $B_r$.
4. Values in brackets indicate observed range.

Sharp (1977): "Do healthy algal cells release organic matter?", most investigators have minimized error sources and presented reliable evidence for answering the question with a yes. It must, however, be remembered that many very different methodological approches have been used to give the answer. There is still (unfortunately) no standard method on these matters.

Recent data, mostly from *in situ* measurements, have been compiled in Table 3.5. The results cover an array of different lakes in temperate areas. To be able to make some comparisons, we have with some hesitation presented the data as p.e.r., being fully aware of all weaknesses inherent in this presentation (see Mague et al. 1980). Otherwise, it would have been impossible to compare mean values from a diel experiment in one lake with a mean of seven depth-integrated values from another lake.

In eutrophic lakes most figures of p.e.r. based on $EOC_n$ are less than 10% and often range between 2 and 6% (Table 3.5). The values from Lake Mendota and Lake Nakanuma are high and perhaps unique. In less productive lakes p.e.r. ranges from 17 to 38%. Large variations within a single lake can be found and can be exemplified using the detailed investigation in slightly eutrophic Lake Vechten, The Netherlands (Blaauboer et al. 1982). In winter, the release was close to zero. High release rates in both relative (55%) and absolute terms were found during a period with a dense and very active metalimnetic population of *Mallomonas caudata*. The data of Blaauboer and co-workers have been transformed to p.e.r., which is repesented in Fig. 3.30 along with absolute values. High release is most pronounced in summer and autumn.

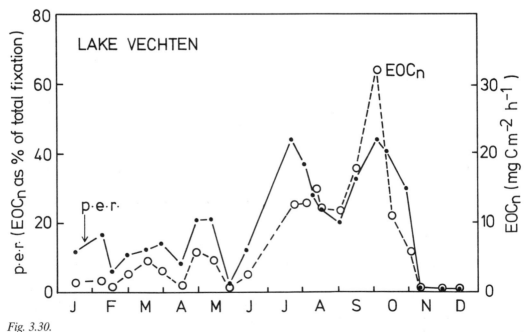

Fig. 3.30.
*Seasonal variations of p.e.r. and absolute $EOC_n$ values from Lake Vechten, The Netherlands. Data recalculated to represent depth-integrated values. From Blaauboer et al. (1982).*

Returning to the general picture, there is clearly a trend of relatively high percent $EOC_n$ values in less productive lakes (Table 3.5). In previous sections we stressed the importance of an ecological data treatment and interpretation. EOC is taken up and metabolized, i.e. assimilated into new bacterial biomass and respired. Especially the high EOC transport often found in eutrophic lakes must here be taken into account (Table 3.4.). Calculation of apparent release ($B_r$ + $B_n$ + $EOC_n$)is only possible from a limited number of investigations. Doing so the general pattern established from $EOC_n$ values is a little blurred although the available data in Table 3.5 do not disprove the general statement. It is, however, striking that the p.e.r. level from four eutrophic lakes is elevated into Fogg's (1983) oligotrophic category. Most studies on EOC only cover shorter periods, which certainly can obscure the results. From the studies in eutrophic lakes covering investigation periods of more than one day or a scatter of few measurements (Hylke, Frederiksborg Slotssø, Mossø, Mendota, Nakanuma and Vechten) it seems likely that the release is generally underrated. One further reason for an underestimation is linked with the dominance of large phytoplankton species in eutrophic as opposed to oligotrophic lakes (Reynolds 1984). In section 3.3.4 it was shown that the rate at which isotopic equilibrium in the internal carbon pools is achieved, is a function of the turnover rate. Larger species have a slow biomass turnover, which can influence the specific activity of the released products and result in underestimation of true release.

Our conclusion is, that viewed over a season taking account of all water depths, the release in eutrophic lakes can prove to be larger than hitherto believed. Presently, only a few samples of quantiative data from studies of a relatively short duration are available to fill into the scenario for EOC presented in Fig. 3.1 (Table 3.6). Measurements from oligotrophic Lake Almind are presented to contrast the more eutrophic lakes. As dissolved organic carbon does not accumulate over longer periods in lakes, the fraction $EOC_n$ (Table 3.6) must be metabolized, and the assimilated fraction added to $B_n$.

### 3.3.7.2. Factors controlling EOC release

Over the years a series of biological and environmental factors have been suggested to control release rates in nature. Among the most cited factors are: Algal species and their physiological state, e.g. senescence, nutrient level and environmental stress factors such as high light intensities.

Species dependent release rates have been studied in algal cultures and *in situ*. It is known that differences among species exist and that low release (p.e.r. $< 10\%$) is normally found at optimum growth conditions (Hellebust 1965, Nalewajko and Marin 1969, Watt 1969, Ignatiades and Fogg 1973, Myklestad 1974, Lee and Nalewajko 1978, Wolter 1982). *In situ* studies are most valuable in an ecological context, but the results are often difficult to interpret due to a mixed phytoplankton population. Knowledge of species specific release rates *in situ* is restricted to situations with total dominance of a single species.

The summer and autumn dominance of *Mallomonas caudata* in the metalim-

Table 3.6. Some examples of pelagial carbon flux measured with the [14]C-method
and differential filtration

| Lake | Period | Algae | Bacterial assimilation (B$_n$) | EOC$_n$ | Units | Reference |
|---|---|---|---|---|---|---|
| Mossø[1] | Mar 24-Apr 24 | 20 | 0.4 | 0.5 | g C m$^{-3}$ | Riemann et al. (1982) |
| Erken[2] | Apr 18-May 7 | 2.3 | 0.004 | 0.9 | g C m$^{-3}$ | Bell & Kuparinen (1984) |
| Frederiksborg Slotssø, enclosure + fish | Jun 2-21 | 23.5 | 10 | 1.2 | g C m$^{-2}$ | Riemann & Søndergaard (1986) |
| Hylke | May 20-Jun 2 | 3.5 | 0.5 | 0.3 | g C m$^{-2}$ | Riemann & Søndergaard (1986) |
| Hylke, control enclosure | Sep 3-22 | 9.2 | 0.3 | 0.6 | g C m$^{-2}$ | Søndergaard (unpubl.) |
| Almind, control enclosure | May 20-Jun 2 | 2.6 | 0.4 | 0.7 | g C m$^{-2}$ | Søndergaard et al. (1986) |
| Almind, enclosures with nutrients | May 20-Jun 2 | 6.1 | 1.1 | 1.0 | g C m$^{-2}$ | Søndergaard et al. (1986) |

[1]) Values from surface incubations
[2]) Mean values for the photic zone

nion of Lake Vechten created such a case (Blaauboer et al. 1982). At these specific
environmental conditons, *M. caudata* had a very high release rate (EOC$_n$ values)
in both absolute and relative (p.e.r. > 50%) terms. Chlorophycean species dominating in the epilimnion at the same time had low release ( > 10%).

Spring blooms in eutrophic lakes are often dominated by few species, often
*Stephanodiscus* sp. or other centric diatoms. Such blooms are, at least in the early
phase, characterized by low or moderate release (EOC$_n$) between 5 and 18% of
the total fixation (Watt 1969, Blaauboer et al. 1982, Riemann et al. 1982, Jensen
1985b), although Bell and Kuparinen (1984) found about 40% in mesotrophic
Lake Erken.

Colonial and filamentous blue-green algae can be dominating during extensive
periods in eutrophic and eutrophicated lakes (section 3.2). No consistent material
on single species is available, but low relative release seems to be common. Low
EOC$_n$ in the eutrophic lakes presented in Table 3.5 is often found concomitant
with dominance of blue-green algae (Lake Ørn, Lake Mossø in October, Frederiksborg Slotssø in September and November, Lake Hylke in September and Lake
Jorzec). As high release ( > 40%) is also found during blue-green algal dominance
(Watanabe 1980, Brock and Clyne 1984, Jensen 1985b) no general conclusion is
possible.

The theory of high relative release rates in oligotrophic waters contrary to low
values in more eutrophic waters (Fogg et al. 1965, Fogg 1983) is primarily based on
marine studies and might be related to species dependent differences and not to

the trophic level as such. Dominance of different species can perhaps offer an explanation to conflicting results, where some authors could support the theory (e.g. Anderson and Zeutschel 1970, Thomas 1971, Chrost 1983), while others could not (Williams and Yentsch 1976, Sellner 1981, Larsson and Hagström 1982).

*Nutrients:* In a seasonal study along a nutrient and productivity gradient in the Southern Bight of the North Sea, Lancelot (1983) produced a series of measurements, which can illustrate the complexities of these questions. In phytoplankton communities dominated by the haptophycean *Phaeocystis pouchetii* and dinoflagellates, a significant negative correlation between p.e.r. (from $EOC_n$ values) and the concentration of mineral nitrogen was found. Approaching depletion of nitrogen *P. pauchetii* released about 70% and dinoflagellates about 45% of the total fixation as EOC (Lancelot 1983). Lancelot suggested a threshold value for this negative relation to be about 10 $\mu$mol N $l^{-1}$ (in the water). At higher concentrations p.e.r. remained rather constant. In a marine diatom population, Lancelot (1983) found no significant correlation between EOC and nitrogen concentrations. These results made her question the general hypothesis of high p.e.r. in oligotrophic areas as such, but rather to focus on algal species and bloom situations, including depletion of nutrients.

Similar studies in lakes have also shown that p.e.r. increase as a function of nutrient depletion and decline (senescence) of a specific population. The spring blooms of *Stephanodiscus hantzschii* in eutrophic Lake Mossø and *S. pusillus* in eutrophic Lake Hylke are examples. In Lake Mossø, p.e.r. (based on $EOC_n$) increased from 3 to 4%, when nutrients were abundant to about 12% at the time when phosphate became undetectable and the algal biomass declined (Fig. 3.31). The decrease of p.e.r. after April 23 occurred when algal species other than diatoms became more abundant (Riemann et al. 1982). In Lake Hylke, Jensen (1985b) found $EOC_n + B_n$ to be 14-19% during population increase and as high as 32-41%, when the diatom population decreased.

The apparent effects of nutrients on release by a natural population dominated by *Fragillaria* and *Dinobryon* have recently been studied in a series of enclosure experiments in oligotrophic Lake Almind (Søndergaard et al. 1986). In all enclosures without addition of nutrients the concentration of orthophosphate was low ($<5$ $\mu$g $PO_4$-P $l^{-1}$) and apparent release of EOC fluctuated between 45 and 60% (Fig. 3.32). In enclosures added N + P *Fragillaria* became dominant and apparent release decreased significantly with time and was about 20%, when the experiments were stopped (Fig. 3.32). These results support the conclusion that some species are sensitive to nutrient depletion, which affects the physiological state and growth of the algae. Furthermore, the results are in agreement with previous studies on algae in culture (Fogg et al. 1965, Ignatiades and Fogg 1973, Watanabe 1980, among numerous).

*Light:* Within a given area and time there is a close relationship between rates of photosynthesis and release. That is, the p.e.r. is almost constant within the photic zone, but can be slightly higher at the surface and at very low light intensities. Elevated p.e.r. at high light intensities has been experienced in several studies (Fogg

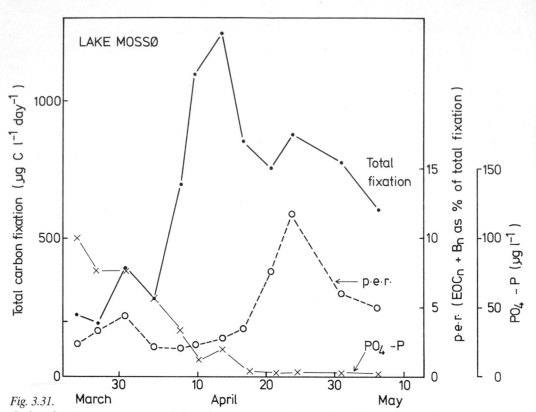

Fig. 3.31.
*Carbon fixation, percent extracellular release and phosphate in Lake Mossø, Denmark, during a spring diatom bloom. Based on data from Riemann et al. (1982) and Søndergaard & Schierup (1982a).*

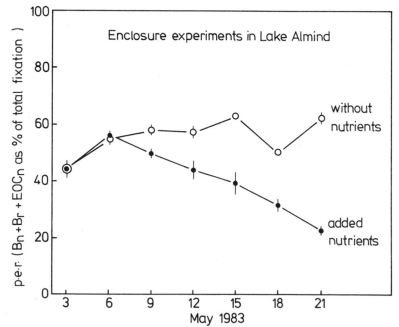

Fig. 3.32.
*Time dependent percent extracellular release in enclosure experiments in Lake Almind, Denmark. Bars are "SD", n = 2-4. In all enclosures the phytoplankton was dominated by Fragillaria and Dinobryon.*
Redrawn after Søndergaard et al. (1986).

et al. 1965, Watt 1965, Thomas 1971, Smith et al. 1977, Mague et al. 1980, Watanabe 1980). From *in situ* measurements in Lake Windermere Watt (1969) observed a relationship between p.e.r. and photoinhibition. The effect was only seen at a photosynthetic inhibition above 60% (Fig. 3.33).

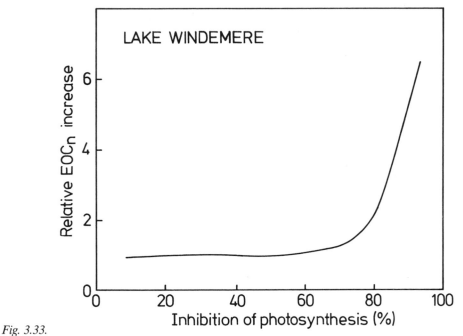

Fig. 3.33.
*Relationship between photoinhibition and increase in relative extracellular release. Based on data from Watt (1969).*

It is characteristic for this review of EOC studies that the literature is rather conflicting if not directly contradictory. The effect of light is no exception. Recent studies have also included vertical release patterns without detecting any significant p.e.r. increase in surface waters (Tilzer and Horne 1979, Blaauboer et al. 1982, Coveney 1982, Lancelot 1983, Søndergaard unpublished). Our conclusion is that a severe photoinhibition can cause elevated p.e.r. However, the relative increase is a function of a decreased total fixation so the quantitative consequences for the carbon flux are minimal. One can also argue along with Harris (1978) that photoinhibition is very often a procedural artefact related to the prolonged exposure to high light intensities in a bottle fixed at the surface. In nature, phytoplankton is freely mixed in most situations and thus exposed to fluctuating light intensities.

From existing evidence it is difficult to make definite conclusions on factors regulating and controlling release of EOC. Basically, the release rates are closely linked with the photosynthetic rates. The amount of EOC available for bacteria is higher in eutrophic lakes than in oligotrophic lakes. Bloom situations followed by depletion of nutrients and a senescent algal population are often linked with high relative release, although it is to some extent species dependent. Søndergaard et al. (1984) experienced a situation with a mixed and declining algal community where the realtive release did not increase.

The ecological importance of EOC should not only be viewed from the standpoint of the carbon budget of the algae, but also from the consumer size viz. bacteria. The importance of EOC to bacterial production is treated in Chapter 4.

### 3.3.8. Composition and decomposition of extracellular products

Experiments with planktonic microalgae in dense cultures have demonstrated release of an array of different organic compounds. Mono-, oligo- and polysaccharides are among the most often identified products, but amino acids, polypeptides, proteins, vitamins, glycollate and products like glycerol, hydroxamate and fatty acids have also been identified. The information on specific EOC products mostly found in algal cultures has previously been reviewed by Fogg (1966, 1971) and Hellebust (1974). In most cases the majority of EOC is still chemically unidentified.

Quantitative and even qualitative information on specific organic compounds released by natural phytoplankton communities are rather limited. In 1965, Hellebust could not identify a large amount (up to 90%) of the EOC released by 23 marine phytoplankters. Basically the situation has not changed, and the EOC classification suggested by Fogg (1966) is still valid and used. He defined two product categories: Type I, includes low molecular weight products (LMW), representing metabolic intermediates and Type II, including high molecular weight products (HMW), most likely metabolic end products. The classification is arbitrary and not further defined, but common practice has over the years defined Type I to be molecules less than 500-700 Daltons.

Three major problems are present in measuring the composition of EOC *in situ*.

1. The use of $^{14}C$ causes rapid labeling of Type I products and a slower $^{14}C$ incorporation into Type II products. This has been demonstrated by Jensen et al. (1985). Thus, in short-time (4-6 hours) incubations the importance of LMW products are overestimated, and the relative distribution of labeled EOC will change with time. After 2 hours' *in situ* incubation in eutrophic Lake Mossø, $EOC_n$ was dominated (87% of the total) by LMW products $< 700$ Daltons (Fig. 3.34). After 5 hours LMW products still dominated (65%), although Type II (here $> 700$ Daltons) became more important. Similar results have been presented in several studies on time dependent release (Steinberg 1978, Iturriaga and Zsolnay 1983, Lancelot 1984) and are related to isotopic dis-equilibrium in EOC precursor pools (section 3.3.4).

2. In nature, algal release and bacterial uptake occur simultaneously, so the $EOC_n$ composition is biased to the more refractory material. The error is dependent on the heterotrophic activity. Such fast changes are illustrated in Fig. 3.35, where $^{14}C$-labeled EOC from a natural phytoplankton population is inoculated with native bacteria. During heterotrophic incubation the contribution of LMW molecules to the total decreased from 61 at time zero to 56 and 48% after 6 and 72 hours, respectively. Even in this experiment the LMW products at time zero

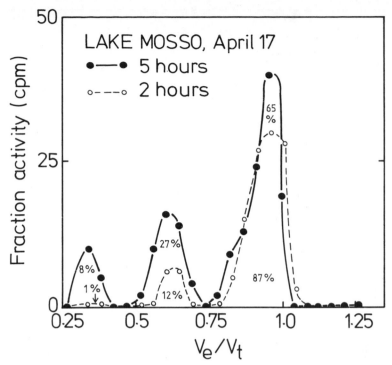

Fig. 3.34.
Gel-chromatography (Sephadex G-50) of $EOC_n$ released by a natural phytoplankton community domi-
nated by Stephanodiscus hantzchii and incubated with $^{14}CO_2$ for 2 and 5 h. Lake Mossø, Denmark.
Redrawn after Søndergaard and Schierup (1982a).

are probably underestimated due to bacterial activity during production of EOC.
The utilization rates depicted from Fig. 3.35 are not *in situ* rates, as the native
bacteria were diluted ten times. Although small molecules have a faster turnover
than larger molecules, the HMW EOC products are also utilized. This is apparent in Fig. 3.35 and has been found in several investigations (Iturriaga and
Hoppe 1977, Iturriaga 1981, Chrost and Faust 1983, Kato and Stabel 1984, Lancelot 1984). The high and fast transport of EOC from algae to bacteria in
eutrophic lakes (Table 3.4) means that $EOC_n$ might only represent about one
third or less of apparent release. The use of antibiotics to suppress bacterial
uptake is one way to investigate the EOC composition without major bacterial
influence. Jensen (1983) did observe some differences in the molecular weight
distribution in samples with and without streptomycin. Although the differences
were not dramatic, the contribution of small molecules was underestimated in
samples without streptomycin.
3. Direct chemical identification of specific EOC products *in situ* without use of a
radioactive tracer is not generally possible, as there is no way to distinguish
between extracellular release and compounds from other sources. Glycollate is
an exception. It only originates as a metabolic intermediate in photorespiration
and thus must be of direct photosynthetic origin.
  Difficulties with chemical identification have made the use of more crude
molecular weight separations by gel-chromatography and ultrafiltration a much

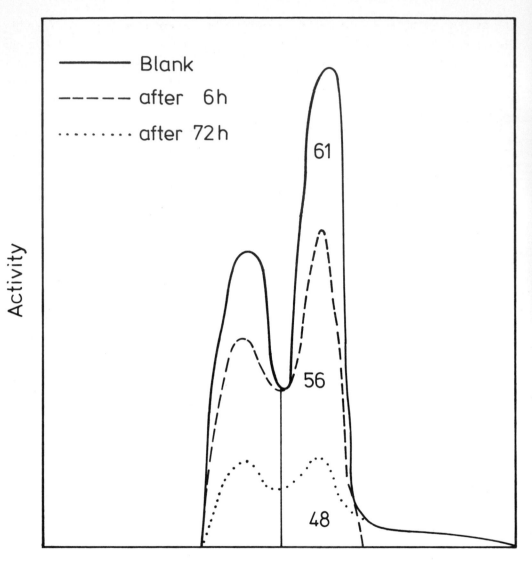

Fig. 3.35.
*Gel-chromatography (Sephadex G-10) of EOC$_n$ from a native marine community exposed to bacterial activity for 0, 6 and 72 h, respectively. The contribution of low molecular weight products ( < 500 Daltons) to the total is inserted for each time. After Iturriaga and Zsolnay (1983).*

used and useful tool in EOC studies (see Fig. 3.34 and 3.35; Nalewajko and Lean 1972, Nalewajko et al. 1976, Chrost and Faust 1980, Søndergaard and Schierup 1982a, Jensen 1983, Lancelot 1984).

The results from these studies covering both oligo- and eutrophic lakes and marine areas are variable. Low molecular weight products are dominating in some cases (Wiebe and Smith 1977, Jüttner and Matuschek 1978, Mague et al. 1980, Iturriaga 1981, Søndergaard and Schierup 1982a, Jørgensen et al. 1983), whereas Nalewajko and Schindler (1976), Watanabe (1980) and Lancelot (1984) found larger molecules to be more frequent. In most cases several molecular weight sizes are present in natural $EOC_n$ pools (Table 3.7). As appears from the variable results it is not possible to make any generalizations at present. This becomes even more pronounced when detailed seasonal results are examined (see Chrost and Faust 1983). From their study we know that the composition of $EOC_n$ changes with season. The changes are probably due to the succession of phyto- and bacterioplankton or in their proportions, and that specific algal species almost exclusively release specific compounds depending on growth and physiological state (Hellebust 1965, Watt 1969, Lancelot 1984). Nalewajko and Lean (1972) have found that larger molecules become predominant when algal cultures age. A similar pattern was observed by Søndergaard and Schierup (1982a), when a spring bloom of *Stephanodiscus hantzschii* in Lake Mossø entered the stationary phase. During the exponential phase HMW-products ($> 10,000$ Daltons) made up about 1%, increasing to 20% in the stationary phase.

Table 3.7. Examples of molecular weight distribution of $EOC_n$ in different lakes and marine areas. Measured by ultra-filtration (UF) and gel-chromatography (GC) and presented as percentages of the total.

| Locality | Method/Remarks | > 10,000 | < 10,000 | > 500 | < 500 | Reference |
|---|---|---|---|---|---|---|
| | | | Daltons | | | |
| | | | % | | | |
| Frederiksborg Slotssø | UF, June, n = 16 | 42 | 28 | | 30 | Søndergaard (unpubl.) |
| Lake Mossø | GC, April, n = 5 | 7 | 25 | | 68 | Søndergaard and Schierup (1982a) |
| Lake Ørn | GC, June 19, n = 3 | 27 | 20 | | 53 | Søndergaard (unpubl.) |
| Lake Hylke | GC, May, n = 8 | 24 | 15 | | 61 | Søndergaard (unpubl.) |
| Lake Almind | GC, May, n = 8 | 17 | 29 | | 54 | Søndergaard (unpubl.) |
| Rhode River estuary | UF, season, n = 17 | 64 | 16 | | 20 | Chrost and Faust (1983) |
| Belgian coast | UF, spring, summer, n = 20 | | 77[1] | | 23 | Lancelot (1984) |
| Lake Constance | GC, summer, n = 7 | | 48[2] | | 52 | Kato and Stabel (1984) |

1) > 500 Daltons
2) > 1500 Daltons

Identification and quantification of EOC released under natural conditions have only been attempted in a few cases. Amino acids, glycollate, mannitol, trehalose, peptides, oligosaccharides and polysaccharides are among the substances most frequently found (Hellebust 1965, Watt 1966, Tanaka et al. 1974, Jüttner and Matuschek 1978, Mague et al. 1980, Watanabe 1980, Hall and Fisker 1983, Jørgensen et al. 1983). Saccharides (mono, oligo, and poly) seem to be most important. Amino acids can comprise from 5 to 32% of $EOC_n$ in natural waters (Mague et al. 1980, Jørgensen et al. 1983, Søndergaard and Jørgensen, unpubl.).

Whether glycollate is an important EOC product is a matter of controversy, which to our opinion is not solved. The interest for glycollate originated from work with algae in culture used to investigate photorespiration. Especially the work on glycollate as the major EOC product carried out by Fogg and co-workers (Nalewajko et al. 1963, Fogg et al. 1965, Watt and Fogg 1966, Watt 1969, Nalewajko et al. 1980) opened for the interest of glycollate in a more ecological context. Indeed, glycollate has been demonstrated as an extracellular product in a number of different algal cultures (Sørensen and Halldal 1977, Bowes and Berry 1972, Colman and Hosein 1980, Stewart and Codd 1981, Soeder and Bolze 1981).

In his 1983 review Fogg stated: "With many algae, in fact, a single substance, glycollic acid, preponderates among the labeled products". Direct proof of glycollate as a dominant extracellular product in natural phytoplankton is, however, sparse, and the above statement does not necessarily apply for natural phytoplankton, but for cultures (Fogg pers. comm.). The only investigations, to our knowledge, presenting evidence for the importance of glycollate in nature are Fogg et al. (1965) and Watt (1966). Indirect proof for release, but not for a quantitative dominance, is the presence of glycollate in water (Al-Hasan et al. 1975, Al-Hasan and Coughlan 1976, Coughlan and Al-Hasan 1977, Wright and Shah 1977, Billen et al. 1980).

The composition of $EOC_n$ is often dominated by molecules larger than glycollate (Table 3.7). Unless glycollate is taken up at a fast rate and is not detected in true proportion, the molecular weight distributions do not unequivocally support glycollate importance. However, it is interesting to find that glycollate in a large number of samples from different areas has much longer turnover times (range: 7 to about 2500 hours, with most above several hundred) than many other LMW substrates (Wright and Shah 1975, Coughlan and Al-Hasan 1977, Billen et al. 1980, Iturriaga and Zsolnay 1981). Long turnover times do not disprove quantitative importance, if concentrations are high (Billen et al. 1980).

Although a direct comparison of metabolic features of glycollate and EOC products is not possible, it is nevertheless striking that glycollate turnover times in most cases is substantially higher than EOC turnover. EOC turnover in uptake kinetic experiments ranged between 10 and 440 hours (mean ~ 100 h) in a Norwegian fjord (Bell and Sakshaug 1980) and with much lower values of 8 to 60 hours in a "eutrophic" lake, even at rather low temperatures (Jensen 1985b). That is one reason to question the importance of glycollate dominance, at least of the EOC products released by the algae in the above studies. Another important reason to dispute the general importance is the discrepancy in respiration values of glycollate and EOC. Glycollate is an energy source for bacteria. The respiratory fraction of total uptake is about 80-90% (Wright and Shah 1975, Billen et al. 1980, Iturriaga and Zsolnay 1981). In contrast most respiration values of EOC are between 16 and 60% (Iturriaga and Hoppe 1977, Bell and Sakshaug 1980, Jensen 1985b). The only exception is the high respiration of LMW products found by Iturriaga (1981) and shown in Fig. 3.22. Glycollate could be a major product in this case. Measurements of glycollate at low natural concentrations (~20 µg l$^{-1}$) are subject to a series of errors, so to our opinion there is no scientific reason to view glycollate as *the* EOC product.

To know the ultimate fate of EOC in the environment is crucial to our understanding of organic carbon budgets and to estimate the importance of EOC to bacterial production. The fate can be divided into three components: 1. The part easily metabolized by bacteria, 2. A part transformed and excreted by the bacteria and 3. A refractory part very slowly decomposed, i.e. entering the more refractory part of the DOC pool.

In a previous section it was shown that immediate transport of EOC to bacteria is not 100% (Table 3.4). In fact, transport can be as low as 4%. If the total pool of EOC is available to bacteria, the bacterial production ($B_p$) should be compared with $B_n + EOC_n$ and not only with $B_n + B_r$. Several authors have assumed the total amount of EOC to be utilized (Larsson and Hagström 1982, Kato and Stabel 1984, Lancelot and Billen 1984). However, incubation of EOC with natural bacteria for several days (3-7) has shown uptake rates to decrease with time and leave approximately 25% of the initial $EOC_n$ in a very slowly metabolized pool (Herbland 1975, Iturriaga and Zsolnay 1983). The decomposition pattern is like those found for the DOC pool (e.g. Søndergaard 1985); asymptotically decreasing rates to approaching zero. Thus, it does not seem appropriate to assume all EOC to be metabolized in short time. Time-courses of decomposition in the above investigations can, however, be biased by depletion of inorganic nutrients. But as the DOC pool does not increase, a steady state between production and utilization is operating on an annual cycle (Cole et al. 1984).

During bacterial utilization, EOC is enriched with HMW-products, partly because these products are utilized at a slower rate than LMW-products and partly due to molecular transformation. With the exception of glycollate it has been shown that several simple sugars and amino acids are transformed and excreted as nominally larger and smaller substances during bacterial decomposition (Dunstall and Nalewajko 1975, Iturriaga and Zsolnay 1981). It is likely the same pattern applies to EOC. Algal lysis products are also transformed to larger molecules.

## 3.4. Phytoplankton lysis

The organic material present in algal biomass can enter the pelagic detritus pathway, if the algae die while still suspended. That algae do die for some reasons and their death rates can be calculated, has most elegantly been shown by Knoechel and Kalff (1978) and Reynolds et al. (1982). The loss process and fate of organic material entering the environment from lysing algae are the subjects of this section.

How algae actually die is not known in detail. However, the most important feature is the loss of cell integrity due to loss of membrane control. The term lysis is here used to describe the events occurring after loss of membrane control.

The lytic process can arbitrarily be divided into shorttime (hours) and long-time (days, weeks) events. Particles ultimately sediment and a mass balance of organic material can be evaluated from intensive sediment trap experiments or decomposi tion studies in closed systems (Jassby and Goldmann 1974, Jones 1976, Fallon and Brock 1979, 1980). Water turbulence and particle transport prevent a quantitative estimation of the time dependent loss, since the age of the sedimenting algae is not known.

Another *in situ* method to investigate the dynamic of algal lysis is the dip-method introduced by Cole (1982) and Cole et al. (1984). The method was developed to avoid very long (weeks') incubations of algae in bottles. The principal outline is as follows (after Cole et al. (1984)): Natural algal populations are labeled with $^{14}$C, harvested on a filter and "gently" killed. The labeled algae are then reintroduced to the natural environment (in a bottle) and the fate of $^{14}$C in particles and in solution is followed with time. To avoid heavy grazing pressure, larger zooplankton (e.g. > 200 µm) are normally removed.

For the purpose of quantification of loss of organic carbon, that is conversion of $^{14}$C to $^{12}$C values it is a pre-requisite that all internal organic pools in the algae are in isotopic equilibrium. Theoretically, the labeling period should cover at least five cell divisions (Jensen 1985a). In practical work with natural populations, such long incubations are to be avoided. Shorter incubation periods of 2-3 cell divisions have proved successful (Cole et al. 1984).

To distinguish between physico-chemical and biological events, the filters are incubated in bottles with sterile water (control) and in bottles with natural lake water populations. Dip-experiments lasting about 4-6 days have proved biologically mediated loss of organic material from the algae to be relatively unimportant (Cole et al. 1984, Hansen et al. 1986), so the event taking place in the control bottles can be viewed as "gross" processes.

Besides problems with a uniform labeling of algal cell constituents, three further problems shall be mentioned. For a detailed discussion the reader should consult Cole (1982): 1. Algal death during the initial labeling period would influence the results, as the particles on the filters would then be a mixture of intact algal cells, and cells having underwent lysis for some time. 2. The killing procedure might create leaching rates not occurring in nature and influence the composition of the leached material. Different procedures have been tested, but the accuracy of the method cannot yet be evaluated (Cole et al. 1984, Hansen et al. 1986). 3. The concentration of particles on a filter create an artificial environment. The diffusion conditions around single cells are altered. Probably, the leached products can reach very high and unnatural levels in the boundary layer and accelerate hetorotrophic activity (Cole and Likens 1979), as bacteria are stimulated to move into microenvironments with high DOC concentrations (Azam and Ammerman 1984). In spite of these problems, the dip-method has proved to be a useful tool in studies of lysis.

The loss of organic carbon from dead algal cells is characterized by a high initial loss rate. The pattern shown for Lake Mossø (Fig. 3.36) is representative for the results found in several eutrophic Danish lakes (Hansen et al. 1986). From 11 to 40% of the $^{14}$C-labeled cell content rapidly entered the environment in a dissolved form (DOC) followed by a decreasing loss rate. Prolonged serial incubations (4-6

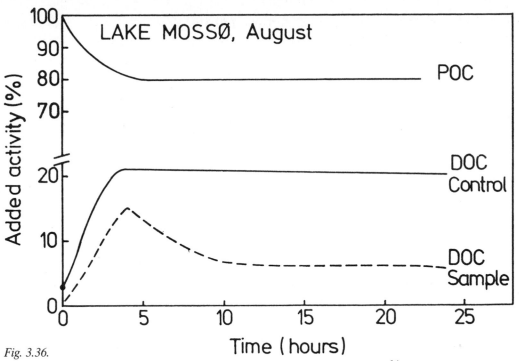

*Fig. 3.36.*
*Time-course of lysis experiment in eutrophic Lake Mossø, August 1982. The $^{14}C$-activity in dead algae (POC) and in leached products in sterile control bottles (DOC-control) and in bottles with a natural microbial community (DOC-sample) was followed. Redrawn after data from Hansen et al. (1986).*

days and weeks in successive 24 h incubations) have confirmed slow loss after about 1-2 days (Cole et al. 1984, Gabrielson and Hamel 1985). The highest initial loss rate of 43% within 24 h were found in situations with dominance of the cyanobacteria *Microcystis* and *Aphanizomenon*. Compared with other taxa the cyanobacteria are among the least resistant to decomposition (Gunnison and Alexander 1975). Probably due to the chemical nature of their cell wall (Fallon and Brock 1979). The loss of organics from the algal cells is not affected by the presence of bacteria. In accordance with the results of Cole et al. (1984) Hansen et al. (1986) observed no temperature dependency during the initial (24 h) leaching phase. It can thus be concluded that the initial lysis is a simple water extraction from cells without membrane control. Cole et al. (1984) found total leaching over several days to be temperature dependent, thus indicating the effect of bacteria.

In the bottles with natural microheterotrophic populations, bacterial utilization of leached DOC was seen as a lowering of DOC compared with the control bottles. From Fig. 3.36 it is depicted that loss rates are initially higher than uptake rates. After about 10 h the bacteria had reduced the DOC level significantly. The actual time of DOC disappearance is temperature dependent. Cole et al. (1984) demonstrated a seasonal cycle in the conversion of particulate detritus to $CO_2$, the ultimate fate of organic carbon undergoing decomposition. In summer the conversion rate averaged 6% day$^{-1}$ for a 6 days' period compared with 0.5% day$^{-1}$ in winter. The bacterial growth efficiency using initial lysis products is about 60% (Cole et al. 1984, Hansen et al. 1986).

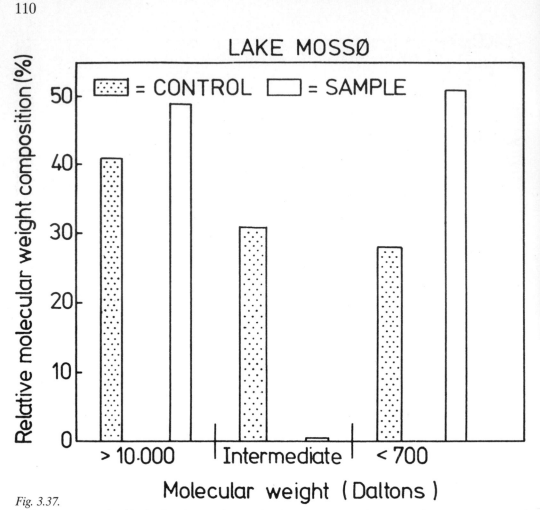

Fig. 3.37.
*Gel-chromatography (Sephadex G-50) of DOC leached from dead blue-green algae (Microcystis and Aphanizomenon) in Lake Mossø, August 1982 after 24 h. Control = algae exposed to sterile lake water. Sample = algae incubated in natural lake water. After data in Krog et al. (1986).*

Lysis products differ among different algal populations. With cyanobacteria as the dominating algal group the molecular weight composition of leaching DOC has almost equal amounts of large, intermediate and small molecules (Fig. 3.37). With diatom dominance, as exemplified by results from Lake Hylke (Fig. 3.38), intermediate sized molecules (2-5,000 Daltons) dominated and could account for about 80% of the total loss.

The chemical composition of the lysis products is not known in detail. In the study performed on diatoms in eutrophic Lake Hylke measurements of cell content revealed polysaccharides and low molecular weight products to dominate during the initial 24 h. No proteins left the dead cells (Krog et al. 1986). This might vary among algal groups. The different patterns of molecules leaching from blue-greens and diatoms indicate species specific products.

Although the composition of lysis products was different in the two situations presented, the bacterial response was similar. All molecular weight groups were assimilated by the bacteria, but preference to the intermediate sized molecules was very distinct (Fig. 3.37 and 3.38). After 24 h of incubation virtually all these molecules were either metabolized or transformed. A relative enrichment of large and small molecules took place. The results from Lake Hylke (Fig. 3.38) showed a bacterial transformation of smaller molecules into larger molecules, which were excreted to the environment.

Considering the available data there is ample evidence (see also Golterman 1960, Fallon and Brock 1980, Gabrielson and Hamel 1985) that immediately after death of an algal cell, a substantial part of the organic constituents is lost to the environment in a dissolved form and readily available as a bacterial substrate. Even death of a minor proportion of a phytoplankton population will create a burst of bacterial substrates influencing the bacterial production as shall be demonstrated in Chapter 4.

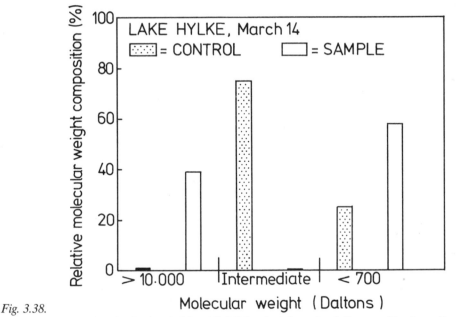

*Fig. 3.38.*
*Gel-chromatography (Sephadex G-50) of DOC leached from dead diatoms (Stephanodiscus) in Lake Hylke, March 1983. See explanation in Fig. 3.24. After data in Krog et al. (1986).*

## Acknowledgement

Several colleagues have made valuable suggestions and linguistic corrections. In particular we would like to acknowledge Dr. J. J. Cole, Institute of Ecosystem Studies, Dr. A. Baily-Watts, Edinburgh, Dr. W. Booth, Dr. J. M. A. Brown, Dr. F. I. Dromgoole, M. Rattray, all University of Auckland, Professor G. E. Fogg, University College of North Wales, Dr. R. Smith, Bedford Institute of Oceanography, Dr. K. Richardson, Danish Institute for Fisheries and Marine Research and Dr. Wayne Bell, University of Maryland. The typing by M. Mikkelsen, A. G. Kreiborg, K. Pedersen, R. Arthur and E. Rosendahl is appreciated.

## 3.5. References

Al-Hasan, R. H. & Coughlan, S. J. 1976. A method for the determination of glycollic acid in the extracellular products of cultures and natural phytoplankton populations. J. exp. mar. Biol. Ecol. 25: 141-149.

Al-Hasan, R. H., Coughlan, S. J., Pand, A. & Fogg, G. E. 1975. Seasonal variations in phytoplankton and glycollate concentrations in the Menai Straits, Anglesay. J. Mar. Biol. Ass. U.K. 55: 557-565.

Anderson, G. C. & Zeutschel, R. P. 1970. Release of dissolved organic matter by marine phytoplankton in coastal and offshore areas of the North-east Pacific Ocean. Limnol. Oceanogr. 15: 402-407.

Anderson, J. P. E. & Domsch, K. H., 1973. Quantificaiton of bacterial and fungal contribution to soil respiration. Arch. Microbiol. 93: 113-127.

Anderson, J. P. E. & Domsch, K. H., 1974. Measurement of bacterial and fungal contribution to respiration of selected agricultural and forest soils. Can. J. Microbiol. 21: 314-322.

Antia, N. J., McAllister, C. D., Parsons, T. R., Stephens, K. & Strickland, J. D. 1963. Further measurements of primary production using a large volume plastic sphere. Limnol. Oceanogr. 8: 166-183.

Arthur, C. R. J. & Rigler, F. H. 1967. A possible source of error in the $^{14}C$ method of measuring primary productivity. Limnol. Oceanogr. 12: 121-124.

Azam, F. & Ammerman, J. W. 1984. Cycling of organic matter by bacterioplankton in pelagic marine ecosystems: Microenvironmental considerations, p. 345-360. In: Fasham, M. J. R. (Ed.), Flow of energy and material in marine ecosystems. Plenum Press.Azam, F. & Hodson, R. E. 1977. Size distribution and activity of marine microheterotrophs. Limnol. Oceanogr. 22: 492-501.

Azam, F. & Holm-Hansen, O. 1973. Use of tritiated substrates in the study of heterotrophy in seawater. Mar. Biol. 23: 191-196.

Azam, F. & Hodson, R. E. 1981. Multiphasic kinetics for D-glucose uptake by assemblages of natural marine bacteria. Mar. Ecol. Prog. Ser. 6: 213-222.

Basham, J. A. & Calvin, M. 1957. The path of carbon in photosynthesis. Englewood Cliffs, N. J.

Bell, R. T. & Kuparinen, J. 1984. Assessing phytoplankton and bacterioplankton production during early spring in Lake Erken, Sweden. Appl. Environ. Microbiol. 48: 1221-1230.

Bell, W. H. 1980. Bacterial utilizaiton of algal extracellular products. 1. The kinetic approach. Limnol. Oceanogr. 25: 1007-1020.

Bell, W. H. 1983. Bacterial utilization of algal extracellular products 3. The specificity of algal-bacterial interaction. Limnol. Oceanogr. 28: 1131-1143.

Bell, W. H. 1984. Bacterial adaptation to low-nutrient conditions as studied with algal extracellular products. Microb. Ecol. 10: 217-230.

Bell, W. H., Lang, J. M. & Mitchell, R. 1974. Selective stimulation of marine bacteria by algal extracellular products. Limnol. Oceanogr. 19: 8323-839.

Bell, W. H. & Mitchell, R. 1972. Chemotactic and growth responses of marine bacteria to algal extracellular products. Biol. Bull. 143: 265-277. Bell, W. H. & Sakshaug, E. 1980. Bacterial utilization of algal extracellular products. 2. A kinetic study of natural populations. Limnol. Oceanogr. 25: 1021-1033.

Berg, K. & Nygaard, G. 1929. Studies on the plankton in the lake of Frederiksborg Castle. Kgl. D. Vid. Selsk. Skr., Naturv.-Math. Afd. 9.R. 1: 227-316.

Berland, B. R. & Maestrini, S. Y. 1969. Study of bacteria associated with marine algae in culture. II. Action of antibiotics substances. Mar. Biol. 3: 334-335.

Berman, T. 1973. Modification in filtration methods for the measurements of inorganic $^{14}$C uptake by photosynthesizing algae. J. Phycol. 9: 327-330.

Berman, T. 1975. Size fractionation of natural aquatic populations associated with autotrophic and heterotrophic carbon uptake. Mar. Biol. 33: 215-220.

Berman, T. 1976. Release of dissolved organic matter by photosynthesizing algae in Lake Kinneret, Israel. Freshwat. Biol. 6: 13-18.

Berman, T. & Gerber, C. 1980. Differential filtration studies of carbon flux from living algae to microheterotrophs, microplankton size distribution and respiration in Lake Kinneret. Microb. Ecol. 6: 189-198.

Berman, T. & Holm-Hansen, O. 1974. Release of photoassimilated carbon as dissolved organic matter by marine phytoplankton. Mar. Biol. 28: 305-310.

Bern, L. 1985. Autoradiographic studies on methyl-$^3$H thymidine incorporation in a cyanobacterium *(Microcystis wesenbergii)* - bacterium association and in selected algae and bacteria. Appl. Environ. Microbiol. 49: 232-233.

Billen, G., Joiris, C., Wijnant, J. & Gillain, G. 1980. Concentration and microbiological utilization of small organic molecules in Scheldt estuary, the Belgian coastal zone of the North Sea and the English Channel. Estuarine Coastal Mar. Sci. 11: 279-294.

Bjørnsen, P.K. 1986. Bacterioplankton growth yield in continous Sea water cultures. Mar. Ecol. Prog. Ser. 30: 191-196.

Blaauboer, M. C. T., Van Kuelen, R. & Cappenberg, Th. E. 1982. Extracellular release of photosynthetic products by freshwater phytoplankton populations, with special reference to the algal species involved. Freshwat. Biol. 12: 559-572.

Bowes, G. & Berry, J. A. 1972. The effect of oxygen on photosynthesis and glycollate excretion in *Chlamydomonas reinhardtii.* Carn. Inst. Year Book 71: 148-158.

Bresta, A.-M., Ursin, Chr. & Jensen, L. M. Submitted. Intercomparison of $^{14}$C-labelled bicarbonate solutions prepared by different institutes for measurements of primary productivity in natural waters and monoalgal cultures. J. Plank. Res.

Brock, T. H. & Clyne, J. 1984. Significance of algal excretory products for growth of epilimnetic bacteria. Appl. Environ. Microbiol. 47: 731-734.

Bölter, M. Liebezeit, G., Wolter, K. & Palmgren, V. 1982. Submodels of a brackish water environment. III. Microbial biomass production and related carbon flux. P.S.Z.N.I. Mar. Ecol. 3: 243-253.

Canter, H. & Lund, J. W. G. 1948. Studies on plankton parasites. 1. Fluctuations in the number of *Asterionella formosa* Hass. in relation to fungal epidemics. New Phytol. 47: 238-261.

Carlucci, A. E., Craven, D. B. & Henrichs, S. M. 1984. Diel production of dissolved free amino acids in waters off Southern California. Appl. Environ. Microbiol. 48: 165-170.

Caron, D. A., Pick, F. R. & Lean, D. R. S. 1985. Chroococcoid cyanobacteria in lake Ontario: Vertical and seasonal distributions during 1982. J. Phycol. 21: 171-175.

Carpenter, E. J. & Lively, J. S. 1980. Review of estimates of algal growth using $^{14}C$ tracer techniques, p. 161-178. In: Falkowski, P. G. (ed.), Primary productivity in the sea.

Chang, V. T.-P. 1980. Zwei neue Synechococcus - Arten aus dem Zürichsee. Schweiz. Z. Hydrol. 42: 247-254.

Chrost, R. J. 1978. The estimation of extracellular release by phytoplankton and heterotrophic activity of aquatic bacteria. Acta Microbiol. Pol. 27: 139-146.

Chrost, R. J. 1981. The composition and bacterial utilization of DOC released by phytoplankton. Kieler Meeresforsch. Sonderh. 5: 325-332.

Chrost, R. J. 1983. Plankton photosynthesis, extracellular release and bacterial utilization of dissolved organic carbon (RDOC) in lakes of different trophy. Acta Microbiol. Pol. 32: 275-287.

Chrost, R. J. & Faust, M. A. 1980. Molecular weight fractionation of dissolved organic matter (DOM) released by phytoplankton. Acta Microbiol. Pol. 29. 79-88.

Chrost, F. & Faust, M. A. 1983. Organic carbon release by phytoplankton: its composition and utilization by bacterioplankton. J. Plank. Res. 5: 477-493.

Cole, J. J. 1982. Microbial decomposition of algal organic matter in an oligotrophic lake. Ph.D. Thesis, Cornell University. 279 pp.

Cole, J. J. 1982. Interactions between bacteria and algae in aquatic ecosystems. Ann. Rev. Ecol. Syst. 13: 291-314.

Cole, J. J. & Likens, G. E. 1979. Measurements of mineralization of phytoplankton detritus in an oligotrophic lake. Limnol. Oceanogr. 24: 541-547.

Cole, J. J., Likens, G. E. & Hobbie, J. E. 1984. Decomposition of planktonic algae in an oligotrophic lake. Oikos 42: 257-266.

Cole, J. J., Likens, G. E. & Strayer, D. L. 1982. Photosynthetically produced dissolved organic carbon: An important carbon source for planktonic bacteria. Limnol. Oceanogr. 27: 1080-1090.

Colman, B. & Hosein, M. L. 1980. The release of glycolate by a freshwater diatom. J. Phycol. 16: 478-479.

Coughlan, S. J. & Al-Hasan, R. H. 1977. Studies of uptake and turnover of a glycollic acid in the Menai Straits, North Wales. J. Ecol. 65: 731-746.

Coveney, M. F. 1982. Bacterial uptake of photosynthetic carbon from freshwater phytoplankton. Oikos 38: 8-20.

Craig, S. R. 1983. Distribution and physiological ecology of picoplankton from Little Round lake, Ontario. Ph.D. Thesis, Queens University, Kingston, Canada.

Craig, S. R. 1984. Productivity of algal picoplankton in a small meromictic lake. Verh. Int. Verein. Theor. Angew. Limnol. 22: 351-354.

Crawford, C. C., Hobbie, J. E. & Webb, K. L. 1973. Utilization of dissolved organic compounds by microorganisms in an estuary, p. 169-180. In: L. H. Stevenson & R. R. Colwell (eds.), Estuarine Microbial Ecology. Univ. S. Carolina.

Cronberg, G. 1982. Phytoplankton changes in Lake Trummen induced by restoration. Folia Limnol. Scand. 18: 1-119.

Cronberg, G. & Weibull, C. 1981. *Cyanodictyon imperfectum,* a new chroococcal blue-green alga from Lake Trummen, Sweden. Arch. Hydrobiol. Suppl. 60: 101-110.

Derenbach, J. R. & Williams, P. J. 1974. Autotrophic and bacterial production. Fractionation of plankton populations by differential filtration of samples from the English Channel. Mar. Biol. 25: 263-269.

Douglas, D. J. 1984. Microautoradiography-based enumeration of photosynthetic picoplankton with estimates of carbon-specific growth rates. Mar. Ecol. Prog. Ser. 14: 223-228.

Dring, M. J. & Jewson, D. H. 1982. What does $^{14}$C uptake by phytoplankton really measure? A theoretical and modelling approach. Proc. R. Soc. London. Ser. B 214: 351-368.

Ducklow, H. W. & Hill, S. M. 1985. The growth of heterotrophic bacteria in the surface waters of warm core rings. Limnol. Oceanogr. 30: 239-259.

Dunstall, T. G. & Nalewajko, C. 1975. Extracellular release in planktonic bacteria. Verh. Int. Verein. Limnol. 19: 2643-2649.

Fallon, R. D. & Brock, T. D. 1979. Decomposition of blue-green algal (cyanobacterial) blooms in Lake Mendota, Wisconsin. Appl. Environ. Microbiol. 37: 820-830.

Fallon, R. D. & Brock, T. D. 1980. Planktonic blue-green algae: Production, sedimentation, and decomposition in Lake Mendota, Wisconsin. Limnol. Oceanogr. 25: 72-88.

Faust, M. A. & Chrost, R. H. 1981. Photosynthesis, extracellular release and heterotrophy of dissolved organic matter in Rhode River estuarine plankton, p. 465-479. In: Neilson, B. F. & Cronin, L. E. (Eds.), Estuaries and nutrients. The Humana Press Inc. Califton, NJ.Findenegg, I. 1943. Untersuchungen über die Ökologie und die Produktionsverhältnisse des Planktons im Kärntner Seengebiete. Int. Rev. ges. Hydrobiol. u. Hydrogr. 43: 368-429.Fogg, G. E. 1952. The production of extracellular nitrogenous substances by a blue-green alga. Proc. Roy. Soc. Lond. Ser. B. 139: 372-397.

Fogg, G. E. 1958. Extracellular products of phytoplankton and the estimation of primary production. Rap. Proc. Verb. Reu. Cons. perm. Int. Expl. Mer. 144: 56-60.

Fogg, G. E. 1966. The extracellular products of algae. Oceanogr. Mar. Biol. Ann. Rev. 4: 195-212.

Fogg, G. E. 1971. Extracxellular products of algae in fresh water. Arch. Hydrobiol. Beih. Ergeb. Limnol. 5: 1-25.

Fogg, G. E. 1975. Biochemical pathways in unicellular plants. In: Cooper, J. P. (ed.), Photosynthesis and productivity in different environments. Cambridge Univ. Press. Cambridge, p. 437-457.

116

Fogg, G. E. 1977. Excretion of organic matter by phytoplankton. Limnol. Ocea-nogr. 22: 576-577.Fogg, G. E. 1983. The ecological significance of extracellular products of phytoplankton photosynthesis. Bot. Mar. 26: 3-14.

Fogg, G. E., Nalewajko, C. & Watt, W. D. 1965. Extracellular products of phytoplankton photosynthesis. Proc. Roy. Soc. Ser. B. 162: 517-534.

Fogg, G. E. & Westlake, D. F. 1955. The importance of extracellular products of algae in freshwater. Verh. Int. Verein. Limnol. 12: 219-232.

Fuhrman, J. A. & Bell, T. M. 1985. Biological considerations in the measurement of dissolved free amino acids in seawater and implications for chemical and microbiological studies. Mar. Ecol. Prog. Ser. 25: 13-21.

Gaarder, T. & Gran, H. H. 1927. Investigations on the production of plankton in the Oslo fjord. Rap. Proc. Verb. Cons. Int. Expl. Mer. 42: 3-48.

Gabrielson, J. O. & Hamel, K. S. 1985. Decomposition of the cyanobacterium *Nodularia spumigena*. Bot. Mar. 28: 23-27.

Gargas, E. 1975. A manual for phytoplankton primary production studies in the Baltic. The Baltic Marine Biologists, Publ. No. 2. Water Quality Institute, Denmark 88 pp.

Gieskes, W. W. C. & Kraay, G. W. 1980. Primary productivity and phytoplankton pigment measurements in the northern North Sea during FLEX 87. Meteor Forsch.-Ergebnisse A22: 105-112.

Gieskes, W. W. C. & Kraay, G. W. 1984. State-of-the-art in the measurement of primary production, p. 171-190. In: Fasham, M. J. R. (Ed.), Flows of energy and materials in marine ecosystems. Theory and practice. Plenum Press, New York.

Goldman, C. R., Steemann Nielsen, E., Vollenweider, R. A. & Wetzel, R. G. 1972. The [14]C light and dark bottle technique, p. 88-91. In: Vollenweider, R. A. (Ed.), A manual on methods for measuring primary production in aquatic environments. IBP Handbook No. 12, Blackwell Sci. Publ. Oxford.Golterman, H. L. 1960. Studies on the cycle of elements in freshwater. Acta Bot. Neerl. 9: 1-58.

Gunnison, D. & Alexander, M. 1975. Resistance and susceptibility of algae to decomposition by natural microbial communities. Limnol. Oceanogr. 20: 64-70.

Gächter, R. & Mares, A. 1979. Comments on the acidification and bubbling method for determining phytoplankton production. Oikos 33: 69-73.

Gächter, R., Mares, A. & Tilzer, M. M. 1984. Determination of phytoplankton production by the radiocarbon method: a comparison between the acidification and bubbling method (ABM) and the filtration technique. J. Plank. Res. 6: 359-364.

Hall, S. L. & Fisher, F. M. 1983. Organic compounds released from a natural population of epibenthic bluegreen algae. J. Phycol. 19: 365-368.

Hammer, K. D. & Brockmann, V. H. 1983. Rhytmic release of dissolved free amino acids from partly synchronized *Thalassiosira rotula* under nearly natural conditions. Mar. Biol. 74: 305-312.

Hansen, L., Krog, G. F. & Søndergaard, M. 1986. Decomposition of lake phytoplankton. 1. Dynamics of short-term decomposition. Oikos 46: 37-44.

Harder, R. 1917. Ernährungsphysiologische Untersuchungen an Cyanophyceen hauptsächlich dem endophytischen *Nostoc punctiforme.* Z. Bot. 9: 145.

Hargrave, B. T. 1969. Epibenthic algal production and community respiration inthe sediments of Marion Lake. J. Fish. Res. Bd. Can. 26: 2003-2026.

Harris, G. P. 1978. Photosynthesis, productivity and growth: The physiological ecology of phytoplankton. Arch. Hydrobiol. Beih. Ergebn. Limnol. 10: 1-171.

Harris, G. P. 1980. The measurement of photosynthesis in natural populations of phytoplankton, p. 129-187. In: Morris, I. (Ed.), The physiological ecology of phytoplankton. Blackwell, Oxford.

Heaney, S. I. 1976. Temporal and spatial distribution of the dinoflagellate *Ceratium hirundinalla* O. F. Müller within a small productive lake. Freshwat. Biol. 6: 531-542.

Hellebust, J. A. 1965. Excretion of some organic compounds by marine phytoplankton. Limnol. Oceanogr. 10: 192-206.

Hellebust, J. A. 1974. Extracellular products. In: Stewart, W.D.P. (ed.), Algal physiology and biochemistry. Blackwell, Oxford. p. 838-865.

Herbland, A. 1975. Utilization par la flore hétérotrophe de la matiére organique naturelle dans l'eau de mer. J. exp. mar. Biol. Ecol. 19: 19-31.

Hickel, B. 1975. Changes in phytoplankton composition since 1894 in two lakes of East-Holstein, Germany. Verh. Int. Verein. Limnol. 19: 1229-1240.

Hobbie, J. E., Daley, R. J. & Jasper, S. 1977. Use of nuclepore filters for counting bacteria by fluorescence microscopy. Appl. Environ. Microbiol. 33: 1225-1228.

Hobbie, J. E. & Rublee, P. 1977. Radioisotope studies of heterotrophic bacteria in aquatic ecosystems, p. 443-476. In: J. Cairns, Jr. (Ed.), Aquatic Microbial Communities. Garland. New York.

Hobson, L. A., Morris, W. J. & Pirquet, K. L. 1976. Theoretical and experimental analysis of the $^{14}$C technique and its use in studies of primary production. J. Fish. Res. Bd. Can. 33: 1715-1721.

Ignatiades, J. A. & Fogg, G. E. 1973. Studies on the factors affecting the release of organic matter by *Skeletonema costatum* (Greville) Cleve in culture. J. Mar. Biol. Ass. (U.K.) 53: 937-956.

Ittekkot, V. 1982. Variations of dissolved organic matter during a plankton bloom: qualitative aspects, based on sugar and amino acid analyses. Mar. Chem. 11: 143-158.

Iturriaga, R. 1981. Phytoplankton photoassimilated extracellular products: heterotrophic utilization in marine environments. Kieler Meeresforsch. 5: 318-324.

Iturriaga, R. & Hoppe, H.-G. 1977. Observations of heterotrophic activity on photoassimilated organic matter. Mar. Biol. 40: 101-108.

Iturriaga, R. & Zsolnay, A. 1981. Transformation of some dissolved organic compounds by a natural heterotrophic population. Mar. Biol. 62: 125-129.

Iturriaga, R. & Zsolnay, A. 1983. Heterotrophic uptake and transformation of phytoplankton extracellular products. Bot. Mar. 26: 375-381.

Jassby, A. D. & Goldman, C. R. 1974. Loss rates from a phytoplankton community. Limnol. Oceanogr. 19: 618-627.

Jensen, L. M. 1983. Phytoplankton release of extracellular organic carbon, molecular weight composition, and bacterial assimilation. Mar. Ecol. Prog. Ser. 11: 39-48.

Jensen, L. M. 1984. Antimicrobial action of antibiotics on bacterial and algal carbon metabolism: on the use of antibiotics to estimate bacterial uptake of algal extracellular products (EOC). Arch. Hydrobiol. 99: 423-432.

Jensen, L. M. 1985a. Carbon-14 labelling patterns of phytoplankon: specific activity of different product pools. J. Plank. Res. 7: 643-652.

Jensen, L. M. 1985b. Characterization of native bacteria and their utilization of algal extracellular products by a mixed-substrate kinetic model. Oikos 45: 311-322.

Jensen, L. M., Jørgensen, N. O. G. & Søndergaard, M. 1985. Specific activity: Significance in estimating release rates of extracellular dissolved organic carbon (EOC) by algae. Verh. Int. Verein. Limnol. 22: 2893-2897.

Jensen, L. M. & Søndergaard, M. 1982. Abiotic formation of particles from extracellular organic carbon released by phytoplankton. Microb. Ecol. 8: 47-54.

Jensen, L. M. & Søndergaard, M. 1985. Comparison of two methods to measure algal release of dissolved organic carbon (EOC) and the subsequent uptake by bacteria. J. Plank. Res. 7: 41-56.

Johnson, P. W. & Sieburth, J. McN. 1979. Chroococcoid Cyanobacteria in the sea: A ubiquitous biomass. Limnol. Oceanogr. 24: 928-935.

Johnson, P. W. & Sieburth, J. McN. 1982. In situ morphology and occurrence of eucaryotic phototrophs of bacterial size in the picoplankton of estuarine and oceanic waters. J. Phycol. 18: 318-327.

Joiris, G., Billen, G., Lancelot, C., Daro, M. N., Mommaerts, J. P., Bertels, A., Bossicart, M. & Hecq, J. H. 1982. A budget of carbon cycling in the Belgian coastal zone: Relative role of zooplankton, bacterioplankton and benthos in the utilization of primary production. Neth. J. Sea. Res. 16: 260-275.

Jónasson, P. M. & Kristiansen, J. 1967. Primary and secondary production in Lake Esrom. Growth of *Chironomus anthracinus* in relation to seasonal cycles of phytoplankton and dissolved oxygen. Int. Rev. ges. Hydrobiol. 52: 163-217.

Jones, J. G. 1976. The microbiology and decomposition of seston in open water and experimental enclosures in a productive lake. J. Ecol. 64: 241-278.

Jüttner, F. & Matuschek, T. 1978. The release of low molecular weight compounds by the phytoplankton in a eutrophic lake. Water Res. 12: 251-255.

Jørgensen, N. O. G., Søndergaard, M., Hansen, H. J., Bosselmann, S. & Riemann, B. 1983. Diel variation in concentration, assimilation and respiration of dissolved free amino acids in relation to planktonic primary and secondary production in two eutrophic lakes. Hydrobiologia 107: 107-122.

Kaplan, L. A. & Bott, T. L. 1982. Diel fluctuations of DOC generated by algae in a piedmont stream. Limnol. Oceanogr. 27: 1091-1100.

Kato, K. & Stabel, H.-H. 1984. Studies on the carbon flux from phyto- to bacterioplankton communities in Lake Constance. Arch. Hydrobiol. 102: 177-192.

Keller, M. D., Mague, T. H., Badenhausen, M. & Glover, H. E. 1982. Seasonal variations in the production and consumption of amino acids by coastal microplankton. Est. Coast. Shelf Sci. 15: 301-315.

Kogure, K., Shimidu, U. & Taga N. 1982. Bacterial attachment to phytoplanktonin sea waters. J. exp. mar. Biol. Ecol. 56: 197-204.

Knoechel, R. & Kalff, J. 1978. An *in situ* study of the productivity and population dynamics of five freshwater plankton diatom species. Limnol. Oceanogr. 23: 195-218.

Krempin, D. W. & Sullivan, C. E. 1981. The seasonal abundance, vertical distribution, and relative microbial biomass of chroococcoid cyanobacteria at a station in southern California coastal waters. Can. J. Microbiol. 27: 1341-1344.

Kristiansen, J. 1980. Ferskvandets Planteplankton (in Danish). In: Danmarks Natur 5: 180-209. Third ed., Politikens Forlag, Copenhagen.

Kristiansen, J. 1985. Occurrence of scale-bearing Chrysophyceae in a eutrophic Danish lake. Verh. Int. Verein. Limnol. 22: 2826-2829.

Kristiansen, J. 1986. Silica-scale bearing chrysophytes as environmental indicators. Br. Phycol. J. (in press).

Kristiansen, J. & Mathiesen, H. 1964. Phytoplankton of the Tystryp-Bavelse Lakes. Primary production and standing crop. Oikos 15: 1-43.

Kristiansen, J., Riemann, B., Jacobsen, Aa., Madsen, O. S. & Sørensen A. 1982. Tystrup Sø 1979-81. Næringssalte - primærproduktion - phytoplankton (in Danish). Report to the Danish Agency of the Environment. 91 pp.

Krog, G. F., Hansen, L. & Søndergaard, M. 1986. Decomposition of lake phytoplankton. 2. Composition and lability of lysis products. Oikos 46: 45-50.

Krogh, A., Lange, E. & Smith, W. 1930. On the organic matter given off by algae. Biochem. J. 24: 1666-1671.

Lancelot, C. 1979. Gross excretion rates of natural phytoplankton and heterotrophic uptake of excreted products in the southern North Sea, as determined by short-term kinetics. Mar. Ecol. Prog. Ser. 1: 179-186.

Lancelot, C. 1983. Factors affecting phytoplankton extracellular release in the southern bight of the North Sea. Mar. Ecol. Prog. Ser. 12: 115-121.

Lancelot, C. 1984. Extracellular release of small and large molecules by phytoplankton in the southern bight of the North Sea. Est. Coast. Shelf Sci. 18: 65-77.

Lancelot, C. & Billen, G. 1984. Activity of heterotrophic bacteria and its coupling to primary production during the spring phytoplankton bloom in the southern bight of the North Sea. Limnol. Oceanogr. 29: 721-730.

Larsson, U. & Hagström, A. 1979. Phytoplankton exudate release as an energy source for growth of pelagic bacteria. Mar. Biol. 52: 199-206.

Larsson, U. & Hagström, A. 1982. Fractionated phytoplankton primary production, exudate release and bacterial production in a Baltic eutrophication gradient. Mar. Biol. 67: 57-70.

Lean, D. R. S. & Burnison, B. K. 1979. An evaluation of errors in the $^{14}C$ method of primary production measurements. Limnol. Oceanogr. 24: 917-928.

Lee, K. & Nalewajko, C. 1978. Photosynthesis, extracellular release and glycolic acid uptake by plankton: Fractionation studies. Verh. Int. Verein. Limnol. 20: 257-262.

Lewin, R. A. 1956. Extracellular polysaccharides of green algae. Can. J. Microbiol. 2: 665-672.

Li, W. K. W. 1983. Consideration of errors in estimating kinetic parameters based on Michaelis-Menten formalism in microbial ecology. Limnol. Oceanogr. 23: 262-279.

Li, W. K. W., Subba Rao, D. V., Harrison, W. G., Smith, J. C., Cullen, J. J., Irwin, B. & Platt, T. 1983. Autotrophic picoplankton in the tropical ocean. Science 219: 292-295.

Lund, J. W. G. 1950. Studies on *Asterionella formosa* Hass. II. Nutrient depletion and the spring maximum. J. Ecol. 38: 15-35.

Lund, J. W. G. & Reynolds, C. S. 1982. The development and operation of large limnetic enclosures in Blelham Tarn, English Lake District, and their contribution to phytoplankton ecology. Progr. in Phycol. Res. 1: 1-65.

Mague, T. H., Friberg, E., Hughes, D. J. & Morris, I. 1980. Extracellular release of carbon by marine phytoplankton: a physiological approach. Limnol. Oceanogr. 25: 262-279.

Maita, Y., Yanada, M. & Rikuta, A. 1974. Rates of net assimilation and respiration of amino acids by marine bacteria. J. Oceanogr. Soc. Japan 30: 1-9.

Marra, J., Landriou, G. & Ducklow, H. W. 1981. Tracer kinetics and plankton rate processes in oligotrophic oceans. Mar. Biol. Lett. 2: 215-223.

McKinley, K. R., Ward, A. K. & Wetzel, R. G. 1977. A method for obtaining more precise measures of excreted organic carbon. Limnol. Oceanogr. 22: 570-573.

Morita, R. Y. 1982. Starvation-survival of heterotrophs in the marine environment. Adv. Microb. Ecol. 6: 171-198.

Morris, I. 1980. Paths of carbon assimilation in marine phytoplankton, p. 139-157. In: Falkowski, P. G. (Ed.), Primary productivity in the sea, Plenum Press.

Morris, I. 1981. Photosynthesic products, physiological state, and phytoplankton growth, p. 83-102. In: Platt, T. (Ed.), Physiological bases of phytoplankton ecology, Canadian Bulletin of Fisheries and Aquatic Sciences, Bulletin 210.

Morris, I., Glover, H. E. & Yentsch, C. S. 1974. Products of photosynthesis by marine phytoplankton: The effect of environmental factors on the relative rates of protein synthesis. Mar. Biol. 27: 1-10.

Morris, I., Yentsch, C. M. & Yentsch, C. S. 1971. Relationship between light carbon dioxide fixation and dark carbon dioxide fixation by marine algae. Limnol. Oceanogr. 16: 854-858.

Myklestad, S. 1974. Production of carbohydrates by marine planktonic diatoms. I. Comparison of nine different species in culture. J. exp. mar. Biol. Ecol. 15: 261-274.

Nalewajko, C. 1977. Extracellular release in freshwater algae as a source of carbon for heterotrophs, p. 589-624. In: Cairns, J. (Ed.), Aquatic microbial communities. Garland, New York.

Nalewajko, C., Chowdhori, N. & Fogg, G. E. 1963. Extrection of glycollic acid and the growth of a planktonic *Chlorella*, p. 171-183. In: Studies on microalgae and photosynthetic bacteria. Jap. Soc. Pl. Physiol. Tokyo.

Nalewajko, C., Dunstall, T. G. & Shear, H. 1976. Kinetics of extracellular release in axenic algae and in mixed algal-bacterial cultures: significance in estimationof total (gross) phytoplankton excretion rates. J. Phycol. 12: 1-5.

Nalewajko, C. & Lean, D. R. S. 1972. Growth and excretion in planktonic algae and bacteria. J. Phycol. 8: 361-366.

Nalewajko, C., Lee, K. & Fay, P. 1980. Significance of algal extracellular products to bacteria in lakes and in cultures. Microb. Ecol. 6: 199-207.

Nalewajko, C. & Marin, L. 1969. Extracellular production in relation to the growth of four planktonic algae and of phytoplankton populations from Lake Ontario. Can. J. Bot. 47: 405-413.

Nalewajko, C. & Schindler, D. W. 1976. Primary production, extracellular release, and heterotrophy in two lakes in the ELA, Northwestern Ontario. J. Fish. Res. Bd. Can. 33: 219-226.

Novitsky, J. A. 1983. Heterotrophic activity throughout a vertical profile of seawater and sediment in Halifax Harbor, Canada. Appl. Environ. Microbiol. 45: 1753-1760.

Nygaard, G. 1949. Hydrobiological studies on some Danish ponds and lakes. Kgl. D. Vid. Selsk. Biol. Skr. 7(1): 1-291.

Nygaard, G. 1958. Furesøens planteplankton (in Danish). In: Berg, K. et al., Furesøundersøgelser 1950-54. Folia Limnol. Scand. 10: 109-113.

Olrik, K. 1973. Phytoplankton from four culturally influenced lakes of the Mølleå system, North Zealand, Denmark. Bot. Tidsskr. 68: 1-29.

Olrik, K. 1978. Cyanophyceae and environmental factors in 15 Danish lakes. Verh. Int. Verein. Limnol. 20: 690-695.

Olrik, K. 1981. Succession of phytoplankton in response to environmental factors in Lake Arresø, North Zealand, Denmark. Schweiz. Z. Hydrol. 43: 6-19.

Overbeck, J. 1979. Dark $CO_2$ uptake-biochemical background and its relevance to in situ bacterial production. Arch. Hydrobiol. Beih. Ergebn. Limnol. 12: 38-47.

Palumbo, A. V., Ferguson, R. L. & Rublee, P. A. 1984. Size of suspended bacterial cells and association of heterotrophic activity with size fractionations of particles in estuarine and coastal waters. Appl. Environ. Microbiol. 48: 157-164.

Payne, W. J. & Wiebe, W. J. 1978. Growth yield and efficiency in chemosynthetic micro-organisms. Ann. Rev. Microbiol. 32: 155-183.

Pearsall, W. H. 1932. Phytoplankton in the English Lakes. 2. The composition of the phytoplankton in relation to dissolved substances. J. Ecol. 20: 241-262.

Peterson, B. J. 1980. Aquatic primary productivity and the $^{14}C$-$CO_2$-method: A history of the productivity problem. Ann. Rev. Ecol. Syst. 11: 359-385.

Platt, T., Rao, D. V. S. & Irwin, B. 1983. Photosynthesis of picoplankton in the oligotrophic ocean. Nature (Lond.) 301: 702-704.

Poindexter, J. S. 1981. Oligotrophy: fast and famine existence. Adv. Microb. Ecol. 5: 63-89.

Rai, H. 1984. Magnitude of heterotrophic metabolism of photosynthetically fixed dissolved organic carbon (PDOC) in Schöhsee, West Germany. Arch. Hydrobiol. 102: 91-103.

122

Reinertsen, H. & Olsen, Y. 1984. Effects of fish elimination on the phytoplankton community of a eutrophic lake. Verh. Int. Verein. Limnol. 22: 649-657.

Reynolds, C. S. 1973. The phytoplankton of Cross Mere, Shropshire. Br. Phycol. J. 8: 153-162.

Reynolds, C. S., Thompson, J. M., Fergusson, A. J. D. & Wiseman, S. W. 1982. Loss processes in the population dynamics of phytoplankton maintained by closed systems. J. Plank. Res. 4: 561-600.

Riemann, B. 1978. Differentiation between heterotrophic and photosynthetic plankton by size fractionation, glucose uptake, ATP and chlorophyll content. Oikos 31: 368-367.

Riemann, B. 1980. Diurnal variations in the uptake of glucose by attached and free-living microheterotrophs in lake water (Lake Mossø, Denmark). Developments in Hydrobiol. Vol. 3: 153-160.

Riemann, B. 1983. Biomass and production of phyto- and bacterioplankton in eutrophic Lake Tystrup, Denmark. Freshwat. Biol. 13: 389-398.

Riemann, B, & Søndergaard, M. 1986. Regulation of bacterial secondary production in two eutrophic lakes and in experimental enclosures. J. Plank. Res. 8: 519-536.

Riemann, B., Søndergaard, M., Schierup, H.-H., Bosselmann, S.,Christensen, G., Hansen, J. & Nielsen, B. 1982. Carbon metabolism during a spring diatom bloom in the eutrophic Lake Mossø. Int. Rev. ges. Hydrobiol. 67: 145-185.

Roberts, R. B., Cowie, D. B., Abelson, P. H., Bolton, E. T. & Britton, R. J. 1955. Studies of biosynthesis in *Escherichia coli*. Publs. Carnegie Instn. 607, p. 1-521.

Rodhe, W. 1958. The primary production in lakes: some results and restrictions of the [14]C method. Rap. Proc. Verb. Cons. Perm. Int. Expl. Mer. 144: 122-128.

Rodhe, W., Vollenweider, R. A. & Nauwerck, A. 1958. The primary production and standing crop of phytoplankton, p. 299-322. In: Buzzati-Traverso, A. A. (Ed.), Perspectives in Marine Biology. Berkeley, University of California Press.

Saks, N. M. 1982. Primary production and release of assimilated carbon by Chlamydomonas provasoli in culture. Mar. Biol. 70: 205-208.

Salonen, K. 1974. Effectiveness of cellulose ester and perforated polycarbonate membrane filters in separating bacteria and phytoplankton. Ann. Bot. Fennici 11: 133-135.

Sand-Jensen, K. & Søndergaard, M. 1981. Phytoplankton and epiphyte development and their shading effect on submerged macrophytes in lakes of different nutrient status. Int. Rev. ges. Hydrobiol. 66: 529-552.

Saunders, G. W. 1972a. The transformation of artificial detritus in lake water. Mem. Ist. Ital. Idrobiol. 29 (Suppl.): 533-540.

Saunders, G. W. 1972b. The kinetics of extracellular release of soluble organic matter by phytoplankton. Verh. Int. Verein. Limnol. 18: 140-146.

Schindler, D. W., Schmidt, R. V. & Reid, R. A. 1972. Acidification and bubbling as an alternative to filtration in determining phytoplankton production by the [14] method. J. Fish. Res. Bd. Can. 29: 1627-1631.

Schleyer, M. H. 1980. A preliminary evaluation of heterotrophic utilisation of a labelled algal extract in a subtidal reef environment. Mar. Ecol. Prog. Ser. 3: 223-229.

Schleyer, M. H. 1981. Microorganisms and detritus in the water column of a subtidal Reef of Natal. Mar. Ecol. Prog. Ser. 4: 307-320.

Seki, H., Nakai, T. & Otobe, H. 1972. Regional differences on turnover rate of dissolved materials in the Pacific Ocean at summer of 1971. Arch. Hydrobiol. 71: 79-89.

Seki, H., Nakai, T. & Otobe, H. 1974. Turnover rate of dissolved materials in the Philippine Sea at winter of 1973. Arch. Hydrobiol. 73: 238-244.

Sellner, K. G. 1981. Primary productivity and the flux of dissolved organic matter in several marine environments. Mar. Biol. 65: 101-112.

Shapiro, J. 1984. Blue-green dominance in lakes: The role and management significance of pH and $CO_2$. Int. Rev. ges. Hydrobiol. 69: 765-780.

Sharp, J. H. 1977. Excretion of organic matter by marine phytoplankton: Do healthy cells do it? Limnol. Oceanogr. 22: 381-399.

Sheldon, R. W. 1972. Size separations of marine seston by membrane and glass fibre filters. Limnol. Oceanogr. 17. 494-498.

Sieburth, J. McN., Smetacek, V. & Lenz, J. 1978. Pelagic ecosystem structure: Heterotrophic compartments of the plankton and their relationship to plankton size fractions. Limnol. Oceanogr. 23: 1256-1263.

Silver, M. W. & Davoll, P. J. 1978. Loss of $^{14}C$ activity after chemical fixation of phytoplankton: error source for autoradiography and other productivity measurements. Limnol. Oceanogr. 23: 362-367.

Smith, D. F. 1974. Quantitative analysis of the functional relationships existing between ecosystem components. I. Analysis of the linear intercomponent mass transfers. Oecologia 16: 97-106.

Smith, D. F. & Higgins, H. W. 1978. An interspecies regulatory control of dissolved organic carbon production by phytoplankton and incorporation by microheterotrophs, p. 34-39. In: Lomit, M. W. & Miles, J. A. R. (Eds.), Springer Verlag.

Smith, D. F. & Wiebe, W. J. 1976. Constant release of photosynthate from marine phytoplankton. Appl. Environ. Microbiol. 32: 75-79.

Smith, D. F., Wiebe, W. J. & Higgins, H. W. 1984. Heterotrophic potential estimates: an inherent paradox in assuming Michaelis-Menten kinetics. Mar. Ecol. Prog. Ser. 17: 49-56.

Smith, R. E. H. 1982. The estimation of phytoplankton production and excretion by $^{14}C$. Mar. Biol. Lett. 3: 325-334.

Smith, R. E. H. & Platt, T. 1984. Carbon exchange and $^{14}C$ tracer methods in a nitrogen limited diatom, *Thalassiosira pseudonana*. Mar. Ecol. Prog. Ser. 16: 75-87.

Smith, W. O. Jr. 1975. The optimal procedures for the measurement of phytoplankton excretion. Mar. Science Com. 1: 395-405.

Smith, W. O. Jr., Barber, R. T. & Huntsman, S. A. 1977. Primary production off the coast of northwest Africa: Excretion of dissolved organic matter and its het

124

erotrophic uptake, Deep-Sea Res. 24: 35-47.

Soeder, C. J. & Boltze, A. 1981. Sulfate deficiency stimulates release of dissolvedorganic matter in synchronous cultures of *Scenedesmus obliquus.* Physiol. Plant. 52: 233-238.

Sommer, U. 1981. The role of r- and k-selection in the succession of phytoplankton in Lake Constance. Acta Oecologia/Oecol. Gener. 2: 327-342.

Spencer, C. P. 1952. On the use of antibiotics for isolating bacteria-free cultures of marine phytoplankton organisms. J. Mar. Biol. Ass. U.K. 31: 97-106.

Steemann Nielsen, E. 1952. The use of radioactive carbon ($^{14}$C) for measuring organic production in the Sea. J. Cons. Perm. Int. Explor. Mer. 18: 117-140.

Steemann Nielsen, E. 1955. The interaction of photosynthesis and respiration and its importance for the determination of $^{14}$C-discrimination in photosynthesis. Physiol. Plant 8: 945-953.

Steinberg, C. 1978. Freisetzung gelösten organische Kohlenstoffs (DOC) verschiedener Molekülgrössen in Planktongesellschaften. Arch. Hydrobiol. 82: 155-165.

Stewart, R. & Codd, G. A. 1981. Glycollate and glyoxylate excretion by *Sphaerocystis schroeteri* (Chlorophyceae). Br. Phycol. J. 16: 177-182.

Storch, T. A. & Saunders, W. 1975. Estimating daily rates of extracellular dissolved organic carbon release by phytoplankton populations. Verh. Int. Verein. Limnol. 19: 952-958.

Strickland, I. D. H. & Parsons, T. R. 1972. A practical handbook of seawater analysis. Fish. Res. Bd. Canada, Bull. 167. Ottawa p. 310

Søndergaard, M. 1980. Adsorption of inorganic carbon-14 to polyethylene scintillation vials - a possible source of error in measures of extracellular release of organic carbon. Arch. Hydrobiol. 90: 362-366.

Søndergaard, M. 1985. On the radiocarbon method: filtration or the acidification and bubbling method. J. Plank. Res. 7: 391-397.

Søndergaard, M. 1986. Photosynthesis of aquatic plants under natural conditions. In: Symoens, J. J. (Ed.), Aquatic Vegetation. Vol. 15: Handbook of Vegetation Science. Dr. W. Junk Publ., Netherlands. In press.

Søndergaard, M., Emmery, L. & Hansen, K. S. 1984. Phytoplankton development and release of extracellular organic products (EOC): Enclosure experiments, p. 35-44. In: Bosheim, S. & Nicholls, M. (Eds.), Interaksjoner mellom trofiske nivoer i ferskvann. Norsk Limnologforening. ISBN 82-990973-3-9.

Søndergaard, M., Riemann, B. & Jørgensen, N. O. G. 1985. Extracellular organic carbon (EOC) released by phytoplankton and bacterial production. Oikos 45: 323-332.

Søndergaard, M. et al. 1986. Pelagic food web processes and community structures in oligotrophic Lake Almind, Denmark. In preparation.

Søndergaard, M. & Schierup, H.-H. 1982a. Release of extracellular organic carbon during a diatom bloom in Lake Mossø: molecular weight fractionation. Freshwat. Biol. 12: 313-320.

Søndergaard, M. & Schierup, H.-H. 1982b. Dissolved organic carbon during a spring diatom bloom in Lake Mossø, Denmark. Water Res. 16: 815-821.

*Anacystis nidulans.* Photochem. Photobiol. 26: 511-518.

Tanaka, N., Nakanishi, M. & Kadota, H. 1974. The excretion of photosynthetic products by natural phytoplankton populations in Lake Biwa. Japan J. Limnol. 35: 91-98.

Theodòrsson, P. & Bjarnasson, J. O. 1975. The acid bubbling method for primary productivity measurements modified and tested. Limnol. Oceanogr. 20: 1018-1019.

Thomas, J. P. 1971. Release of dissolved organic matter from natural populations of marine phytoplankton. Mar. Biol. 11: 311-323.

Tilzer, M. M. & Horne, A. J. 1979. Diel patterns of phytoplankton productivity and extracellular release in ultra-oligotrophic Lake Tahoe. Int. Rev. ges. Hydrobiol. 64: 157-176.

Venrick, E. L., Beers, J. R. & Heinbokel, J. F. 1977. Possible consequences of containing microplankton for physiological rate measurements. J. exp. mar. Biol. Ecol. 26: 55-76.

Vollenweider, R. A. (ed.) 1972. A manual on methods for measuring primary production in aquatic environments. IBP Handbook No. 12, 2nd ed. Blackwell Sci. Publ. Oxford, 225 p.

Watanabe, Y. 1980. A study of the excretion and extracellular products of natural phytoplankton in Lake Nakanuma, Japan. Int. Rev. ges. Hydrobiol. 65: 809-834.

Waterbury, K. B., Watson, S. W., Guillard, R. R. & Brand, L. E. 1979. Widespread occurrence of a unicellular, marine, planktonic cyanobacterium. Nature. (Lond.) 277: 293-294.

Watt, W. D. 1966. Release of dissolved organic material from the cells of phytoplankton populations. Proc. Roy. Soc. Ser. B. 164: 521-551.

Watt, W. D. 1969. Extracellular release of organic matter from two freshwater diatoms. Ann. Bot. 33: 427-437.

Watt, W. D. & Fogg, G. E. 1966. The kinetics of extracellular glycollate production by *Chlorella pyrenoidosa.* J. exp. Bot. 17: 117-134

Welschmeyer, N. A. & Lorenzen, C. J. 1984. Carbon-14 labelling of phytoplankton carbon and chlorophyll a carbon: Determination of specific growth rates. Limnol. Oceanogr. 29: 135-145.

Wesenberg-Lund, C. 1904. Studier over de danske søers plankton. Specielle del (in Danish). Gyldendal, København.

Wesenberg-Lund, C. 1912. Furesøstudier (in Danish). Kgl. D. Vid. Selsk. Skr., Naturv. Math. Afd. 8, III, 1: 1-208.

Wetzel, R. G. 1983. Limnology. 2nd ed. Saunders College Publ. Philadelphia. 767 p.

Wetzel, R. G. & Likens, G. E. 1979. Limnological analysis. Saunders Co. Philadelphia. 35 pp.

Wiebe, W. J. & Smith, D. F. 1977. Direct measurement of dissolved organic carbon release by phytoplankton and incorporation by microheterotrophs. Mar. Biol. 42: 213-223.

Williams, P. J. LeB. 1981. Incorporation of microheterotrophic processes into the classical paradigm of planktonic food web. Kieler Meeresf. Sonderh. 5: 1-28.

Williams, P. J., Berman, T. & Holm-Hansen, O. 1972. Potential sources of error in the measurement of low rates of planktonic photosynthesis and excretion. Nature (Lond.) 236: 91-92.

Williams, P. J. LeB. & Yentsch, C. S. 1976. An examination of photosynthetic production, excretion of photosynthetic products and heterotrophic utilization of dissolved organic compounds with reference to results from a subtropical sea. Mar. Biol. 35: 31-48.

Wolter, K. 1980. Untersuchungen zur Exsudation Organischer Substanz und deren Aufnahme durch natürliche Bakterienpopulationen. Dissertation (Ph.D. thesis), Universität Kiel. 127 p.

Wolter, K. 1982. Bacterial incorporation of organic substances released by natural phytoplankton populations. Mar. Ecol. Prog. Ser. 7: 287-295.

Wright, R. T. 1972. Some difficulties in using [14]C-organic solutes to measure heterotrophic bacterial activity, p. 199-217. In: L. H. Stevenson and R. R. Colwell (Eds.), Estuarine microbial ecology. Univ. S. Carolina.

Wright, R. T. & Hobbie, J. E. 1965. The uptake of organic solutes in lake water. Limnol. Oceanogr. 10: 22-28.

Wright, R. T. & Shah, N. M. 1975. The trophic role of glycolic acid in coastal seawater. I. Heterotrophic metabolism in seawater and bacterial cultures. Mar. Biol. 33: 175-183.

Wright, R. T. & Shah, N. M. 1977. The trophic role of glycolic acid in coastal seawater. II. Seasonal changes in concentration and heterotrophic use in Ipswich Bay, Massachusetts, USA. Mar. Biol. 43: 257-263.

Yetka, J. E. & Wiebe, W. J. 1974. Ecological application of antibiotics as respiratory inhibitors of bacterial populations. Appl. Environ. Microbiol. 28: 1033-1039.

# Chapter 4. BACTERIA

by Bo Riemann and Morten Søndergaard
with contribution by Niels O. G. Jørgensen (section 4.5.)

## 4.1. Introduction

Knowledge on the ecological importance of aquatic bacteria has increased dramatically during the last few years. The development of new methods has enabled ecologists to measure bacterial biomass as well as the flux of organic matter passing through the bacteria. A huge amount of recent literature concludes that (1) bacterial biomass is a significant proportion of the total biomass (Hobbie et al. 1977, Azam and Fuhrman 1984), (2) free-living bacteria are well adapted for growth in a dilute environment, and their production rates often exceed rates of particle-associated bacteria (Azam and Hodson 1977), and (3) carbon requirement of natural populations of bacteria accounts for of 10-60% of the carbon fixed by the phytoplankton (Fuhrman and Azam 1980, Riemann 1983). This recently developed scenario for the ecological significance of aquatic bateria contrasts with more traditional thinking, which states that (1) the bacterial biomass is insignificant compared to total biomass, (2) pools of utilizable dissolved organic matter are too dilute to support significant growth, and (3) most free-living bacteria are dormant (Wangersky 1977, Stevenson 1978).

The classical method to estimate the number of bacteria was enumeration of colony-forming units on agar plates. For a number of reasons (van Es and Meyer-Reil 1982), only 0.0001-0.1% of the total population grows on such plates. Recent microscopical improvements and new staining procedures have given evidence for the existence of a significant bacterial biomass in both freshwater and marine environments (Francisco et al. 1973, Zimmermann and Meyer-Reil 1974, Hobbie et al. 1977).

The organic matter in lakes is chemically diverse and physically heterogenous. In order to estimate the total flux of various chemical constituents into bacteria, it is necessary to determine concentrations of all utilized constituents as well as their uptake rates. This has been attempted for selected, dissolved components (Andrews and Williams 1971, Crawford et al. 1974, Williams and Yentsch 1976, Bölter et al. 1982); however, it is not possible to measure flux of particulate organic carbon.

A number of recent methodological improvements and new techniques have enabled the aquatic ecologist to estimate the bacterial secondary production in natural environments. Most of the new procedures comprise incorporation of various radiotracers into cellular macromolecules like proteins (Jordan and Likens 1980, Cuhel et al. 1981, Hansen 1984), RNA (Karl 1979, 1982) or DNA (Fuhrman and Azam 1980, Moriarty and Pollard 1981). Quantification of the number of dividing cells has also been used to estimate growth rates of bacteria (Hagström et al. 1979).

These methodological improvements have changed our way of thinking (Hobbie and Williams 1984) and created a new scenario for the ecological importance of aquatic bacteria (Azam and Ammermann 1984). However, much work is still needed to present quantitative evaluations.

Presently, bacterial ecological research has provided a frame for major events controlling the biomass and productivity of aquatic bacteria. Considering that the discussions concerning the reliability of ecological methods are still going on, some details on selected methods are presented as a basis for understanding the data on biomass and production rates which follow.

## 4.2. Determination of cell numbers, biovolume and carbon content

Enumeration of bacterial cells is traditionally carried out by means of 1) plate count procedures or 2) direct counting procedures. Plate count procedures include growth of the bacterial assemblages on agar plates, and after incubation for certain periods of time the developed colonies are counted. Results obtained from plate counts consistently underestimate the microbial community when compared with direct counting procedures (Zobell 1946, Jannasch and Jones 1959), and the majority of ecological literature in aquatic environments is now based on direct count procedures. However, isolation of bacterial strains is still an important tool for various scientific purposes, especially numerical taxonomy and physiological identification (Witzel et al. 1982).

The direct counting procedures include staining of the bacteria and subsequent microscopical enumeration with transmitted light microscopy. Several different dyes have been used, e.g. erythrosin (Sorokin 1969), aniline blue and Ziehl's carbol fuchsin (Casida 1971). Most recently, however, epifluorescence microscopy with fluorochromes has now proved to be most useful in visualizing bacteria from various aquatic environments. Probably the most common dyes are the acridine-based fluorochromes; acridine orange (AO) (Hobbie et al. 1972, Hobbie et al. 1977), and euchrysine 2GNX, 3,6-diamino-2,7-dimethyl-9, methyl acridinium chloride (E-2GNX) (Jones 1974, Jones and Simon 1975). Both dyes predominantly stain nucleic acids.

Another commonly used dye is 4'6-diamidino-2-phenylindole (DAPI), which produces a bright blue fluorescense with DNA (Porter and Feig 1980).

A number of comparisons have been carried out with respect to choice of filters, dyes, volume of sample, lamps etc. (Francisco et al. 1973, Jones 1974, Jones and Simon 1974, Daley and Hobbie 1975, Hobbie et al. 1977). Polycarbonate filters are preferable to ordinary membrane filters, as many of the small bacteria are retained in the matrix of the latter and therefore escape counting. It is necessary to stain the polycarbonate filters in Irgalan black (Hobbie et al. 1977). A short staining time of 10-30 min. is sufficient. The concentration of AO should be about 0.01% (final concentration), and 1-2 minutes' exposure of the bacterial cells to the dye is sufficient to produce thorough staining of the bacteria. To achieve an even distribution of the bacteria on the filter, the surface area of the filters and the volume of water are critical. An uneven distribution has been observed when small volumes (<5 ml) are filtered (Jones and Simon, 1975). We normally use 0.2-1 ml sub-samples diluted to about 5 ml with 0.2 μm filtered distilled water and rinse the 25 mm (diameter) polycarbonate membranes twice, using about 3 ml of distilled water. The damp filter may be used directly or stored dry in the dark at room temperature for months, at least 2 months. J. Overbeck (pers. comm.) has stored filters for years without any significant reduction in the fluorescence. Further details on microscopic equipment (lamps, filters, magnification) are described in Hobbie et al. (1977) and Daley (1979).

Determination of cell dimensions are required to calculate cell volume and, subsequently, the biomass. Both Scanning Electron Microscopy (SEM) and epifluorescence microscopy have been used to measure bacterial cell sizes (e.g. Hagström and Larsson 1984, Fuhrman 1981, Krambeck et al. 1981). Examples of mean cell volumes from freshwater and marine environments vary from 0.015 to 0.462 $\mu m^3$ (Table 4.1). Part of these differences in cell size is probably due to the use of various techniques, although it is claimed that results obtained from SEM may be seriously underestimated compared with epifluorescence (Fuhrman 1981). Seasonal, diel and a variety of abiotic and biotic factors may control the cell volume, yet only few reports have been published. Krambeck et al. (1981) reported only subtle diel changes in the mean cell volume during two diel studies in Lake Pluss-See, and Hagström and Larsson (1984) reported diel changes in cell length between ~0.6-0.9 μm during one study off the Swedish west coast.

Table 4.1. Mean cell volumes of bacteria from variety of natural aquatic environments.

| Locality | Mean cell volume ($\mu m^3$) | Reference |
|---|---|---|
| freshwater | 0.040-0.240 | Salonen (1977) |
| freshwater | 0.015-0.130 | Krambeck et al. (1981) |
| freshwater | 0.042-0.048 | Riemann et al. (1982a) |
| freshwater | 0.083-0.279 | Bell et al. (1983) |
| freshwater | 0.052-0.462 | Pedrós-Alió and Brock (1983) |
| freshwater | 0.09 | Riemann (1983) |
| seawater | 0.081-0.145 | Fuhrman (1981) |
| seawater | 0.015-0.146 | Riemann et al. (1984) |
| seawater | 0.087-0.279 | Watson et al. (1977)[x] |
| seawater | 0.09 | Ferguson and Rublee (1976) |

[x]single cells of **Hyphomicrobium** were > 1 $\mu m^3$

An important drawback of measuring cell volume in the epifluorescence microscope is that the limit of resolution is about 0.2 μm, close to the sizes of many aquatic bacteria. Even though black and white photographs of the epifluorescence picture may improve accuracy of sizing, only small changes in cell diameter, which may be difficult to detect, cause large changes in the calculated volume. In contrast, most SEM equipments have a resolution around 0.01 μm (Krambeck et al. 1981). A more precise measurement of cell size can thus be made at the expense of possible shrinkage phenomena due to handling of the cells (Montesinos et al. 1983, Krambeck et al. 1981) during the more tedious preparation procedure of samples.

The final step to calculate the bacterial carbon biomass involves the use of a conversion factor, since a direct measurement of the carbon content of natural bacterial assemblages is difficult due to presence of often large amount of detrital particles of the same size as bacteria. Proposed values vary from 0.08 to 0.56 pg C μm$^{-3}$ (Ferguson and Rublee 1976, Watson et al. 1977, Bowden 1977, Bratbak and Dundas 1984, Bratbak 1985, Bjørnsen 1986). Several of these estimates are based on *E. coli,* which is not important in natural environments, and moreover, as pointed out by Bratbak and Dundas (1984), proper corrections for intracellular water have not been made. Several authors have used a conversion factor of 0.121 pg C μm$^{-3}$ (Watson et al. 1977, Fuhrman and Azam 1980, Eppley et al. 1981, Riemann and Søndergaard 1984a, Riemann et al. 1984). However, recently Bratbak and Dundas (1984) presented values corrected for intracellular water for *Bacillus subtilis* and *Pseudomonas pirata* to be about 0.22 pg C μm$^{-3}$). Both occur in natural environments. Bjørnsen (1986) reported a conversion factor, obtained from batch cultures of estuarine and freshwater bacteria, of 0.35 pg C μm$^{-3}$.

Possibly, the true conversion factor for natural bacterial assemblages should be higher than 0.121 pg C μm$^{-3}$ (Bratbak 1985, Bjørnsen 1986). Whether or not a revised conversion factor should be exactly 0.22 or even higher or is variable is not clear, since this estimate is based on limited material.

The ecological consequences using the recently proposed factor (0.22 pg C μm$^{-3}$) is that many published values of both biomass and production rates of aquatic bacteria should be about doubled. This certainly stresses the need for a detailed evaluation of the conversion factor to calculate carbon biomass from cell volumes.

## 4.3. Methods to determine growth rates of natural bacterial assemblages

In the chemostat under constant temperature and nutrient conditions, bacteria may be grown under balanced growth conditions. During balanced growth, a number of different approaches may be used to characterize the growth, since cellular constituents like DNA, RNA, and protein remain constant in relation to cell numbers or biomass.

In nature, balanced growth conditions are probably never fulfilled. Changes in temperature and nutrient regime as well as a selective grazing pressure are rapid and may create rapid changes in the growth of bacteria. During such unbalanced growth conditions the relationships between macromolecular synthesis and cell numbers or biomass vary and depend on the growth rate (Måløe and Kjeldgård 1966). In natural bacterial assemblages, different strains, species and specimens might grow at different rates. Thus, determination of the growth may require a detailed examination of growth rates and macromolecules from various parts of the population; a tedious and awkward examination, which is a more or less impossible project in an ecological context.

From the above it appears that since the growth rate of natural bacterial assemblages cannot be defined, no measurements of macromolecular synthesis can be used to define completely the growth of the assemblage. As a consequence, measurements of single constituents in the synthesis of macromolecules give an inaccurate estimate of the growth of the population. How far off we are from true values for growth depends on the method applied, e.g. the cellular RNA content may vary more than the cellular DNA and protein content, so estimates of DNA and protein synthesis may be preferable.

### 4.3.1. Choice of method

During the past decade a number of new methods have been developed to determine growth of natural populations of bacteria. A number of prerequisites must be accepted for each specific method, and a detailed analysis of these prerequisites should preceed the choice of the proper method.

Below, several methods to determine bacterial growth rates are discussed. This section is not intended to be a thorough review of published literature. It is rather our personal evaluation based on literature and on our laboratory and field experiments.

### 4.3.2. $^3$H-thymidine incorporation into DNA

This procedure utilizes the observation that bacteria take up and incorporate exogenous thymidine into DNA and that unlike protein or RNA, DNA is only synthesized for growth and does not turn over within cells. The amount of $^3$H-thymidine incorporated is used to calculate cell production by means of a conversion factor. A number of assumptions are made, and below the following are discussed: (1) concentration of exogenous thymidine, (2) incorporation into cellular macromolecules, (3) dilution by *de novo* synthesis, (4) termination of incubations and extractions of macromolecules, (5) the use of factors to convert thymidine incorporated to cell production, (6) non-bacterial $^3$H-thymidine uptake.

### 4.3.2.1. Concentration of exogenous thymidine

The addition of exogenous thymidine should be high enough to achieve maximal labeling of DNA, but kept low enough to reduce the risk for uptake by phytoplankton and protozoa (Grivell and Jackson 1968, Fuhrman and Azam 1980). During eight experiments in lakes, $^3$H-thymidine flow into total TCA precipitate increased with increasing $^3$H thymidine added (Fig. 4.1). When more than 9.6 nM $^3$H-thymidine was added, a constant incorporation rate in each individual lake was obtained, independent of lake productivity. When less than 9.6 nM $^3$H-thymidine was added, lower incorporation rates were observed. Thus, in eutrophic Lake Lillesø (F in Fig. 4.1), incorporation of $^3$H-thymidine into cold TCA was 150 DPM, when 5 nM $^3$H-thymidine was added, compared with 1,865 DPM when 9.6 nM was

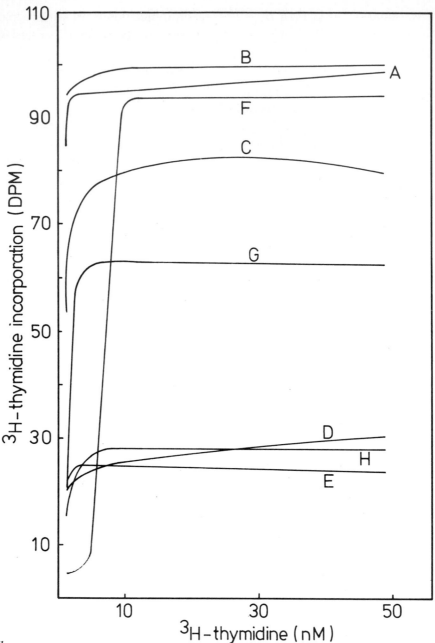

Fig. 4.1.
*The effects of increasing concentration of ³H-thymidine on the rate of ³H-incorporation (DPM × 20) in cold TCA insoluble material measured for water samples from various Danish lakes. Redrawn from Riemann et al. (1982).*

added. In some of the other lakes, such large differences in the incorporation were not found. Experiments from marine environments suggested that between 1.7 and 11 nM ³H-thymidine was enough to maximally label the extracellular and intracellular thymidine pools (and hence DNA) (Fuhrman and Azam 1980), and often 5 nM ³thymidine is used (Fuhrman and Azam 1982). It appears that in lakes additions of about 10 nM $^3$H-thymidine are sufficient. If lower concentrations are used, tests should be made to evaluate whether such low concentrations of $^3$H-thymidine produce lower incorporation rates into DNA.

### 4.3.2.2. Incorporation of $^3$H-thymidine into cellular macromolecules

The biosynthesis of nucleotides in the cells follows two pathways: (1) the *de novo* route and (2) the salvage pathway. A number of metabolic reactions may occur, including catabolism and further degradation of thymidine, which cause a transport of labeled material to both DNA, RNA and protein fractions (seeMoriarty 1986 and Fig. 4.2).

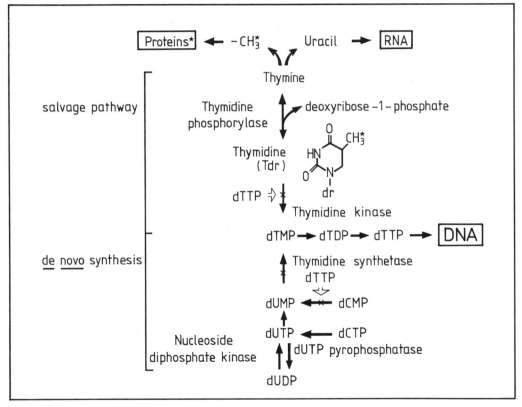

*Fig. 4.2.*
*Pathways of the thymine and thymidine metabolism. dTMP, dTDP, dTTP = thymidine mono-, di- and triphosphate respectively; dUMP, dUDP, dUTP = deoxyuridine mono-, di- and triphosphate, respectively; dCMP, dCTP = deoxycytisine mono- and triphosphate, respectively (redrawn after Moriarty 1986).*

Previously, attempts have been made to differentiate incorporation of $^3$H-thymidine into various macromolecules by means of a modified Schmidt-Tannhäuser procedure (Fuhrman and Azam 1982). After incubation with $^3$H-thymidine, the water sample is divided into three fractions: *(1)* extraction in 5% TCA (final concentration) for 5 min. (this fraction is assumed to contain DNA, RNA and proteins); *(2)* extraction in 0.5 N NaOH (final concentration) at 60°C for 1 h, then acidification and chilling (this fraction is assumed to contain DNA and proteins, RNA is assumed to be hydrolyzed); (3) extraction in 5% TCA at 95°C for 1 h, then chilling (this fraction is assumed to contain only protein). The amount of $^3$H in DNA, RNA and protein fractions can then be calculated by subtraction. Sometimes a "carrier solution" (unlabeled DNA) has been used to improve retention efficiency of $^3$H-labeled macromolecules (Fuhrman and Azam 1982). However, in freshwater samples we found no such effects (Riemann and Søndergaard 1984a).

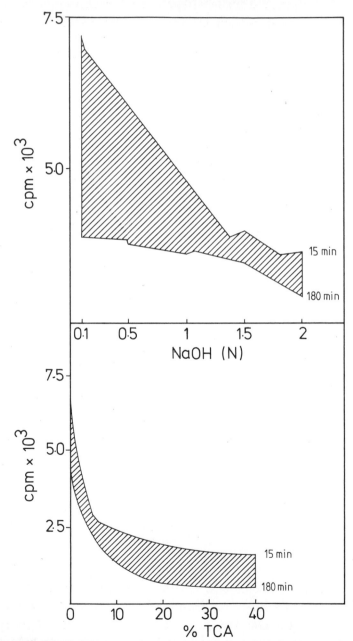

Fig. 4.3.
*The effects of NaOH strength and hot TCA strength and incubation time on the $^3$H-thymidine incorporated into macromolecules. Samples from eutrophic Frederiksborg Slotssø. Redrawn from Riemann (1984).*

By means of the above fractionation procedure, Hollibaugh et al. (1980) reported that 5-20% of total TCA acid insoluble label was resistant to hydrolysis by 95°C hot TCA (step 3 in the procedure described above) and Fuhrman and Azam (1982) concluded from marine environments that 65-80% of the total TCA-insoluble material was DNA. Using lake water samples, the radioactivity in DNA (as a percentage of total TCA insoluble material) decreased with incubation time, probably due to intracellular degradation of thymidine by thymidine phosphorylase (Riemann et al. 1982a). Furthermore, even during short incubations (5-20 min.), radioactivity in DNA was appreciably lower than the 65-80% reported from seawater samples. These discrepancies, and the apparently rapid degradation of thymidine in the freshwater samples, raised the question whether the applied fractionation procedure was in fact adequate (Riemann 1984). Using lake water, changes in concentrations of NaOH and TCA as well as in the incubation time resulted in marked changes in the amount of label in the various fractions (Fig. 4.3). These hyperbolic curves suggest that hydrolysis of the specific macromolecules depend on the experimental design. They also suggest that no distinct separation can be expected from such fractionation procedures. Further evidence for these ideas comes from experiments using RNase and DNase to check if any of the three macromolecular fractions had any cross-contamination of DNA and RNA (Riemann 1984). All three macromolecular fractions were measured for radioactivity. Parallel samples were then treated with RNase or DNase for 3 and 18 hours, respectively, and radioactivity in chilled, acified TCA insoluble material was measured. The difference between the first measurement of radioactivity and the one after RNase or DNase experiment was interpreted as material which was hydrolyzed by the enzymes (presumably RNA and DNA, respectively). An example of such an experiment is presented in Fig. 4.4. RNase-hydrolyzed material was only found in the total TCA insoluble fraction (assumed to contain RNA + DNA + protein), whereas DNase-hydrolyzed material was found in the first two fractions (where DNA is expected to occur), but the protein fraction contained also material which was hydrolyzed by DNase. When the amount of radioactivity calculated from the standard subtraction procedure (cold TCA fraction - base fraction = RNA, base fraction - hot TCA fraction = DNA) was compared with the results obtained from the RNase and DNase experiments (Table 4.2), the major part of the radioactivity was found in the expected fractions, as the results obtained from the two methods were not significantly different. The somewhat lower values in "calculated" radioactivity may be caused by the presence of some DNase-hydrolyzed material in the protein fraction which leads to an underestimation of the calculated DNA.

In view of the theoretical distribution of $^3$H in various compartments of bacterial cells after $^3$H-thymidine uptake, precaution has been taken to ensure purifications of the macromolecules and that measurements are made on $^3$H incorporation into DNA alone. Considering, however, that most often radioactivity in DNA constitutes more than 60% of the radioactivity in the total TCA insoluble material and that some of the remaining 40% may also be DNA material, it is now recom-

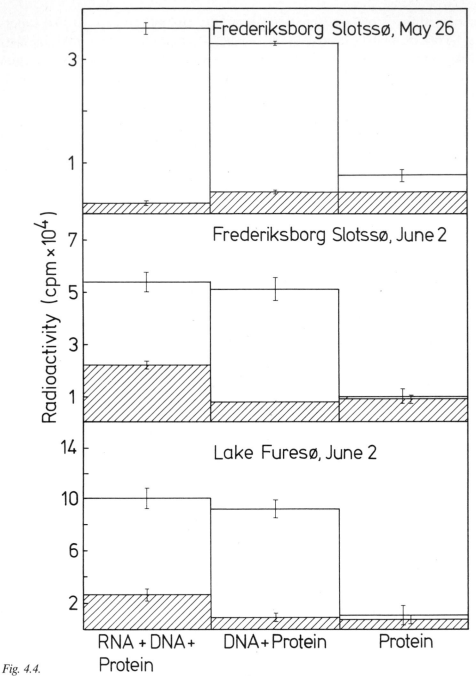

Fig. 4.4.
*The effects of RNase (A) and DNase (B) on the $^3$H-thymidine incorporated into three macromolecular
fractions from eutrophic Frederiksborg Slotssø. Redrawn from Riemann (1984).*

mended to use the radioactivity in the total TCA insoluble material, with only occasional checking of the distribution of $^3$H activity in the various macromolecular fractions (Riemann 1984). Moriarty (1986) reached the same conclusion, but cautioned that longer incubation periods may result in labeling of other macromolecules than DNA.

Table 4.2. [3]H-thymidine incorporated into DNA and RNA estimated from standard procedure by means of macromolecular fractionations and from RNase and DNase experiments (results in cpm, and % of [3]H in cold TCA precipitate. Numbers in brackets are SE, n = 3. From Riemann 1984).

| | Calculated cpm | % | Enzym. determinations cpm | % |
|---|---|---|---|---|
| DNA | 26,314 (1,870) | 78 (5) | 32,000 (1,650) | 95 (8) |
| RNA | 551 (190) | 10 (3) | 323 (96) | 6 (2) |

### 4.3.2.3. Dilution by de novo synthesis

The incorporation rate of exogenous [3]H-thymidine into DNA may be diluted by *de novo* synthesis of deoxythymidine monophosphate (dTMP). Fuhrman and Azam (1980) mentioned the problem and concluded that any correction for unlabeled thymidine and its phosphorylated derivates would increase their production estimates. Moriarty and Pollard (1981) suggested a thymidine isotope dilution procedure to correct for isotope dilution due to de novo synthesis of dTMP and also from exogenous sources.

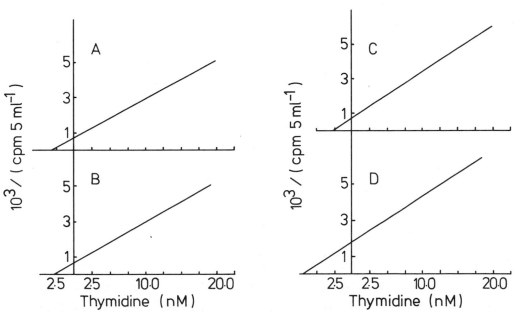

Fig. 4.5.
*Four examples of plots of the reciprocal of [3]H-thymidine incorporated into purified DNA from eutrophic Lake Ørn. Additions of [3]H-thymidine corresponded to 2.5 nM. (A) 8 a.m., (B) 12 a.m.), (C) 16 p.m., (D) 20 p.m. Redrawn from Riemann (1984).*

Briefly, a number of water samples are incubated with a constant amount of $^3$H-thymidine to which increasing amounts of non-$^3$H-labeled (cold) thymidine are added. The $^3$H activity is measured in DNA or in total TCA insoluble material, and plots are made of the reciprocal of the amount of $^3$H incorporated against the concentration of unlabeled thymidine added. The negative intercept on the abscissa refers to dilution of the added thymidine (Moriarty and Pollard 1981). This procedure was used in freshwater (Riemann et al. 1982a). However, varying the range of added cold thymidine from 25-250 nM, non-linear plots were often obtained. Later the amount of cold thymidine was reduced (Riemann 1984), and in most of the experiments straight lines were produced. Four examples of the isotope dilution procedure are presented in Fig. 4.5. In all four examples, purified DNA was used (purified by means of the conventional macromolecular fractionation procedure), and the negative intercept on the abscissa indicated no dilution of the added thymidine by exogenous or endogenous sources in Fig. 4.5. A, B and C. (2.5 nM $^3$H-thymidine was added, so dilution is indicated only when the negative intercept reached more than 2.5 nM). In Fig. 4.5 D, more than a two-fold dilution of the added thymidine was found. Similar results were found by Moriarty (1984) from seawater samples. Thus, it appears that dilution of $^3$H-thymidine into DNA is not often important in pelagic environments and that results obtained from the isotope dilution experiments would not be higher (probably lower, see below) than those based on addition of one single concentration of $^3$H-thymidine. In sediment samples, Moriarty and Pollard (1981) demonstrated large dilutions. However, it was not verified whether these were caused by external or internal pools of thymidine or any phosphorylated derivates. In fact, it might be suspected that the external dilution was important compared with the external dilution in offshore marine and freshwater environments and that the large dilutions obtained from sediments could be reduced by additions of more $^3$H-thymidine. In Fig. 4.5 D a 2-3 fold dilution was found. However, since the Fuhrman and Azam (1982) approach already assumes some dilution (roughly a factor of 4.2), a 2-3 fold dilution obtained from the isotope dilution procedure would not suggest further dilution from external and internal sources. Even though Pollard and Moriarty (1984) demonstrated that growth rates of *Alteromonas undina* (grown in aerobic chemostat) obtained from isotope dilution experiments were not significantly different from growth rates calculated from direct counts, interpretations of the isotope dilution curves are still debatable when natural bacterial assemblages are measured: sometimes non-linear curves are found. The whole analytical procedure is very time-consuming, and the statistical uncertainty for the intercept on the abscissa is large. Moreover, most of the experience from pelagic freshwater and also marine systems appear to demonstrate that dilution of thymidine by de novo synthesis is not important when excess amounts of exogenous thymidine are used. In marine environments, about 5 nM $^3$H-thymidine seems to be sufficient, although only few published data support the choice of this value. In the eutrophic lakes studied, addition of about 10 nM $^3$H-thymidine is required to overcome external dilutions and to reduce de novo synthesis.

### 4.3.2.4. Termination of incubations and extractions of macromolecules

Terminations of water samples incubated with $^3$H-thymidine follow a rapid cooling to about 0-5°C, and additions of equal volumes of 10% ice-cold trichloracetic acid (TCA) is two-fold: (1) it stops further incorporation of $^3$H-thymidine into DNA and (2) releases $^3$H-thymidine taken up by the cells but not incorporated into macromolecules (this material may account for more than 50% of total cellular uptake (Hollibaugh et al. 1980, Riemann unpubl. data)). Most often the bacteria are not visibly harmed (by epifluorescence microscopy) by a 5-30 min. extraction in 5% ice-cold TCA as used by Fuhrman and Azam (1980) and Bell et al. (1983). The above procedure has certain limitations, e.g. when large volumes of sample material should be cooled, no precise termination of the experiment can be done. Moreover, a large amount of ice or other cooling facilities are needed, and this may be impractical in the field. Instead, addition of formalin has been used (Bell et al. 1983, Riemann 1984) and extraction of macromolecules directly from the filters containing the labeled material. In Fig. 4.6, slight decreases (~5%) in $^3$H activity in TCA insoluble material were observed during storage of samples from Lake Norviken, Sweden and from two Danish lakes. We have later improved the procedure by cooling the samples when long storage periods are needed before macromolecular extractions are performed. This appears to prevent any loss of radioactivity for days.

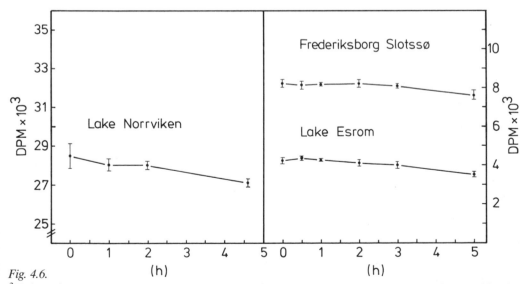

*Fig. 4.6.*
*$^3$H-thymidine incorporation into DNA obtained from terminating the incubation with ice-cold TCA (Time 0) and from terminating the incubation with formalin for various periods of time before filtration. From Riemann (1984) and Bell et al. (1983).*

Formalin fixed samples may be extracted in ice-cold TCA before or after filtration through membrane filters. A simple and reproducible procedure involves filtrations of ice-cooled formalin treated samples by means of a Ten Place Filter Holder

(Bio-Rad). The heavy steel filtration funnels cooled to $<2°C$ prior to use, maintain the temperature below 3-5° during filtration. After filtration the filters are rinsed thoroughly with 5% ice-cold TCA. Small volumes of about 1 ml TCA are most efficient, and filter edges should also be rinsed. When more than 10 ml ice-cold TCA, added in small portions, was drawn through the filters without delay, similar results were found compared with the results obtained from samples extracted prior to filtration (Riemann 1984). It is our experience that reproducibility is better when macromolecular extractions are carried out on the filters in the chilled steel funnels, probably because the temperature may rise above 3-5°C when the TCA extraction is carried out prior to filtration. Especially when 10 ml water samples or more are extracted in equal volumes of 10% TCA, reductions in the $^3H$-activity may occur (Moriarty 1985).

### 4.3.2.5. The use of factors to convert thymidine incorporated into cell production

When the amount of thymidine incorporated into DNA is determined, cell production may be calculated by multiplying results with a conversion factor (Fuhrman and Azam 1980). This factor is based on a considerable number of assumptions (e.g. that the DNA content of bacteria is constant and that thymidine accounts for 25% of the total number of bases incorporated into DNA). A large number of published values concerning these assumptions are present from laboratory cultures of bacteria. These bacteria may not be important in nature, and they are often grown under conditions which seldom occur in natural environments.

Table 4.3. Some published factors to convert rates of thymidine incorporation into rates of bacterial cell production. (M) = marine environments, (F) = freshwater environments.

|  | Conversion factor (x $10^{18}$) |
|---|---|
| Fuhrman and Azam (1980) | 0.2-1.3 (M) |
| Fuhrman and Azam (1982) | 1.7-2.4 (M) |
| Riemann et al. (In press) | 1.1 (SD 0.4) (M) |
| Ducklow and Hill (1985b) | 2.8-6.2 (M) |
| Kirchman et al. (1982) | 1.9-68 (F + M) |
| Bell et al. (1983) | 1.9-2.2 (F) |
| Riemann (1984) | 0.9-7.0 (F) |
| Riemann (1985) | 2.0-7.2 (F) |
| Lovel and Konopka (1985) | 2.2 (F) |
| Murray and Hodson (1985) | 5.8-8.7 (F) |

Attempts have been made to evaluate the conversion factor in natural bacterial assemblages. Fuhrman and Azam (1980) compared the incorporation of $^3H$-thymidine with direct observations of increases in number of bacteria in presumably predator-free (3 or 1 μm filtrates) seawater samples. Values range from 1.7 to 2.4 x $10^{18}$ cells produced per mole thymidine incorporated. Kirchman et al. (1982) used another approach. They diluted the bacterial inoculum with filter sterilized water from which the inoculum originated. In Table 4.3 are listed conversion factors from a number of marine and freshwater environments. Excluding the lowest value from Fuhrman and Azam (1980), because this was calculated exclusively on literature data and has not yet been experimentally confirmed, nearly a 10-fold variation is present. Such a variation may reflect changes between the different environments; however, use of different methods may also affect the results. Thus, Moriarty (1986) suggested that if the bacterial growth rate was stimulated by the dilution or filtration step, then a higher proportion of $^3$H-thymidine may be incorporated into DNA compared with natural, undisturbed samples in which some of the isotope might be incorporated into other cellular materials. This artefact may bias published values of conversion factors, however, to which extent it explains the 10-fold variation in Table 4.3 is not clear. On the other hand, the conversion factors present in Table 4.3 suggest that true rates of bacterial production may be off by a factor of 10, exclusively dependent on the choice of the conversion factor. A detailed examination of factors regulating the conversion factor is urgently needed. At present, it is important to evaluate these factors when new habitats are investigated. Recently, Riemann et al. (in press) reported on evaluation of the conversion factor in coastal marine environments. Various media of natural origin were used as substrates, and temperature and generation time ranged 6-30° and 1 - > 200 h, respectively. The average conversion factor was 1.1 (SE 0.05, n = 63), and no significant changes were found between the conversion factors obtained from the various growth rates, media, or temperatures. The calculation of the conversion factor followed Fuhrman and Azam (1980), which, in contrast to the Kirchman et al. (1982) does not assume exponential growth and that all cells are active. The 1.1 x $10^{18}$ conversion factor is close to the theoretical factor (Fuhrman and Azam 1980), and is without the marked unsystematic changes reported by Kirchman et al. (1982); however, as pointed out by Riemann et al. (in press), this conversion factor can only be used in coastal marine environments. Both the Fuhrman and Azam (1980) procedure and the Kirchman et al. (1982) approach may be used to establish a conversion factor, but careful examinations should be considered with respect to filtration and dilution artefacts, and furthermore it should also be considered that 3.0 or 1.0 μm filtrates of natural water samples cannot always be defined as predator-free, as eukaryote grazers may still be present (Fuhrman and McManus 1984, Andersen and Fenchel 1985).

### 4.3.2.6. Specificity of $^3$H-thymidine uptake (incorporation)

Microorganisms may take up exogenous $^3$H-thymidine; however, the subsequent incorporation into DNA requires the presence of the enzyme thymidine kinase. Using the conventional $^3$H-thymidine incorporation procedure, it is assumed that (1) all active bacteria incorporate exogenous thymidine, and (2) other organisms like algae and protozoa do not incorporate thymidine.

Concerning the first point, some bacteria may not incorporate $^3$H-thymidine. Ramsay (1974) reported that two *Pseudomonas* species could not be labeled, and Pollard and Moriarty (1984) found no labeling of the DNA in two other *Pseudomonas* species and only little DNA labeling in *Alcaligenes aquamarinus*. However, most other bacteria are believed to incorporate $^3$H-thymidine into DNA (Pollard and Moriarty 1984, Moriarty 1985). Supporting indirect evidence comes from studies using microautoradiographic examinations of natural bacterial assemblages (Fuhrman and Azam 1982, Riemann et al. 1984, Marcussen et al. 1984), but more direct evidence is needed to evaluate the extent of this problem.

Concerning point 2, most algae do not contain thymidine kinase (Moriarty 1985). Fuhrman and Azam (1982) and Bern (1985) demonstrated by means of microautoradiography that algae did little or no incorporation of $^3$H-thymidine. In contrast many protozoa have thymidine kinase (cited in Moriarty 1986). It is, however, not likely that protozoa would have transport mechanisms efficient enough to take up exogenous thymidine in nM concentrations during short time incubations (minutes). So, even though other organisms do contain thymidine kinase and have the ability to incorporate exogenous $^3$H-thymidine into DNA, the experimental design seems specific in labeling only bacterial macromolecules.

### 4.3.3. Frequency of dividing cells (FDC)

This method is probably the most elegant of published procedures to determine growth rates of bacteria. It involves enumeration of dividing cells and measurement of the *in situ* temperature. The growth rate is then calculated from relationships between FDC and the temperature. Neither any manipulations of the water samples in question, nor incubations are needed. The bacteria are killed by a fixative, and the number of dividing and non-dividing cells are counted in the epifluorescence microscope.

The immediate simple construction of the method and of the simple performance make it an attractive tool in many studies; however, there are drawbacks and debatable problems which limit its application to a broader scientific audience. Below the following topics are discussed: (1) Determination of FDC, (2) relationships between FDC, the growth rate and the temperature, and (3) the importance of inactive cells.

#### 4.3.3.1. Determination of FDC

A primary assumption to calculate the growth rate is an accurate enumeration of the dividing cells. In the literature, dividing cells are characterized by an invagination of the cell wall, but a clear zone between the daughter cells means that the cells have divided and should not be included in the FDC (Hagström et al. 1979, Newell and Christian 1981). Considering, however, (a) the resolution of the epifluorescence microscope, (b) that mean volumes of aquatic bacteria may be as low as 0.020 $\mu m^3$, and (c) that the number of dividing cells most often is below 10% of the total number of cells (Hagström et al. 1979, Newell and Christian 1981, Larsson and Hagström 1982, Riemann 1983, Hagström and Larsson 1984), large uncertainties may be suspected in the exact enumeration of truly dividing cells. Nevertheless, the coefficients of variation (CV) range between 10% and 29% for trained observers (Riemann 1983, Hagström and Larsson 1984). To achieve such low CV values, it is important to optimize the preparation of sample (e.g. staining procedure, low background etc.), and if very small cells dominate, it is worthwhile evaluating the microscopic counts with numbers of dividing cells enumerated from scanning electron microscopy. The counting procedure is certainly time consuming, and the necessary eye concentration often limits the number of samples. If computer-aided image analyses could be developed to take over the manuel enumeration, the FDC procedure would probably receive much more general interest in future.

### 4.3.3.2. Relationships between FDC, the growth rate, and the temperature

Based on empirical relationships, the growth rate ($\mu$) of the bacterial assemblage is calculated from the FDC and the temperature. Hagström et al. (1979) compared the growth rates of nutrient-enriched batch cultures of bacteria isolated from marine environments (growth rates were determined from measurements of optical density) with FDC. A linear relationship was established for 5, 10 and 15°C. Further measurements from batch and continuous cultures were added (Hagström and Larsson 1984), and new calculations gave the following relationships:

$$0°C; FDC = 11.0\mu + 7.2$$
$$5°C; FDC = 60.2\mu + 5.5$$
$$10°C; FDC = 70.0\mu + 3.8$$
$$15°C; FDC = 36.3\mu + 3.5$$
$$20°C; FDC = 21.2\mu + 0.5$$

Newell and Christian (1981) and Hanson et al. (1983) calibrated FDC with $\mu$ and ln $\mu$ from nutrient-enriched samples of natural bacterial assemblages and found a better fit of the data using ln $\mu$, when growth rates were $< 0.1$ h$^{-1}$ and that the range of deviation of predicted or measured $\mu$ was lower (10-14%) than the deviations calculated from the Hagström et al. formula (20-76%). Whether or not a straight-line fit or a natural-logarithm function should be used is probably not important, at least when growth rates are $<0.1$ h$^{-1}$. Riemann (1983) plotted temperature (°C) versus specific growth rates (h$^{-1}$) for various FDC values (data from Hagström et al. 1979). Points with the same FDC were connected through 5, 10 and 15°C and extrapolated to 20°C. This procedure ensures a direct FDC relationship to the accurate temperature (Fig. 4.7). From these curves it appears that when temperature decreases, a given FDC value corresponds to a decreased growth rate. When the growth rate is kept constant, a lowering of the temperature would increase FDC. Hagström and Larsson (1984) thus reported a doubling of FDC from 5.3 to 10.3%, when the temperature decreased from 20 to 5°C. However, two other experiments did not produce significant changes in FDC when temperatures decreased from 15 to 10 or 5°C. More data are certainly needed, and the present

144

discussion about linear or logarithmic relationships between µ and FDC raises the question whether data from nutrient-enriched mixed cultures and even batch cultures using natural populations can be used to predict growth of natural, undisturbed samples from FDC data. Even short time incubations (minutes, hours), using natural assemblages, may create rapid changes in cell size (Newell and Christian 1981) and in activity (Ferguson et al. 1984). Thus, a rapid succession may favour growth of bacteria which have FDC/µ-relationships different from bacteria in undisturbed samples. Relationships between FDC and µ need to be further specified in seawater, but almost nothing is published from freshwater. When the growth rates of freshwater bacterial assemblages have been determined from FDC, the empirical formulas from marine environments have been used (Pedrós-Alió and Brock 1982, Riemann 1983). A comprehensive evaluation of these fundamental relationships are urgently needed from freshwater systems.

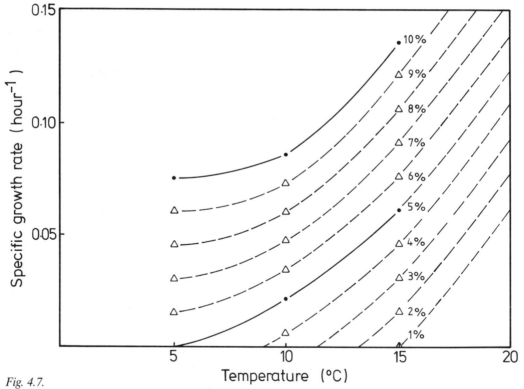

*Fig. 4.7.*
*A plot of temperature (°C) versus specific growth rate (h⁻¹) for various FDC (%) values. Data from Hagström et al. (1979).*

### 4.3.3.3. The importance of inactive cells

Although the problems discussed in the previous section are the most crucial ones, additional problems arise, considering that a variable part of natural populations of bacteria is inactive (dormant?) (Faust and Correl 1977, Stevenson 1978, Meyer-Reil 1978, Fuhrman and Azam 1982, Riemann et al. 1984, Marcussen et al. 1984). Considering that a cell, at least in theory, could stop the growth in the midst of division, then true FDC percentages and the subsequent calculated production rate would be overestimated. On ther other hand, assuming that all dividing cells are in fact active, then FDC and the bacterial biomass may be corrected for inactive cells by means of microautoradiography. Thus, Riemann et al. (1984) found that 20-80% of the bacteria were active, and the corrected production rates, calculated from FDC, were as an average 28% higher than those predicted from the standard FDC procedure. Whether or not these errors in the calculated production rates reflect that the time from constriction to division is not constant at very low growth rates (inactive or apparently inactive cells), as found for *E. coli* (Woldringh 1976), is not clear.

### 4.3.4. Sulphate uptake and incorporation into protein

Considering that the variation of protein per cell as a function of growth rate appears to be small (Måløe and Kjeldgaard 1966), measurements of bacterial protein synthesis is an interesting tool to describe growth rates of natural bacteria. A direct determination of carbon transport into proteins is hampered by additional transport into other cellular compartments. Alternatively, Monheimer (1972, 1974) proposed measuring uptake of S (as $SO_4$), which is assumed to be assimilated in direct proportion to C (Roberts et al. 1955). Jassby (1975) refined the procedure to be used as a direct estimate of bacterial production, and it has been used to measure bacterial growth in relation to phytoplankton production in lakes (Jordan and Likens 1980, Pedrós-Alió and Brock 1982) and rivers (Campbell and Baker 1978).

Briefly, the procedure involves addition of carrier-free $^{35}S$-$SO_4$ to water samples. Incubations are terminated by adding $Na_2SO_4$ (to greatly dilute further $^{35}S$-$SO_4$ uptake), and the samples are filtered through polycarbonate filters (Jordan et al. 1978). The radioactivity incorporated into particulate material is measured by means of conventional scintillation counting procedures. To relate $^{35}S$-$SO_4$ uptake to total $SO_4$ uptake it is necessary to determine the total $SO_4$ content of the water, usually by means of turbidimetric methods (e.g. Lazrus et al. 1968, Tabatabai 1974). From total uptake of $SO_4$, bacterial production estimates can be computed from C:S ratio which may vary from about 50 to 500 (Roberts et al. 1955, Monheimer 1972, 1974). Most often a C:S ratio of 50 or 100 is suggested to be used in natural environments (Jassby 1975, Jordan and Likens 1980, Pedrós-Alió and Brock 1982). As an alternative, Cuhel et al. (1981) proposed to use a C:S ratio in proteins instead of using a C:S ratio based on total cellular S uptake. Apparently the C:S ratio in protein is more constant.

When production rates of aquatic bacteria are determined, the $^{35}S$-$SO_4$ uptake should be carried out in darkness, since uptake in light is carried out by a number of microorganisms (Monheimer 1972). Even in darkness, phytoplankton take up sulphate (Monheimer 1978). Attempts to correct for algal sulphate uptake is not valid, since the C:S ratios vary from 1:150 to 1:10,000 (Monheimer 1981). In consequence, to reduce or eliminate the possible interference from phytoplankton, a detailed size differentiation must be carried out in an attempt to define a size fraction without phytoplankton (see also chapter 3).

The bacterial uptake of S and the subsequent incorporation into protein may come from S-containing compounds of DOM rather than $SO_4$. The S-containing amino acids methionine, cystein and glutathione are found in low concentrations in freshwater (Jørgensen et al. 1983). Assuming that uptake rates of these amino acids are similar to measured uptake rates of other amino acids, this source of S appears not to be important for freshwater bacteria.

The $^{35}S$-$SO_4$ procedure cannot be used under anaerobic conditions, in which $SO_4$ is reduced to $H_2S$. Moreover, if significant amounts of thiosulphate are present, uptake of $SO_4$ may be reduced due to competitive uptake of thiosulphate (Roberts et al. 1955).

### 4.3.5. Dark fixation of $CO_2$

The anaplerotic $CO_2$ dark fixation regenerates metabolic carbon intermediates which are lost from the TCA cycle, and out of five known enzyme systems catalyzing the $CO_2$ fixation, pyruvate and phosphoenolpyruvate appear to be most important (Wood and Stjernholm 1962). From studies in the Rybinsk reservoir, Kutnetsov and Romanenko (1966) reported that dark $CO_2$ uptake accounted for about 6% of the bacterial biomass, however, Overbeck (1979) found that this percentage varied from 0.4 to 30% in Lake Pluss-See. Jordan and Likens (1980) and Riemann and Søndergaard (1984) compared the results from dark $CO_2$ fixation with results from other procedures ($^{35}S$-$SO_4$ uptake, total carbon balance and $^3H$-thymidine incorporation) and found that $CO_2$ dark fixation overestimated the production of the bacteria. Riemann et al. (1982b) reported that a considerable proportion of the measured dark fixation was found as extracellular organic carbon during a spring diatom bloom in eutrophic Lake Mossø. In consequence, the $^{14}C$-activity in the "bacterial" size fraction was not only due to anaplerotic uptake by the bacteria, but probably also to assimilation of extracellular organic carbon products released by the algae. Like the $^{35}S$-$SO_4$ procedure, a size fractionation is required to differentiate between algae and bacterial dark uptake.

### 4.3.6. Increase in cell numbers, cell biomass or ATP

This method is based on observations of cell growth in 3 or 1 µm filtrates. Bacteriovores are assumed to be removed by the filtration procedure, and growth of the bacteria is determined from direct counts of cells (Fuhrman and Azam 1980) or from changes in the ATP content (Sieburth et al. 1977). A direct estimate of biomass may be carried out from the counts when the average cell volume is determined.

The direct counts from such filtrates are specific. The bacteria and the experimental design is simple. It is, however, important to check changes in the cell volume, since marked increases in cell size have been reported (Newell and Christian 1981, Riemann, in prep.). In other situations, changes in cell size were not found (Fuhrman and Azam 1980). These different responses may be due to release of organic substrates during the filtration procedure. Thus, the amount of particulate material and the sensitivity of e.g. phytoplankton cells to filtration stress may control the extent to which the bacteria are nutrient enriched or not. Furthermore, growth of bacteria $>1$ or 3 µm and of bacteria attached to larger particles are ignored.

The measurement of bacterial biomass production from ATP analyses is an extremely sensitive procedure, but ATP may come from other non-bacterial sources that pass through the filters applied. Moreover, C:ATP ratios may vary between organisms and are influenced by nutrient conditions (Karl et al. 1978).

### 4.3.7. [3]H-adenine incorporation into DNA and RNA

This procedure includes a simultaneous measure of both DNA and RNA synthesis, and careful size fractionations have to be carried out, since [3]H-adenine is assimilated by both bacteria and algae (Karl 1981, 1982). Originally, the procedure was designed for RNA synthesis alone (Karl 1979), but later modified for DNA as well (Karl 1981). The procedure involves incubation of water samples with (2-[3]H)-adenine to label the pools of cellular ATP and dATP (precursors of RNA and DNA, respectivly). After incubations, nucleic acids are isolated and measured for radioactivity. The size and radioactivity of the cellular ATP pool are also determined. Calculations of the DNA and RNA synthesis can thus be made from the incorporation rates of [3]H into the nucleic acids as well as into the precursor pools.

The method has been proposed to measure growth rates of entire microbial assemblages (Karl and Winn 1984). However, in an ecological context several limitations arise, since both autotrophic and heterotrophic processes are involved in the measured biosynthesis of nucleic acids. Nevertheless, in studies concerning production and turnover rates of the total microbial RNA and DNA content, the Karl and Winn procedure is interesting. Theoretically, the procedure appears to be directly applicable to freshwater systems, however, at least to our knowledge, only few experimental evidences have yet supported this statement.

### 4.3.8. Determination of bacterial growth by means of different methods

When results obtained from different methods are compared, the following two concepts should be kept in mind: 1) Each method measures different aspects of growth, e.g. aspects of "cell activity", cell division or cell growth, so the information obtained from mixed bacterial populations by means of various techniques are not expected to give the same information, and 2) when comparisons of results are made, the "true" value is not known. In other words, two methods yielding the same quantitative information do not absolutely confirm that true, e.g. growth rates, have been obtained, and results obtained from a third method should not be disregarded because they differ from the other two methods. Are comparisons then of any value? The answer is yes, when several conditions are considered.

For example, the measured growth rates may be compared with major routes of input and output of organic compounds to the bacteria. Although this requires an enormous amount of information in natural environments, it is possible. In such situations, measurements of growth can be discussed in relation to a detailed description of the function of more or less complete ecosysems and not just the results obtained from another method. In other examples, it might be obvious that one method is particularly difficult to apply, e.g. in sediments, under anaerobic conditions or in seawater/freshwater samples.

At present, few comparisons have been carried out. In most cases only two methods are compared (Newell and Fallon 1982, Pedrós-Alió and Brock 1982, Bell et al. 1983). However, a few examples are published where results from several procedures are compared (Riemann and Søndergaard 1984a,b). For some of the methods (e.g. $^3$H-thymidine incorporation versus FDC) results are different by sometimes two orders of magnitude, whereas at best results are within the same order of magnitude (Newell and Fallon 1982, Riemann et al. 1984).

In the previous sections, several techniques to estimate growth rates of natural bacterial assemblages have been discussed. These techniques have been selected on basis of either established or possible applications to pelagic freshwater communities. We did not intend to make a comprehensive review on methods, rather did we want to demonstrate that several methods can be used to measure bacterial production in eutrophic lakes. None of the methods can yet be used as standard techniques. Nevertheless, we have focused on the $^3$H-thymidine incorporation procedure as it fulfils most of the requirements as a specific, quantitative method, which is easy to use in routine studies. There are still uncertainties, e.g. concerning the conversion factor, to calculate the number of cells produced per mole thymidine incorporated, however the majority of recent literature suggests a positive attitude regarding a general application to many pelagic freshwater systems.

In the following sections various factors affecting bacterial growth rates are discussed based on results obtained from various techniques. Here, differences in results much smaller than orders of magnitude have been classified as significant (Newell and Fallon 1982, Pedrós-Alió and Brock 1982, Riemann and Søndergaard 1984a, 1986). In these studies, it is assumed that possible methodological errors are systematic.

## 4.4. Bacterial production rates in eutrophic lakes (diel, vertical and seasonal changes)

### 4.4.1. Diel changes

In order to calculate annual, monthly or weekly production rates of aquatic bacteria, changes during shorter time scales (days, hours) must be evaluated. Nine examples of diel changes in bacterial production are presented in Fig. 4.8. There were only subtle and no systematic diel changes in the production rates. Rates among different lakes varied more than twenty-fold from below 0.05 to more than 1 mg C m$^{-3}$h$^{-1}$, however, rates may exceed 5 mg C m$^{-3}$ h$^{-1}$ (Riemann 1983, Riemann and Søndergaard 1986). In six of the nine studies, production rates increased shortly after sunrise. No specific time of the day or night gave significantly different results (Riemann and Søndergaard 1984a). The diel changes in the bacterial production rates may be ascribed to changes in the activity of phyto- and zooplankton. Thus, it has been demonstrated that phytoplankton release of low-molecular-weight photosynthesis products initiates just after sunrise (Iturriaga

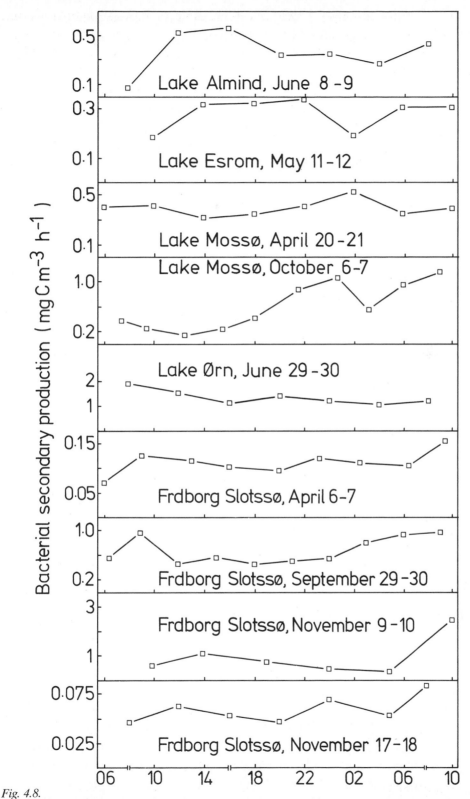

Fig. 4.8.

*Nine examples of diel changes in bacterial production measured by $^3$H-thymidine into DNA. Redrawn from Riemann and Søndergaard (1984a).*

1981) or in some cases with lower phytoplankton biomass, release of labile organic compounds from grazing zooplankton may also be important (Lampert 1978, Riemann et al. 1986, Riemann and Søndergaard 1986, see section 4.5 for further details) Other processes, like sudden wind-induced mixing of deep water bodies, diel rain-induced increased in allochthoneous materials by inlets etc. can also be important, so predictions of diel rhythms in the bacterial secondary production are not possible. In order to estimate diel mean values it is, therefore, necessary to measure production several times. Thus, Riemann and Søndergaard (1984a) calculated, from 13 diel studies that diel mean values would have 95% confidence limits (C.L.) of 42% (% of the mean value), when 3 analyses were carried out. By means of 5 diel samples, C.L. was 16% of the mean. These C.L. estimates were based on measurements of $^3$H-thymidine incorporation. The other methods (FDC and dark $CO_2$ assimilated), all gave higher C.L. values.

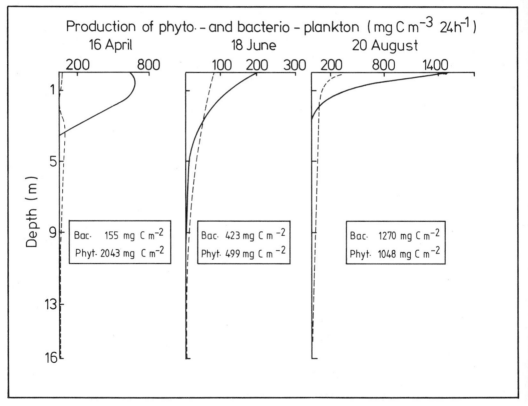

Fig. 4.9.
*Three examples of the vertical distribution of phytoplankton primary production (solid lines) and bacterial secondary production (determined by means of FDC) from eutrophic Lake Tystrup. Redrawn from Riemann (1983).*

### 4.4.2. Vertical changes
Three examples of vertical changes in the bacterial secondary production are presented from the Danish eutrophic Lake Tystrup (Fig. 4.9). In contrast to the

depth distribution of the phytoplankton primary production, the bacterial production continued to the bottom, although rates decreased continuously downwards, probably because of lower temperatures and/or lower concentrations of phyto- and zooplankton. During 16th April the temperature was 6.5°C, and the bacterial gross secondary production accounted for about 8% of the organic matter produced by the phytoplankton based on integrated production values per $m^2$. During 18th June the temperature differences in the vertical profile were about 6°C with 18.5°C in the surface layers. Primary production was 499 mg C $m^{-2}$ compared with 423 mg C $m^{-2}$ assimilated by the bacteria. During 20th August the bacterial gross production, per $m^2$ basis, was 1.2 times the phytoplankton primary production. It appears that during the spring period with low temperatures the bacterial secondary production was controlled by temperature (although grazing might have played a role, but grazing was not measured), since primary production is in fact the highest value compared with those obtained in the two other vertical profiles. During August (Fig. 4.9) the bacterial production was higher than the primary production. During such periods phytoplankton primary production is not sufficient to balance bacterial heterotrophy; however, other processes like photosynthesis of macrophytes and/or epiphytes may contribute. Furthermore, "sloppy feeding" by zooplankton and algal lysis can also be important in the transport of labile organic compounds to the bacteria.

### 4.4.3. Seasonal changes

Most studies concerning daily/diel changes in the bacterial growth in natural plankton communities have hitherto been restricted to time scales less than 48 hours. The reason is that several processes occur in bottles after a while. These processes, often called "bottle effects", may completely change the plankton community, and even though large bottles have been used (up to 10 liters), several unpredicted changes in the production and biomass of the bacteria may occur.

Recently, the growth of natural populations of bacteria were measured during periods of 12-21 days in large, clear plastic enclosures (1 1/2 m in diameter, 3-4.5 m deep and 0.1 mm thick), (Riemann and Søndergaard 1986). In order to induce changes in the growth of the planktonic community, some of the enclosures were manipulated with nitrogen and phosphorus and pelagic, planktivorous fish, and the growth pattern of bacteria, phytoplankton and zooplankton were followed with a sampling frequency of 3-4 days.

In two eutrophic lakes, the bacterial secondary production was compared with the phytoplankton primary production and biomass (Figs. 4.10a,b). In Lake Hylke, the biomass and primary production of the phytoplankton as well as the bacterial secondary production were identical in the open lake and in the enclosures without fish. Addition of fish initiated increases in the phytoplankton production and biomass to a higher level than found in the lake and in the enclosures without fish. After a maximum, a decrease in both production and biomass occurred, caused by phosphate depletion. The addition of fish furthermore increased production rates of the bacteria, and in contrast to the culmination and decrease of the phytoplankton biomass and production, the bacterial production continued to increase and reached, at the end of the experiment, the highest value obtained from all the enclosures and the lake. This increase in the bacterial production was probably the result of lysis of phosphate-depleted algal cells and the release of large quantities of organic products readily available to bacterial assimilation (section 4.5).

The addition of both fish and nutrients resulted in a more or less continuous increase in both phytoplankton biomass and production, and at the end of the period the biomass was nearly ten times higher than the levels found in the lake and in the enclosures without fish.In all the enclosures, and in the open lake, the amount of extracellular organic products released from the phytoplankton(EOC) parallelled the changes found for primary production and biomass (Riemann and Søndergaard 1986).

In the second example, the bacterial production was measured in Frederiksborg Slotssø during 21 days as well as in plastic enclosures. The phytoplankton biomass and production decreased to values of 9 μg chlorophyll liter$^{-1}$ and 0.4 g C m$^{-2}$ day$^{-1}$ at the end of the period in the enclosures without fish (Fig. 4.10b). The enclosures

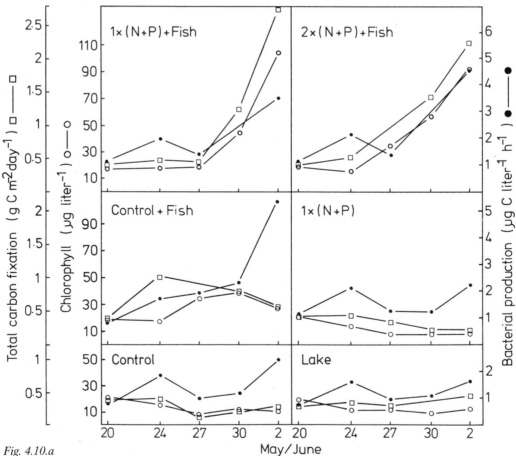

*Fig. 4.10.a*
*Daily changes in phytoplankton total carbon fixation, phytoplankton chlorophyll content and bacterial secondary production in eutrophic Lake Hylke (a) and eutrophic Frederiksborg Slotssø (b) as well as in experimental enclosures manipulated with nutrients (N and P) as well as with pelagic planktivorous fish. Redrawn from Riemann and Søndergaard (1986).*

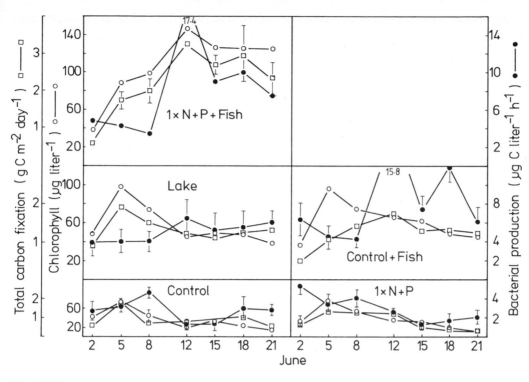

*Fig. 4.10.b*
*See text for fig. 4.10.a.*

with added fish and the lake developed similarly. In the enclosures with added fish and nutrients, the phytoplankton biomass and production reached a level about 120 mg chlorophyll liter$^{-1}$ and 2.5-3 g C m$^{-2}$ day$^{-1}$ at the end of the period. The bacterial production increased in the enclosures with added fish compared with the rates in the enclosures without fish.

The results from the two experiments provided an opportunity to examine the bacterial secondary production as a function of phytoplankton growth and senescence. The best correlation was found between the bacterial production and the phytoplankton production ($r = 0.55$); however, the correlation between the bacterial production and the phytoplankton biomass ($r = 0.50$) or the bacterial net assimilation of EOC ($r = 0.42$) were nearly the same. These rather low correlation coefficients probably demonstrate a complex relationship of simultaneously occurring events like changes in phytoplankton biomass due to lysis, the amount and lability of released EOC, activity from zooplankton, and maybe activities from fish.

Summarizing these observations, marked seasonal changes in the growth of natural populations of bacteria may occur. These changes are controlled by a variety of events occurring on several trophic levels as well as in the nutrient and temperature regime. These events change from place to place and during day and season. Often the bacterial secondary production in eutrophic environments is controlled by EOC released from phytoplankton. However, small changes in the phytoplankton biomass via cell lysis or via intensive grazing by zooplankton may also create marked changes in the growth rate of freshwater bacteria.

Two examples of annual changes in bacterial secondary production in eutrophic Lake Tystrup and Lake Mendota are presented in Fig. 4.11. During winter and early spring (November-April/May) production rates were low in both lakes, and almost all production took place in the summer period. In Lake Tystrup, annual bacterial net production accounted for 61 g C m$^{-2}$, whereas in Lake Mendota annual production determined by means of $^{35}$-SO$_4$ uptake varied between 145 and

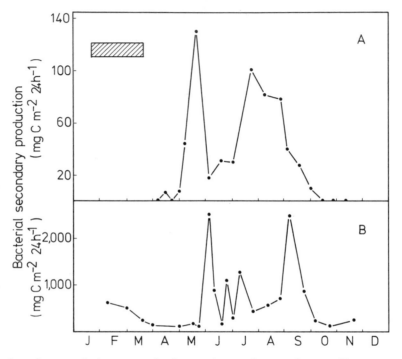

*Fig. 4.11.*

*Two examples of seasonal changes in the bacterial secondary production. Upper curve: data from eutrophic Lake Tystrup determined by means of FDC. Lower curve: data from eutrophic Lake Mendota determined by means of $^{35}$-SO$_4$ uptake. From Riemann (1983) and Pedrós-Alió and Brock (1982).*

205 g C m$^{-2}$ year$^{-1}$ (Pedrós-Alió and Brock, 1982). Annual estimates of bacterial secondary production from a number of lakes which differ with respect to primary production are presented in Table 4.4. Neglecting methodological difficulties, annual values of the bacterial secondary production varied from about 6 g C m$^{-2}$ year$^{-1}$ in oligotrophic Lake Mirror to more than 300 g C m$^{-2}$ year$^{-1}$ in eutrophic lakes.

Considering the limited material and the uncertainties in several of the estimates of the annual bacterial production, there is still a clear and expected correlation between the phytoplankton primary production and the bacterial secondary production. On the other hand, Table 4.4 could not support evidence for an increased bacterial utilization of the primary production in eutrophic lakes. Probably, better material from other oligotrophic lakes than just the result from Lake Mirror could have supported this idea.

Table 4.4. Annual estimates of bacterial gross production (determined by various methods and assuming a 60% growth efficiency) and phytoplankton primary production.

| Locality | Bacterial gross production (g C m⁻²year⁻¹) | Phytoplankton primary production (g C m⁻²year⁻¹) | Reference |
|---|---|---|---|
| Lake Mirror | 6-16 | 37 | Jordan and Likens (1980) |
| Crooked Lake | 92+ | 125+ | Lovell and Konopka (1985a) |
| Little Crooked Lake | 111+ | 160+ | Lovell and Konopka (1985b) |
| Lake Tystrup | 102 | 227 | Riemann (1983) |
| Lake Mendota 1979 | 148-337 | 654 | Pedrós-Alió and Brock (1982) |
| Lake Mendota 1980 | 180-342 | 802 | Pedrós-Alió and Brock (1982) |

+Production period was from April to November

Summarizing this evidence, the bacterial secondary production in pelagic freshwater environments constitutes an important role in the carbon metabolism. Even in oligotrophic lakes the bacteria play a significant role, and in many eutrophic lakes more than 50% of the annual phytoplankton primary production is channelled through pelagic bacteria.

## 4.5. Bacterial assimilation of organic compounds

### 4.5.1. Introduction

In the oxic pelagic environment, bacteria are usually chemoheterotrophic and utilize reduced organic compounds for both energy metabolism and synthesis of new cell material. A number of biological processes release organic compounds into the ambient water, from which they are assimilated by the bacteria (see Chapter 2 for sources of dissolved organic matter). Natural aquatic bacteria have proven capable of taking up and assimilating a large spectrum of organic molecules, including not only amino acids, fatty acids, monosaccharides and other building blocks of biological tissues (Poltz 1972, Bölter 1981, Simon 1985), but also various organic acids (Wright 1970), organic nitrogen compounds like purines and putrescine (Lee and Cronin 1982, Höfle 1984) and even hydrocarbons in petro-polluted areas (Cooney et al. 1985).

In addition to dissolved organic compounds, bacteria may also utilize particulate organic matter. However, before being assimilated, larger components, such as structural carbohydrates and proteins, must be degraded to smaller molecules through the action of hydrolytic exoenzymes, released by the bacteria (Hoppe

1983). Exoenzymes are apparently active only when associated with bacterial cells or for a short period after being released, since most exo-enzymatic activity, at least in the sea, has been measured only in the presence of bacterial cells (Rego et al. 1985). Exoenzymatic decomposition of high molecular-weight organic matter may be a very efficient process; thus, proteins > 100,000 Daltons have been found readily to be taken up by pelagic bacteria (Hollibaugh and Azam 1983, Ammermann et al. 1984).

Dissolved organic matter in natural waters to a large extent consists of biologically unattractive and relatively non-degradable compounds (Schnitzer and Khan 1972). Organic compounds that can be assimilated by bacteria make up a maximum of 20% of the pool of dissolved organic matter (DOM), as discussed in Chapter 2. In freshwaters, the amount of DOM, measured as dissolved organic carbon (DOC), typically ranges from 1 to 10 mg $l^{-1}$ (Søndergaard 1984), which means that from 0.2 to 2 mg C $l^{-1}$ is directly available to the bacteria. Since the bulk of this labile fraction mainly consists of simple carbohydrates, amino acids and organic acids with an average molecular weight of roughly 100, the total concentration of low molecular-weight DOC varies from about 2 to 20 μM. These concentrations are very low relative to the pool of free organic compounds in bacterial cytoplasm. For example can the concentration of intracellular free amino acids be higher than 1 mM (Stanley and Brown 1976). Thus, simple diffusion cannot be used by bacteria to take up organic molecules; instead they possess active transport systems to transfer organic substances into their cells.

Knowledge of the rates and mechanisms of uptake and utilization of natural dissolved organic substances was scarce until new techniques appeared within the past two decades. When radioactive tracers of the most abundant naturally-occurring organic molecules became available, it was demonstrated that bacteria actually assimilate and metabolize a number of different organic compounds. These radiotracer studies indicated that uptake was due to active transport (i.e. working against a concentration gradient at the expense of energy) and that this transport was to a large extent specific for different molecules (Overbeck 1979). The specificity appears to be caused by a number of binding proteins located in the cell membrane, and that each combine with specific molecules (Morita 1984). More recently, the radiotracer technique has also been used to measure natural *fluxes* of DOC, that is, the amount of substance turned over per unit of time. This was made possible by the development of sensitive methods to determine *in situ* concentrations of specific organic compounds. The measured assimilation rates of individual organic species have given valuable information on bacterial activity and metabolism, and on mineralization rates of organic materials caused by bacterial degradation.

In the following, uptake and assimilation are used as synonymous terms to describe transport of organic substances into the bacteria. Since the assimilated compounds are utilized both for metabolism (respiration) and for biosynthesis (incorporation), *gross* uptake/assimilation denotes respiration + incorporation, while *net* uptake/assimilation denotes incorporation only.

### 4.5.2. Determination of turnover rates of dissolved organic compounds

Natural concentrations of dissolved organic substances reflect a balance between production and assimilation. Concentration changes alone can seldom be used as a measure of assimilation rates. Since a number of auto- and heterotrophic processes contribute dissolved organic compounds at varying rates to the water, the assimilation of these compounds does not necessarily cause a reduction in their concentration. However, natural pools of organic substances can be reduced if assimilation exceeds input. This was observed for free amino acids in a water sample collected beneath the ice in a eutrophic Danish lake (Fig. 4.12). In this case the amino acid pool declined after a 24-h incubation period. A substantial production of amino acids took place initially, since the expected concentration decrease due to assimilation (measured with radiotracers of the amino acids) differed significantly from the measured concentration change. This experiment shows that actual assimilation of an organic substance is not always reflected in a corresponding concentration decrease. To determine an actual flux, the turnover rate of the organic substance must be measured.

Turnover of a dissolved organic compound can be determined from assimilation of a radiotracer of the compound, assuming that assimilation of the labeled tracer is identical to assimilation of the non-labeled compound. Depending on the specific isotope, both incorporation and respiration of the compound can be determined. For example, using a $^{14}C$ or a $^{3}H$ label, respiration can be calculated from the produced $^{14}CO_2$ or $^{3}H_2O$, respectively.

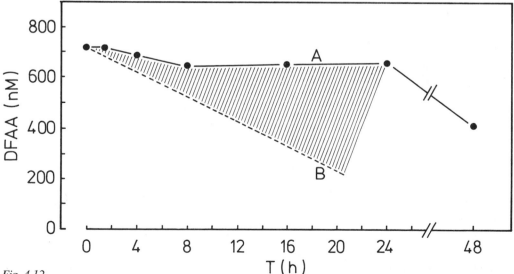

Fig. 4.12.
Concentration of dissolved free amino acids (DFAA) in a water sample from the Danish Lake Knudsø. Water was collected under the ice and incubated in the dark in the laboratory at in situ temperature (2°C). The concentration of DFAA (graph A) was followed for 48 h. Bacterial assimilation of DFAA was measured with $^{14}C$-amino acid tracers. The expectred concentration decline due to assimilation is indicated by graph B. The shaded area represents the production of DFAA. Modified from Jørgensen and Søndergaard (1984).

Basically, two different procedures can be applied when determining turnover rates with radiotracers: (1) a single-substrate approach, and (2) a multi-concentration approach.

### 4.5.3. Single-substrate approach

The turnover of an organic compound can be determined by adding a tracer amount of labeled compound and measuring its assimilation by bacteria. A true tracer concentration is one that causes an insignificant increase of the natural concentration; an increase less than 1% is often an operational value. When the amount of tracer assimilated by the bacteria per unit of volume and time is measured, the turnover time can be determined. For instance, if 2% of the added tracer is assimilated in 1 h, the turnover time of the substance is 50 h. Since only one labeled substrate it used, this procedure has been characterized as the single-substrate approach, originally introduced by Williams and Askew (1968). The turnover time of various organic compounds, mainly glucose and amino acids, have been measured in a number of aquatic environments. Some examples from eutrophic lakes are given in Table 4.5.

The considerable range of turnover times in Table 4.5 does not necessarily indicate corresponding variations in the bacterial activity, since different organic compounds may have variable importance as substrate for the bacteria. Also, the turnover times are determined from only the net and not the gross assimilation of some of the compounds. As respiration of various organic substrates have been found to vary from about 30 to 70% of the gross assimilation (to be discussed later), turnover times based on a net assimilation are not representative, but may be indicative, of the actual cycling of the compounds.

### 4.5.4. Multi-concentration approach: Michaelis-Menten saturation kinetics

As an alternative to measuring only the turnover of the organic substrates, the kinetics of the uptake can be characterized using a multi-concentration approach.

Table 4.5. Single-substate approach for turnover of dissolved organic in eutrophic lakes

| Locality | Compound | Turnover time | Period | Reference |
|---|---|---|---|---|
| Lake Kinneret, Israel | Glucose[a] Amino acids[a] | 34-168 20-152 | Feb/Mar Jun/Mar | Berman and Gerber (1980) |
| Lake Mossø, Denmark | Glucose[b] | 20-5000 | April | Riemann et al. (1982b) |
| Kalgaard Sø, Knudsø, Skanderborg Sø, Denmark | Amino acids[b] | 0.46-52.4 | March | Jørgensen and Søndergaard (1984) |
| Lake Constance, FRG | Amino acids[a] | 28-449 | Apr/May | Simon (1985) |

[a]net assimilation (incorporation), respiration not included
[b]total (gross) assimilation (incorporation + respiration)

This procedure is based on the Michaelis-Menten equation for enzyme reactions and assumes that a maximum reaction rate (saturation) can be obtained. According to the Michealis-Menten equation, the reaction (uptake) rate V at a substrate concentration S can be determined when the maximum rate $V_{max}$, and the substrate concentration at which the rate equals $\frac{1}{2} V_{max}$ (denoted $K_t$, half-saturation constant), are known:

$$V = \frac{S \cdot V_{max}}{K_t + S}$$

This Michaelis-Menten equation has proven useful for characterization not only of enzymatic processes, but also of uptake kinetics of organic compounds by natural populations of bacteria. When the concentration of different organic compounds is increased experimentally in similar water samples, different types of uptake responses by the bacteria are generally found. In many cases the uptake kinetics can reflect the natural concentration levels. For example, if substance A normally occurs at low concentrations, it is advantageous for bacteria to have a high affinity for A, corresponding to a low $K_t$ in this concentration range. But since the concentration of A seldom reaches high levels, the uptake of A may saturate at rather low levels, implying a small $V_{max}$. The opposite may apply to a substance B that generally occurs at higher concentrations than A. In order for assimilation to

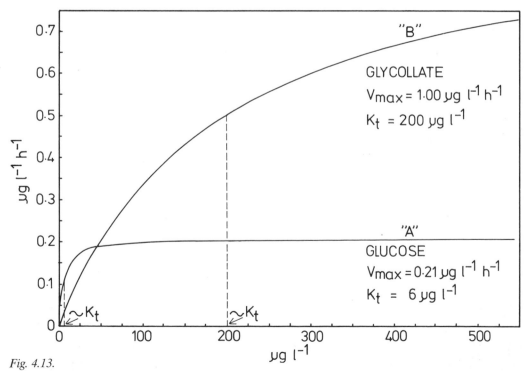

Fig. 4.13.
*Uptake kinetics of glucose and glycollate at a depth of 3 m in Gravel Pond, Mass., USA, August 1978. The indicated $K_t$ values are actually $K_t + S_n$ values, as the natural concentration ($S_n$) of glucose and glycollate are included in the $K_t$'s. Using actual $K_t$ values slightly different graphs would have resulted, but the principal difference would not have been affected. "A" and "B" refer to the text. From Wright (1970).*

occur over a wider concentration range, bacteria assimilate low levels of B more slowly than they assimilate A, but assimilation of B saturates at a higher level than for A. As a consequence, higher values of both $K_t$ and $V_{max}$ characterize the uptake of B. An example of uptake kinetics for compound type A and B is given in Fig. 4.13, which shows the uptake of glucose ("A") and glycollate acid ("B") in a freshwater pond. The natural concentrations of glucose and glycollate were not measured in this particular experiment, but glucose usually occurs in much lower concentrations than glycollate in natural waters (El-Hasan et al. 1975, Münster 1984). Uptake by bacteria probably causes the low concentrations of glucose, since this compound appears to be very attractive to most microorganisms. As a result, the bacteria have to compete for small amounts of glucose, favouring organisms with low $K_t$ values.

The constants $V_{max}$ and $K_t$ of the Michaelis-Menten equation can be determined, if the equation is linearized according to Wright and Hobbie (1966). The natural substrate concentration $S_n$ in most studies is unknown due to analytical difficulties in measuring natural low levels of dissolved substances. Thus, $S_n$ and $K_t$ cannot always be separated, and $K_t + S_n$ rather than $K_t$ is typically determined:

$$\frac{t}{f} = \frac{S_a}{V_{max}} + \frac{K_t + S_n}{V_{max}}$$

In this equation, $S_a$ is the experimentally added substrate concentration, t is the incubation time, and f is the fraction of substrate taken up by the bacteria. t/f therefore corresponds to the turnover time at a specific substrate concentration. In Fig. 4.14 a linearized plot of a Michaelis-Menten equation is illustrated for uptake of glucose in a Czechoslovakian lake. When $S_a = 0$, the y-intercept of the graph represents the turnover time at $S_n$, i.e. the turnover time of the natural substrate. $V_{max}$ can be calculated as the reciprocal of the slope, while the x-intercept gives the $K_t + S_n$ concentration.

Uptake kinetics of various organic compounds, mainly glucose, amino acids and organic acids, have been studied in different freshwater habitats. In the selected examples in Table 4.6, $V_{max}$ varies over two orders of magnitude for glucose and amino acids. This variation may not reflect characteristic differences between the lakes. Similar large changes in $V_{max}$ have also been measured on a seasonal basis in a single lake, e.g. Lake Constance (Simon 1985) and in Lake Cristalino (Rai and Hill 1982).

Changes in $V_{max}$ must be associated with the metabolic activity of the bacterial population. Therefore, $V_{max}$, especially of glucose, has been used to characterize the "heterotrophic potential" of the bacteria. It must be emphasized that this is a potential and not an exact measurement of bacterial activity, since the actual assimilation rate is generally below $V_{max}$. Jørgensen and Søndergaard (1984) found that the actual assimilation rate of amino acids in three Danish lakes ranged from 12 to 87% of $V_{max}$. However, $V_{max}$ may be a valuable tool for estimating bacterial activity. For example, in Lake Constance and Lake Cristalino (Table 4.6), values of $V_{max}$ peaked simultaneously with the concentration of chlorophyll in the water,

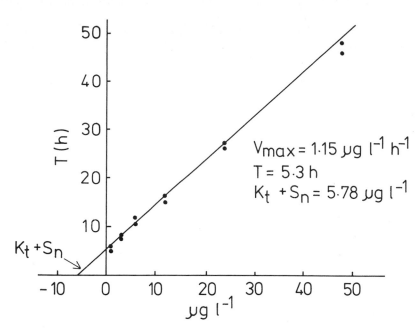

*Fig. 4.14.*
*Linearized plot of glucose uptake kinetics in the Slapy reservoir at the Vltava River, Czechoslovakia, 24 July 1976. t/f represents the turnover time of the added substrate ($S_n$). The y-intercept of the graph at $S_a = 0$ corresponds to the turnover time T of the natural substrate. See text for further details. From Straskrabová and Fuksa (1982).*

indicating that organic substances were released from the phytoplankton and readily assimilated by the bacteria. Notice in Table 4.6 that high values of $V_{max}$ often coincide with short turnover times of the organic substances, also indicating an enhanced bacterial activity.

Respiration of organic substrates taken up by the bacteria must be included in the assimilation rates to estimate the total uptake. This is clearly illustrated for aspartic acid in Dairy Pond, (Table 4.6) where $V_{max}$ increased significantly, and the turnover time decreased, when respiration was taken into account. The simultaneous change of $K_t + S_n$ is unexpected, since respiration should not influence $K_t$ nor $S_n$.

The ecological significance of $K_t + S_n$ concentrations are difficult to interpret, as the half-saturation constant $K_t$ and the natural substrate concentration $S_n$ often cannot be determined separately. $K_t + S_n$ has been used to estimate $K_t$, as this cannot be larger than $K_t + S_n$. However, if $S_n$ is large relative to $K_t$, this estimation of $K_t$ is doubtful. Unfortunately, $S_n$ has been measured in only a few studies. In Danish lakes, Jørgensen and Søndergaard (1984) measured $S_n$ pools of free amino acids and related them to $K_t + S_n$ concentrations. $K_t$ was found to vary independently of $S_n$ and made up from 12 to 897% of $S_n$, and from 11 to 79% of $K_t + S_n$. On the average, $K_t$ was two times larger than $S_n$ and made up 65% of $K_t + S_n$. It is probably advantageous for bacteria to have values of $K_t$ higher than $S_n$ concentrations, as the uptake in that substrate range is very efficient (fast increase of the uptake rate for an increase in concentration, see Fig. 4.13). But the data from the

Table 4.6. Multi-concentration approach for turnover of dissolved organics

| Locality | Compound | $V_{max}$ $\mu g\ 1^{-1}\ h^{-1}$ | $K_t + S_n$ $\mu g\ 1^{-1}$ | Turnover time (h) | Time | Reference |
|---|---|---|---|---|---|---|
| Lake Erken, mesotrophic, Sweden | Glucose[a] Acetate[a] | 180 190 | 3 7 | 17 37 | October | Wright and Hobbie (1966) |
| Dairy Pond, North Carolina, USA | Aspartic acid[a] Aspartic acid[b] | 28 72 | 92 39 | 131 70 | March | Hobbie and Crawford (1969) |
| Upper Klamath Lake, Oregon USA | Glycine[b] | 116 | 12.3 | 296 | March | Burnison and Morita (1974) |
| Lake Grevelingen, saline lake, Holland | Aspartic acid[b] | 407 | 4.8 | 11.7 | June | Sepers (1981) |
| Lago Cristalino, Central Amazon, Brazil | Glucose[a] | 9600 | 1300 | No data | October | Rai and Hill (1982) |
| Lake Tuusulanjärvi, eutrophic Finland | Glucose[a] | 2460 | 12.7 | 5.2 | July | Tamminen (1982) |
| Fraser River, Canada | Glucose[a] | 280 | 0.28 | 98 | February | Albright (1983) |
| Lake Skanderborg, eutrophic, Denmark | Glutamic acid[b] Ornithine[b] | 8880 948 | 7.97 7.81 | 0.64 8.24 | March | Jørgensen and Sønder- gaard (1984) |

[a] net assimilation, respiration not included
[b] gross assimilation (incorporation + respiration)

Danish lakes illustrate that $K_t$ of bacteria is not necessarily larger than $S_n$; hence, bacteria do not always take up dissolved organics at optimum conditions. Variation of the relation between $K_t$ and $S_n$ is probably caused by rapid changes of $S_n$ and not of $K_t$. This is supported by observations of rapid short-term changes in free amino acid concentrations in lake water where 5-fold diel changes of $S_n$ were measured (Jørgensen 1986). Comparisons of measured and kinetically determined $K_t + S_n$ concentrations have not always given consistent results, however, as discussed later.

As an alternative to a chemical measurement of $S_n$ in kinetic studies to determine $K_t$, Wright and Hobbie (1966) attempted to estimate $S_n$ by diluting the natural water sample with substrate-free water. These experiments were not generally successful, as the bacteria had more reduced uptake rates than expected from the degree of dilution. This was probably due to changes of the micro-environments of the bacteria. Also, the water applied for the dilution may not have been free of organic substances.

### 4.5.5. Variations of Michaelis-Menten uptake kinetics

Uptake of dissolved organic compounds by natural assemblages of bacteria does not always obey simple Michaelis-Menten kinetics. Non-linear curves for uptake of glucose have been observed in some cases (Wright and Hobbie 1966, Azam and Hodson 1981, Nissen et al. 1984). Wright and Hobbie suggested that this non-linearity was caused by diffusion, probably into phytoplankton, at high substrate concentrations; but the studies by Azam and Hodson and Nissen et al. indicate that multiphasic uptake kinetics, at least of glucose, may also lead to a non-linear uptake. The implication of multiphasic kinetics is that $K_t$ changes (increases) with the substrate concentration. Whether such multiphasic systems commonly occur in bacteria is as yet unknown, but such a substrate dependent change in $K_t$ would allow the bacteria to benefit from occasionally high substrate concentrations, as the transport system would not saturate.

Calculation of the half-saturation constant, $K_t$, from chemically measured $S_n$ concentrations and kinetically determined $K_t + S_n$ concentrations have in some cases given inconsistent results. Burnison and Morita (1974) and Gocke et al. (1981) found that measured $S_n$ concentrations of amino acids and glucose were larger than calculated concentrations of $K_t + S_n$. This discrepancy may be caused by the chemical procedure for measuring $S_n$, or the radiotracer technique used for determination of the uptake kinetics. Chemical treatment prior to analysis may change natural pools of dissolved organic compounds (Garrasi et al. 1979). Alternatively, the radiotracer added in an uptake experiment may not be representative of the actual pool of the organic substance, as it may be in an equilibrium with a similar, absorbed, but exchangeable pool (Gocke et al. 1981). However, in similar kinetic experiments, Jørgensen and Søndergaard (1984) found good agreement between $S_n$ and $K_t + S_n$ of free amino acids in both sea and freshwater samples. Further kinetic studies will determine the validity of the Michaelis-Menten approach.

Originally the Michaelis-Menten equation was intended for characterization of enzyme reactions with well-defined values of $K_t$ and $V_{max}$. Therefore, how does the Michaelis-Menten equation apply to natural assemblages of bacteria if each species has its own kinetic constants? Williams (1973) analyzed this problem theoretically, using a computer program. He concluded that if (1) the whole population adheres to the kinetic equation, it can be treated as a single species, and (2) if $S_a$ concentrations close to $S_n$ are used, the turnover time would be only 25% in error, assuming a 10-fold variation of $K_t$, and an $S_n$ concentration one third of $K_t$. The variation of $K_t$ in natural populations of bacteria has not been determined, but in most environments, the majority of bacteria may be expected to have rather similar kinetic constants, reflecting the ambient substrate concentrations. Otherwise they would not be able to compete equally for available substrate pools. The findings by Williams therefore seem to lend support to the use of the Michaelis-Menten equation for natural populations of bacteria, if low $S_a$ concentrations are used.

### 4.5.6. Bacterial utilization of organic substrates and isotope dilution experiments

All chemoheterotrophic bacteria use organic substances for synthesis of new cell components and for production of energy, mainly as ATP. The ATP produced during respiration is necessary for sustaining metabolic processes and for maintaining vital activities of the cells, such as membrane potentials.

Which molecules are used for energy production (respiration) and which are used for biosynthesis (incorporation)? Most organic molecules can enter both routes. For example, via the Embden-Meyerhof pathway and the tricarboxylic acid (TCA) cycle, glucose is respired during ATP production, but it may also be used for synthesis of aliphatic and aromatic amino acids and pentoses (Levy et al. 1973). Similarly, assimilated amino acids can be used for ATP production after deamination, but they can also serve as building blocks in the synthesis of proteins, nucleotides, cell wall components and chlorophylls (Levy et al. 1973).

The quantitative significance of respiration of organic compounds assimilated by bacteria was first examined by Hobbie et al. (1968) and Hobbie and Crawford (1969). Earlier, utilization of dissolved organic molecules was typically determined from the incorporation of a $^{14}$C-labeled compound into bacteria which were retained on a filter after suitable incubation periods. If a substantial portion of the organic substance and its radioisotope were respired during the incubation and therefore converted to $^{14}CO_2$, the radioactivity in the bacteria on the filters would underestimate the gross assimilation. Hobbie and Crawford (1969) observed that a significant portion of different organic substances were respired by bacteria. From 18 to 32% of glucose and 8 to 61% of the amino acids were respired in a North Carolina (USA) pond. If these respiration percentages were not included in the assimilation rates, too long turnover times would result.

Today, respiration of organic substances has been determined in several aquatic environments. Some examples are given in Table 4.7. The respiration values show that the largest portion of glucose, fructose and of most non-acidic amino acids is generally used for biosynthesis. The acidic amino acids and glycollate are incorporated to a lesser extent, probably because they are directly related to intermediates in a respiratory pathway. For example, the TCA cycle intermediate a-ketoglutaric acid is formed after deamination of glutamic acid. Low respiration of a compound may indicate that it is essential for the cell's biosynthesis and that it is not produced by the organism itself. This is known to be true of several amino acids (Morita 1984). Conversely, a high respiration might indicate that the bacteria do not need a certain compound in biosynthesis.

When a radiotracer, typically $^{14}$C, is used to determine the assimilation of an organic compound, $^{14}CO_2$ produced during the incubation may not always represent actual respiration since the specific activity of the isotope may change due to variation of the intracellular concentration of the compound, as pointed out by Billen et al. (1980). This possible isotope dilution effect was studied by King and Berman (1984) who conclude that observed changes of the respiration of glucose assimilated by a *Vibrio* strain was most likely caused by dilution of the glucose iso-

Table 4.7. Respiration percentages of various organic compounds

| Locality | Compound | Respiration %* | Reference |
|---|---|---|---|
| Dairy Pond, North Carolina, USA | Glucose | 18-32 | |
| | Acetate | 38 | |
| | Non-acidic amino acids | 8-40 | Hobbie and |
| | Glutamic aicd | 61 | Crawford (1969) |
| | Aspartic acid | 47-60 | |
| Upper Klamath Lake, Oregon USA | Non-acidic amino acids | 15-51 | Burnison and |
| | Glutamic acid | 40-63 | Morita (1974) |
| | Aspartic acid | | |
| Pamlico River estuary, North Carolina, USA | Glucose | 9-17 | |
| | Non-acidic amino acids | $27.5 \pm 11.3$ | Crawford et al. |
| | Glutamic acid | 42-57 | (1974) |
| | Aspartic acid | 45-55 | |
| Rhode River, Maryland, USA | Glutamic acid | 30-43 | Carney and Coll-well (1976) |
| Lake Grevelingen, Holland | Aspartic acid | 75-87 | |
| | Protein hydrolysate | 31-54 | Sepers (1981) |
| Lake Mossø, Denmark | Glucose | 24-39 | Riemann et al. (1982b) |
| Eel Pond, Mass., USA | Glucose | 39-43 | Bell (1984) |
| Lake Esrom, Denmark | Glutamic acid | 57 | |
| | 4 non-acidic amino acids | 43-50 | Jørgensen (1986) |
| Lake Almind, Denmark | Fructose | 34 | Jørgensen, unpubl. |
| Frederiksborg Slotssø, Denmark | Glycollate | 80-90 | Nordby and Søndergaard (unpubl.) |

*percentage of gross uptake

tope or its intermediates inside the cells. An increased respiration percentage therefore may illustrate an actual increase, as well as a smaller dilution of the isotope. Since the intracellular specific activity of a compound increases with the incubation time, until an equilibrium is achieved, the isotope dilution problem is most serious in short-term experiments. Thus, a very short incubation time would give infinitely low respiration.

The effect of possible isotope dilution on measured respiration percentages is difficult to evaluate. In Table 4.7, the respiration of glucose ranges from 9 to 43%; but does this variation reflect an isotope dilution rather than an actual variation of the mineralization rate of glucose? Time-course studies of the respiration in some cases may clarify this problem, but not necessarily. This can be illustrated by the respiration of three amino acids in water samples from a Danish lake (Fig. 4.15).

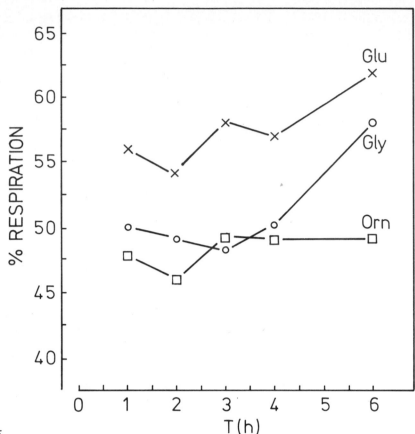

*Fig. 4.15.*
*Respiration of glutamic acid (glu), glycine (gly) and ornithine (orn) in similar dark-incubated water samples from Lake Esrom 11 May 1981. Respiration (determined with $^{14}C$-tracers of the amino acids) is shown as percentage of the gross assimilation. From Jørgensen (1986).*

After 4 h of incubation, the respiration of glutamic acid and glycine increased. But from the course of the respiration it cannot be determined whether a reduced intracellular isotope dilution or an actual increase of the respiration caused this change. In favour of the latter, Sepers (1981) did not observe an isotope dilution effect in time-course studies of bacterial respiration of protein hydrolysates and amino acids in a Dutch lake.

The use of long incubation times in an attempt to examine a possible isotope dilution may introduce another source of error. If the studied compound is being produced continuously, the *extracellular* specific activity decreases during the assimilation of the compound, causing an underestimation of the actual assimilation. Generally, no more than 10% of the added isotope should be removed from the ambient water to minimize a possible extracellular dilution of the isotope. When radiotracers are used for measuring incorporation and respiration of organic compounds, the assumption of an isotope equilibrium must always be carefully considered. Among recent papers discussing this subject, the following should be mentioned: Wiebe and Smith (1977); Smith and Horner (1981); Dring and Jewson (1981); King and Berman (1984); Jensen et al. (1985).

In the previous section (4.4), methods for measuring total bacterial production were discussed. Assimilation of organic compounds can in some cases also be used as a measure of bacterial growth. While the assimilation of carbohydrates such as glucose mainly has been used as an indicator of the level of bacterial activity ("heterotrophic potential"), incorporation of specific amino acids into proteins has proven useful for estimating protein synthesis. Since proteins make up about 50% of the biomass of bacterial cells, the total bacterial growth can be estimated. Kirchman et al. (1985) used incorporation of the amino acid leucine which constitutes a rather constant portion of 8.8% in bacterial protein. They found that incorporation of leucine gave growth rates comparable to those obtained from incorporation of $^3$H-thymidine into DNA. In addition to estimates of the bacterial growth, the rate of protein synthesis also indicates the energy consumption of the cells, as practically all energy used for synthesis of macromolecules, at least in *E. coli,* is used for protein synthesis (Ingraham et al. 1983).

### 4.5.7. Organic substances as substrates for bacteria in eutrophic lakes

Studies of carbon fluxes in freshwater environments have usually demonstrated a close relationship between phytoplankton production and bacterial activity (Tanaka et al. 1974, Blaauboer et al. 1982, Simon 1985, Søndergaard et al. 1985). Therefore, in eutrophic lakes with high phytoplankton productivity, a large bacterial production with a correspondingly large assimilation of dissolved organics would be expected. Despite a higher production of dissolved organic material, e.g. from phytoplankton in eutrophic lakes, larger pools of labile organic compounds are generally not found due to an increased bacterial assimilation. This has been found to be true for common organic compounds like glucose and amino acids (Gocke et al. 1981, Jørgensen et al. 1983). Measurements of pools and turnover rates of organic acids and amino acids along a planktonic activity gradient in the English Channel also supports this relation between concentrations and assimilation rates (Billen et al. 1980).

The higher assimilation rates in eutrophic waters typically result in a faster cycling of the organic compounds. Thus, in lakes during periods of high bacterial production, turnover times of natural pools of organic compounds may be few hours or even below one hour (Table 4.6, Allen 1971). Exact measurements of such fast turnover times are often difficult to perform. As an illustration, the assimilation of glucose in the Pamlico River estuary in September 1969 (Crawford et al. 1974) had a turnover time of only 14 min. This means that mixing of the tracer with the water sample, incubation and sample handling should be completed within 1.4 min, since a maximum of 10% of the isotope should be assimilated in true tracer experiments. If a longer incubation time is used, the glucose assimilation might be underestimated due to external isotope dilution, as discussed in the previous section.

Concentrations of most labile dissolved organic compounds generally cannot be related to the trophic status of a lake, as stated above, but the amplitude of temporal concentration changes appears in many cases to be related to the level of production of the lake. In an oligotrophic Danish lake, an 18-fold change in the concentration of free amino acids was measured within a period of 5 days, while smaller and more damped variations of the amino acid pool occurred in eutrophic lakes (Jørgensen, in press). Even more marked differences in assimilation of the amino acids were found. In the oligotrophic lake, a 90-fold variation in assimilation occurred during a three week period, but in the eutrophic lakes, only a 20-fold variation was measured. A similar relationship between primary productivity and cycling of glucose was found by Riemann et al. (1982b) in a eutrophic lake. The amount of glucose varied from 30 to 600 nM before start of the spring bloom, but during the bloom the concentration fluctuated around 20 nM. The turnover time demonstrated a similar pattern.

The greater variation of both natural concentrations and assimilation rates in oligotrophic lakes and in more productive lakes during periods of low phytoplankton growth clearly suggest that input of dissolved organic matter from phytoplankton has a regulatory effect on the growth of pelagic bacteria. In less productive waters, minor but successive blooms of different phytoplankton species can give rise to occasional high concentrations of dissolved organic matter. However, the low bacterial biomass that commonly occurs in oligotrophic waters does not immediately respond to a sudden, large pool of organic substances. Not until the bacteria have increased in number, and thus assimilation rate, does the ambient concentration of organic substances decline. If a substantial growth of bacteria occurs, very low concentrations of organic compounds may result. Thus, large concentration amplitudes may be observed within short periods of time.

In contrast to oligotrophic lakes, the high primary production of phytoplankton in eutrophic lakes provides a larger contribution of organic substances to the bacteria. Although this contribution changes over time, e.g. due to varying growth rates and composition of the phytoplankton, changes in the concentration of organic compounds are often damped due to efficient assimilation by a large population of bacteria, relative to oligotrophic lakes. However, in eutrophic as well as in oligotrophic lakes, both pools and assimilation rates of organic substrates fluctuate since the activity of the bacterial populations vary, i.e. the proportion of active cells (Riemann et al. 1982b, Simon 1985) and since the need for or benefit from specific organic compounds depends on the co-availability of other organic species and inorganic nutrients, as both carbon- and nitrogen-containing compounds are required for cell growth. Hence, responses of bacteria to changing concentrations of different organic compounds are complex and as yet largely unpredictable.

### 4.5.8. Turnover of dissolved organic compounds in lakes: Variations in concentrations of organic compounds relative to biological activity

Sampling in natural waters at short intervals has demonstrated that concentrations of most low molecular-weight organic substances fluctuate widely. In lakes, from 3 to 9-fold diel changes in the pool of free amino acids have been measured

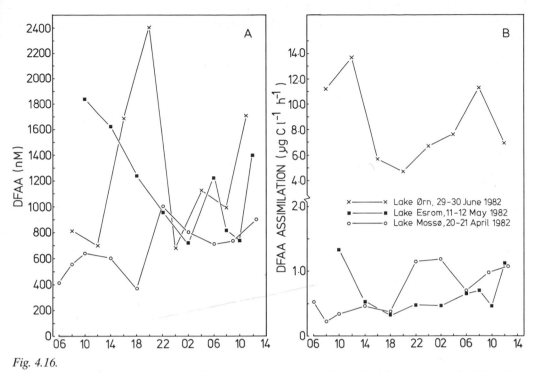

Fig. 4.16.
*Diel changes in concentration (A) and gross assimilation (B) of dissolved free amino acids (DFAA) in three Danish lakes. From Jørgensen (1986).*

(Jørgensen et al. 1983, Jørgensen 1986). In Fig. 4.16 representative diel changes of free amino acids in three Danish lakes are shown. The variations did not demonstrate any obvious diel trends. Similar short-term concentration changes have been measured for simple carbohydrates and amino acids in the sea (Meyer-Reil et al. 1979, Jørgensen 1982, Mopper and Lindroth 1982). The observed variations have been related to biological activities, e.g. nocturnal migration of zooplankton (Mopper and Lindroth 1982) or zooplankton grazing (Riemann et al. 1986), but the concentrations revealed obvious diel rhythms in only a few cases. In another study of Danish lakes, one of the lakes had higher pools of free amino acids in the mornings over a three week period. This coincided with the diel minimum bacterial production rates, suggesting that a reduced bacterial assimilation increased the amino acid pools (Jørgensen, in press). Two other lakes in the study, however, showed no correlations between biological activity and amino acid pool size.

The absence of a diel rhythm in concentration changes of organic substances does not imply that the production of the substances does not vary regularly. Thus, phytoplankton releases more extracellular compounds during photosynthesis than during darkness (Hellebust 1971). Diel feeding patterns in the grazing of zooplankton (Haney 1985, Lampert and Taylor 1985) probably contribute more organic compounds to the water in the night than in the day. However, such pulses of organic compounds, if they occur, could easily be reduced by bacterial assimilation, and thus might not be seen. Dissolved organic compounds released during microbial degradation of particulate and dissolved organic matter (Zygmuntowa 1981, Amano et al. 1982) may also mask a possible diel pulse in the production rate of organic compounds. Furthermore, the bacteria themselves may introduce diel concentration changes of organic compounds, as short-term variations of bacterial growth rates commonly occur (Riemann and Søndergaard 1984).

In addition to diel variations, different levels (i.e. average concentrations measured over a longer period of time, e.g. a week) of dissolved organic compounds commonly occur in aquatic environments. Thus, higher concentration levels of simple carbohydrates and amino acids have been reported during phytoplankton blooms. These increased concentrations may occur during both the active growth phase and the stationary phase of the algae (Burnison and Morita 1974, Myklestad 1977, Brockmann et al. 1979), but more often they coincide with the decline of the phytoplankton (Wangersky 1959, Gardner and Lee 1975, Haan and Boer 1979, Riley and Segar 1979, Ochiai et al. 1979, Jørgensen, in press). This suggests that decomposition of phytoplankton, by either autolysis or microbial degradation, is quantiatively a more important source of organic substances than release during photosynthesis. However, some substances like glycollic acid apparently are released only during photosynthesis (El-Hasan et al. 1975). Also, the composition as well as the concentration of the organic compounds, e.g. of sugars (Ittekkot 1982) and amino acids (Brockmann et al. 1979) may change during a bloom.

In most investigations, variations in pool size of organic molecules have been analyzed only in surface waters. In shallow lakes, these variations may be representative of the whole water column. In deeper lakes with a pronounced summer stratification, concentrations of dissolved organic compounds usually decrease with depth; but if anoxic conditions develop, increased concentrations may occur in the hypolimnion, probably due to the lack of suitable terminal electron acceptors for complete bacterial mineralization of the organic compounds (Starikova and Korzhikova 1969, Ochiai and Ukiya 1981, Jørgensen 1984).

### 4.5.9. Natural variations in actual and potential assimilation rates

The changing pools of dissolved organic substances in lakes may be expected to influence the bacterial assimilation rate, since the organic substances serve as a source of nutrition for the bacteria. Bacterial assimilation of organic compounds actually does show short- and long-term variations. Diel variations of assimilation rates unfortunately have been examined in only a few studies. In marine, offshore areas, up to 3-fold diel changes of the turnover rate (not assimilation rate) of glucose and amino acids may occur (Williams and Yentsch 1976, Ferguson and Sunda 1984). But in eutrophic lakes, more pronounced fluctuations have been observed. In studies of Danish lakes, 4 to 5-fold diel fluctuations of the assimilation rate of amino acids were typically measured, but occasionally up to 10-fold diel changes occurred (Jørgensen et al. 1983, Jørgensen 1984, 1986, in press). The variation of the assimilation in some of the lakes is shown in Fig. 4.16B. Some conclusions about assimilation of free amino acids in lakes can be drawn from these studies:

- no diel trends in the assimilation rate normally occur;

- if the concentration has significant short-term changes, the assimilation rate also changes, but these changes do not typically coincide;
- the assimilation rate is generally more variable than the concentration changes.

Although free amino acids constitute a minor portion of the low molecular-weight substances and only sustain part of the metabolism of the bacteria, their assimilation pattern, or lack of specific patterns, may apply to other organic compounds. The cycling of glucose and fructose, among the most common simple carbohydrates in natural waters (Liebezeit 1980, Ittekott et al. 1981), support this assumption. For example, diel variation of turnover rates of both glucose and fructose resembled those of the amino acids in lakes (Jørgensen unpublished results). Similarly, Meyer-Reil et al. (1979) observed substantial short-term fluctuations of glucose and fructose assimilation in the sea. These fluctuations did not demonstrate any diel trends. An important implication of the frequent inconsistency between pool sizes and assimilation rate of dissolved organic molecules is that bacterial activity cannot be predicted from the concentration of organic compounds available to the bacteria, as emphasized by Billen et al. (1980).

The apparently random variations of the bacterial assimilation rate of organic compounds with respect to changes in the concentration of the compounds can have several causes. The bacterial population may be too small to reduce a sudden burst of organic substances. The compounds may not be utilized by the bacteria, either because more enzymes have to be induced to metabolize the compounds (Edenborn and Litchfield 1985), or because other organic compounds are preferred, e.g. for biosynthesis. Changes in the proportion of bacteria attached to particles relative to free-living bacteria may also influence assimilation rates. Attached bacteria have been observed to have higher uptake rates of organic compounds than free-living bacteria, but conversely, the attached bacteria appear to have the lowest substrate affinity (Bright and Fletcher 1983, Simon 1985). Finally, heterotrophic requirements of different strains of bacteria may vary. In sea water, Wambeke and Bianchi (1985) identified two different bacterial communities, one which was capable of degrading macromolecules, and another which had a high affinity for small molecules. Similar heterotrophic "subpopulations" may develop in freshwaters.

In contrast to short-term variations in *actual* assimilation rates of organic compounds, the *potential* maximum uptake rate, $V_{max}$, has been found to vary only a little in eutrophic lakes. Thus, Overbeck (1979) and Straskrabova and Fuksa (1982) observed only minor changes in $V_{max}$ of glucose within 24-h study periods in European lakes. This may indicate that adaption to new concentration regimes (optimization of $V_{max}$) is a slow process, or that there was no need for changing $V_{max}$, because the concentration range in which the existing kinetic constants $K_t$ and $V_{max}$ operated caused an optimum assimilation. In studies of short-term variations in the assimilation rate and in substrate concentrations, the sampling procedure itself may introduce differences, if the same body of water is not being sampled throughout the whole period. In lakes, currents and internal waves may cause circulation and mixing of different bodies of water (Krambeck 1974). One way to circumvent this problem is to use enclosed bodies of water (limnocorrals). To ensure close-to-natural conditions, the enclosures must be sufficiently large to allow vertical migration of zooplankton, if *in situ* events are to be examined.

Like diel changes, vertical changes in the assimilation of organic compounds have been observed in lakes. In stratified Gravel Pond, Mass., USA, Wright (1970) observed that $V_{max}$ of glucose, acetate and glycollate decreased with depth, probably illustrating a coupling between photosynthetic activity (release of organic compounds) and bacterial activity. In contrast to Gravel Pond, low values of $V_{max}$ of glucose typically occurred in the photic zone in the similarly stratified Lake Pluss-See (Overbeck 1979). Maximum $V_{max}$-values were measured just below the photosynthetic zone, and then lower $V_{max}$ at deeper depths. In Lake Pluss-See, the low $V_{max}$ in the photic zone suggests that exudates from active algae were of less importance to the glucose assimilation than cell lysis and decomposition of sinking cells at larger depths.

In vertical profiles, not only $V_{max}$, but also the actual assimilation rate may change. Although a relatively constant assimilation of amino acids was observed in the thermocline of a stratified lake (Jørgensen 1984), additional studies have demonstrated significant short-term vertical changes in the assimilation of both amino acids and monosaccharides (Jørgensen, unpublished results). These changes did not reveal any significant trends. Variable assimilation rates in vertical profiles may in part be caused by heterogenous densities of bacteria, e.g. due to feeding in bacteria-grazing zooplankton with diel patterns of migration (Lampert and Taylor 1985, Bjørnsen et al. 1986). Heterotrophic microflagellates may also change concentrations of both bacteria and organic substances. Thus, when marine flagellates graze bacteria, the ambient concentration of free amino acids has been observed to increase, probably due to release by the flagellates (Andersson et al. 1985). In freshwater, heterotrophic flagellates may also be important (Sorokin and Pavaljeva 1972, Sheer et al. 1982, Riemann 1985).

Apparently, short-term changes in bacterial assimilation rates generally cannot be related to individual biological processes, such as phytoplankton release of EOC or zooplankton grazing; but different *levels* of assimilation, e.g. of glucose, often have been associated with major changes of the planktonic productivity (Overbeck 1974, Hobbie and Rublee 1977). Thus, the *potential* maximum uptake, $V_{max}$, of glucose in Amazonian lakes and of amino acids in Lake Constance coincided with the maximum concentration of chlorophyll (Rai and Hill 1982, Simon 1985). Similarly, there was a good correlation between primary production and $V_{max}$ of glucose in the upper 10 m water column in Lake Pluss-See (Overbeck 1979). In Upper Klamath Lake, Oregon, USA, Burnison and Morita (1974) measured the highest $V_{max}$ of amino acids during a bloom of cyanobacteria. These coincidences of phytoplankton maxima and large capacities of bacteria for uptake of organic compounds suggest that the phytoplankton release significant amounts of organic substances. Therefore, the *actual* flux of dissolved organic matter may also be expected to correlate with the phytoplankton production. In an annual study of cycling of glucose in the Kiel Fjord, assimilation actually was observed to

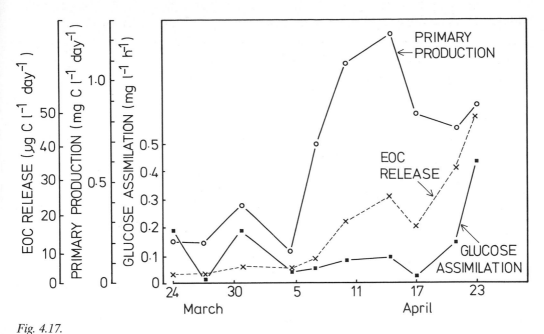

Fig. 4.17.
*Primary production, release of extracellular organic carbon (EOC) from the phytoplankton and gross assimilation of glucose in the Danish Lake Mossø in March and April 1980. From Riemann et al. (1982b).*

vary with primary production (Bölter 1981). Also, Riemann et al. (1982b) measured the highest glucose assimilation in Lake Mossø simultaneously with the maximum exudate release, but during decline of the phytoplankton (Fig. 4.17). However, release of exudates do not always parallel assimilation of organic compounds. In sea water from the coast of Maine, USA, annual maximum rates of exudate release and amino acid assimilation coincided only occasionally (Keller et al. 1981) . Obviously, some inconsistencies between rates of phytoplankton production, exudate release and bacterial assimilation of dissolved organic compounds exist. The observations by Riemann et al. and Keller et al. suggest that high assimilation rates of organic compounds may not only be caused by release of exudates from healthy algae; apparently decomposition of algal cells during decline of the phytoplankton may be another important source of organic compounds (see section 4.6).

### 4.5.10. Assimilation of organic compounds relative to bacteria and phytoplankton production

A large spectrum of dissolved organic compounds are available to bacteria in natural waters, but which ones are quantitatively important? Assimilation of specific organic compounds relative to total bacterial production (carbon uptake) has only been estimated in a few cases. In some Danish lakes, both assimilation of amino acids and bacterial production have been determined (Table 4.8). The assimilation data indicate, as expected, that free amino acids are of variable impor-

Table 4.8. Assimilation of organic compounds relative to total production of bacteria and phytoplankton. For the comparison, assimilation and production rates have been converted to similar fluxes of carbon. Assimilation given as gross values, except where other produces are indicated.

| Locality | Assimilated compound | Assimilation vs. bacterial production % | Assimilation vs. phytoplankton production % | Reference |
|---|---|---|---|---|
| York River estuary, USA[a,d] | Amino acids | ND | 1 | Hobbie et al. (1968) |
| English Channel[a,e] | Amino acids | ND | 6 | Andrews and Williams (1971) |
|  | Glucose |  | 1.5 |  |
| Upper Klamath Lake, USA[b] | Amino acids | ND | 10 | Burnison and Morita (1974) |
| Various Holstein lakes, FRG[b] | Glucose | ND | 2-10 | Overbeck (1979) |
| Pamlico River estuary[a] | Amino acids | ND | 1-10 | Crawford et al. (1974) |
| Californian coast[a] | Amino acids | ND | 1-10 | Williams et al. (1976) |
| Florida Straits, USA[a] | Amino acids | ND | 16 | Williams and Yentsch (1976) |
| Kiel Fjord, FRG[a] | Glucose | ND | 0.2-83 | Bölter (1981) |
| Lake Grevelingen and Lake Haringliet, Holland[b,c] | Amino acids & protein hydrolysates | 2.8-3.7 | ND | Sepers (1981) |
| Lake Mosso, Denmark[a,1] | Glucose | ND | 0.07-1.3 | Riemann et al. (1982) |
| Lake Mosso & Fr. Slotsso, DK[a,2,4] | Amino acids | 18 & 84 | 2 & 9 | Jørgensen et al. (1983) |
| Gulf of Mexico, USA[b,d] | Amino acids | ND | 7 | Ferguson and Sunda (1984) |
| Lake Almind, Denmark[a,2] | Amino acids | 8 | 13 | Jørgensen (1984) |
| Lake Almind, Esrom & Ørn, DK[a,2] | Amino acids | 6 & 6 & 25 | 2 & 9 & 12 | Jørgensen (1986) |
| 5 different lakes, DK[a,2,3] Lake Almind, Hylke | Amino acids | 5-31 | 5-11 | Søndergaard et al. (1986) |
| & Fr. Slotsso, Denmark[a,3] | Amino acids | 93 & 29 & 37 | 11 & 7 & 10 | Jørgensen (in press) |

a: Assimilation based on actual assimilation rates
b: Assimilation based on $V_{max}$-rates
c: Assimilation related to mineralization, not bacterial production
d: Incorporation only
e: Respiration only
ND: No data

Bacterial productivity was calculated from:
1: CO2-uptake,
2: Frequency of dividing cells (FDC),

tance to the bacteria; from 6 to 93%, but commonly below 30%, of the gross carbon requirement could be sustained from assimilation of amino acids. The contribution from amino acids of over 80% to bacterial consumption appears high, considering that several other organic substances (simple carbohydrates, organic acids) are available. But large assimilation rates of specific compounds do occur, as indicated by the glucose assimilation in Kiel Fjord (Bölter 1981; Table 4.8). However, methodological uncertainties in determining actual bacterial growth (see section 4.3) may also influence estimates of the relation between assimilation of organic compounds and total bacterial production. Sepers (1981) related bacterial production and assimilation of amino acids and hydrolyzed proteins to the mineralization rate ($O_2$ consumption) in two Dutch lakes (Table 4.7). In these lakes, the contribution from amino acids was estimated to be about 3% of the bacterial production, which is slightly less than the values measured in the Danish lakes.

In contrast to bacterial production, phytoplankton growth has been compared with assimilation of dissolved organic compounds in several environments (Table 4.8). Assimilation of naturally occurring amino acids (based on both net and gross rates) has been found to correspond to 1 to 16% of the primary production. Similar values were obtained whether the assimilation was based on the actual assimilation rates or the potential maximum rate, $V_{max}$. This is surprising, since $V_{max}$ of amino acids in lake and sea water can be from 1 to 100-fold larger than the actual assimilation rate (Crawford et al. 1974, Jørgensen and Søndergaard 1984). Possibly, kinetic constants calculated for an average of all amino acids may smooth out deviating uptake kinetics which might occur for individual amino acids. For glucose, bacterial assimilation has been determined to amount from less than 1 to 10% of the planktonic primary production. An exception is Kiel Fjord in which glucose assimilation constituted up to 83% of the phytoplankton production in the winter. If assimilation of the total pool of simple saccharides which in natural waters mainly include glucose and fructose (Liebezeit 1980) is considered, the data in Table 4.8 suggest that monosaccharides and amino acids are of similar importance to the bacteria.

The results in Table 4.8 also provide insight into decomposition processes. Uptake of glucose and amino acids apparently make up from 1 to 20% of the primary production, based on the flux of carbon. Despite the fact that proteins make up about 50% of the biomass of most phytoplankton species (Parsons et al. 1961) and glucose is another major portion of phytoplankton biomass, only a small amount of organic matter of phytoplankton seems to be degraded to simple molecules and mineralized by pelagic bacteria, as the data by Sepers (1981) also indicate. Apparently, a substantial portion of protein and carbohydrates escapes bacterial assimilation in the water column. Probably it is metabolized by other pelagic organisms, e.g. zooplankton, or its sinks to the bottom to be decomposed in the sediment. Regarding the decomposition of proteins, Hollibaugh and Azam (1983) have shown that even large peptides are easily degraded and assimilated by pelagic bacteria. Therefore, more protein amino acids may actually be metabolized in the water column than the data in Table 4.8 suggest.

Table 4.9. Examples of phytoplankton primary production (PP), bacterial assimilation of EOC ($B_n$) and bacterial net production ($B_p$)

| Locality | Trophic status | PP | $B_n$ | $B_p$ | Units | $\dfrac{B_n}{B_p}$ (%) | Reference |
|---|---|---|---|---|---|---|---|
| Lake Hylke:[1] | eutrophic | | | | | | |
| Control | | 3.3 | 0.2 | 1.8 | g C m$^{-2}$ | 13 (May) | Riemann & Søn- |
| Fish | | 9.4 | 1.0 | 2.9 | g C m$^{-2}$ | 35 (May) | dergaard (1986) |
| Fish + nutrients | | 11.9 | 1.6 | 2.9 | g C m$^{-2}$ | 55 (May) | do. |
| Frederiksborg | | 24.6 | 0.9 | 6.6 | g C m$^{-2}$ | 13 (June) | do. |
| Slotssø | eutrophic | | | | | | |
| Lake Mendota | eutrophic | 650 | [2] | - | g C m$^{-2}$ | ~14 (year) | Brock & Clyne (1984) |
| Lake Erken | mesotrophic | 3,200 | 34 | 24-34 | mg C l$^{-1}$ | ~100 (spring) | Bell & Kuparinen (1984) |
| Mirror Lake | oligotrophic | 40 | 4 | 6-16 | g C m$^{-2}$ | ~33 (year) | Cole et al. (1982) and Cole (1985) |
| Lake Almind | oligotrophic | 2.9 | 0.6 | 0.5 | g C m$^{-2}$ | ~100 (May) | Søndergaard et al. (unpubl. data) |

1) Results from enclosure experiments
2) Exact data not given

Some evidence suggests that phytoplankton in addition to bacteria may take up dissolved organic compounds at in situ concentrations. Previously, it has been shown that algae might assimilate amino acids, mainly when other nitrogen sources were depleted (North and Stephens 1971, Stephens and North 1971). In 1984, Moll demonstrated that under natural growth conditions, phytoplankton in Lake Michigan, USA, assimilated glucose and amino acids. Similarly, benthic microalgae have been found capable of reducing naturally low concentrations of amino acids (Admiraal et al 1984). Whether assimilation by algae is *truly significant* to the total assimilation of organic compounds in lakes has as yet to be determined.

## 4.6. Phytoplankton as a bacterial carbon source

In a recent analysis on the fate of planktonic primary production in lakes, Forsberg (1985) suggested grazing and sedimentation to be of minor importance. Metabolic losses in the water column were in four of five cases considered responsible for more than 70% of the loss.

Carbon from phytoplankton is a potential substrate supporting bacterial production. Gross production of the bacterioplankton in eutrophic lakes vary from about 20 to 70% of the phytoplankton primary production (Table 4.9) and can during shorter periods increase to 85% (Riemann and Søndergaard 1986).

Bacterial substrates must be in a dissolved form before they are taken up, so algal carbon has to enter the DOC pool to become available. Release of EOC and leaching of DOC from lysing algae are two potentially important processes to provide bacterial substrates. They were both included in Forsberg's metabolic loss category. A third possibility is that utilization of DOC released via zooplankton activity grazing can damage algae, and DOC is released. Leaching from fecal material is another possibility, and zooplankton loses DOC by direct excretion (Lampert 1978). Recently, decomposition of algae in oligotrophic Mirror Lake was treated in detail by Cole (1985).

The importance of EOC to bacterial production was recognized in earlier studies and hypothesized as the dominant carbon and energy source by Williams (1981) in his review on pelagic carbon cycling in marine ecosystems. More comprehensive quantitative comparisons between bacterial net EOC uptake ($B_n$) and total bacterial production ($B_p$) could only be made when more reliable methods on both processes became available in the period 1978-1981. Comparative seasonal studies are, however, still in need.

In section 3.3.6 it was concluded that, theoretically, bacterial gross production versus total EOC uptake is a better comparison than bacterial net production ($B_p$) versus bacterial net EOC uptake ($B_n$). However, in the following section we will outline a comparison based on values actually measured, that is $B_n$, $EOC_n$ and $B_p$ as previously defined. The two main reasons to do so are: 1) Low bacterial *in situ* growth yields are still to be supported by more experimental evidence, and the realiability of $^3$H-thymidine incorporation as an estimate of net production is also uncertain. 2) Apparent release of EOC may be a substantial underestimation of true release (section 3.3.4). The use of bacterial growth yield values around 0.2 would decrease a calculated importance of EOC based only on $B_n$ and $EOC_n$. An underestimation of EOC by a factor 2-3 would almost outbalance the decrease and bring the importance back to the levels reached by a $B_n/B_p$ comparison. These uncertainties must be kept in mind in the forthcoming comparison.

Release of EOC might occur with a diel pulse, where light initiates a release which in turn enhances bacterial production rates (Riemann and Søndergaard 1984a, Fuhrman et al. 1985). In a series of diel experiments in Danish lakes of different trophic levels the bacterial net uptake of EOC could support from less than

Table 4.10. Results on primary production, extracellular release and bacterial production in enclosure experiments in eutrophic Lake Hylke, Denmark, 20 May-2 June, 1983[1])

| Enclosure | Algae | $B_n$ | $EOC_n$ | $B_p$ | $\dfrac{B_n}{B_p}$ | $\dfrac{B_n + EOC_n}{B_p}$ |
|---|---|---|---|---|---|---|
| | | g C m$^{-2}$ | | | % | |
| Control | 3.0 | 0.2 | 0.1 | 1.8 | 13 | 15 |
| Fish | 7.8 | 1.0 | 0.6 | 2.9 | 35 | 47 |
| Fish + nutrients | 9.7 | 1.6 | 0.6 | 2.9 | 55 | 67 |

1) From data in Riemann & Søndergaard and Søndergaard (unpubl.)
2) Uptake of $EOC_n$ calculated with a growth yield of 0.6

38 to about 80% of the bacterial net production measured by means of thymidine incorporation (Søndergaard et al. 1985). Values below 38% occurred during late autumn senescence of cyanobacteria.

Both oligotrophic and eutrophic lakes have been investigated with respect to the importance of EOC, and most available results including both $B_n$ and $B_p$ estimates are presented in Table 4.9. The range is very wide, covering situations where 13 to 100% of the bacterial production can be explained by EOC uptake. From previously presented results it can be expected that EOC can pay a significant contribution to the organic carbon cycling, also in eutrophic lakes. Investigations covering this subject are scarce, and it is not possible to evaluate any significant differences between oligotrophic and eutrophic lakes.

In marine areas high EOC contribution (50%) to annual bacterial production was found by Larsson and Hagström (1982) at two stations in the Baltic. Bacterial production was attributed entirely to EOC by Joiris et al. (1982) in an annual carbon budget off the Belgian coast and also for a spring bloom situation in the same area (Lancelot and Billen 1984). The latter studies used $B_n$ estimates from a kinetic-method (Lancelot 1979) and $B_p$ estimates from single substrate uptake experiments, so a direct comparison with other studies using differential filtration and more direct bacterial production estimates is not possible.

In their study on the importance of EOC, Larsson and Hagström (1982) used $B_n + EOC_n$ values in the comparison with $B_p$. One good reason to do so is the seeming lack of seasonal variation in the DOC pool. All or most released EOC must ultimately be utilized. The consequence to use $B_n + EOC_n$ is an elevated EOC importance. The effect in a eutrophic situation with rather high EOC transport rates is demonstrated in Table 4.10. The $B_n : B_p$ values here increased 20 to 30%.

From Table 4.9 it is apparent that situations exist where only a minor part of the bacterial prouction can be explained by the measured EOC release. Thus, in eutrophic Frederiksborg Slotssø bacterial substrates from phytoplankton lysis or zooplankton mediated DOC release were more important in June 1984. A series of enclosure studies carried out in Lake Hylke can illustrate the dynamic changes, which affect pelagic carbon flow. A detailed description of the experiments was given by Riemann and Søndergaard (1986), and the phytoplankton composition and succession was treated by Søndergaard et al. (1984). A brief summary is given in section 4.4.

At the start of the experiments the growth of the phytoplankton was controlled by macrozooplankton grazing and remained so in the enclosure without fish (control) and in the enclosure only added nutrients. The addition of fish removed the grazing by zooplankton and phytoplankton biomass increased for about a week until phosphate was depleted. In the enclosures with both fish and nutrients added the phytoplankton biomass continued to increase until the experiments ended 2 June (Fig. 4.18 above). As expected, phytoplankton primary production showedsimilar changes as the chlorophyll concentrations. In the three enclosures presented here, bacterial net production ($B_p$, estimated by thymidine incorporation)

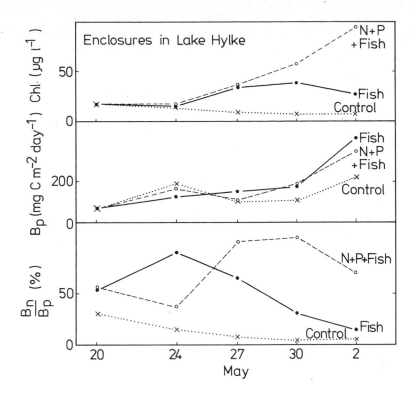

*Fig. 4.18.*
*Development in phytoplankton biomass (above), bacterial net production (middle) and the percentage of the bacterial net production explainable by EOC assimilation (bottom). Based on data in Riemann and Søndergaard (1986)*

from May 20 to May 30 gradually increased from about 80 to 150 mg C m$^{-2}$ day$^{-1}$ (Fig. 4.18, middle). This rather constant situation was followed by a dramatic increase in the enclosures added fish and fish + nutrients. For the entire period the integrated bacterial net production was 1.8, 2.9 and 2.9 g C m$^{-2}$ in the control, fish and fish + nutrients, respectively. Although very different patterns in the phytoplankton development occurred under the different treatments, these resulted in a rather similar bacterial response.

In the enclosure with a continuous phytoplankton growth, the importance of EOC as a bacterial substrate was high throughout the entire period (Fig. 4.18, bottom), averaged 55% of the bacterial carbon requirement and increased to 67% when EOC$_n$ was included (Table 4.10). The importance of EOC was also high in the enclosure with added fish as long as the phytoplankton biomass increased. When the biomass reached a constant value around May 27, B$_n$:B$_p$ decreased and was below 10% at June 2 (Fig. 4.18). For the entire period B$_n$:B$_p$ averaged 35% (Table 4.10). In the control enclosure both the algal biomass and the EOC importance decreased with time.

As macrozooplankton was removed due to fish predation (S. Bosselmann, pers. comm.) and the EOC importance decreased in the enclosure added fish, the only possible explanation of the increased B$_p$ is an input of DOC from lysing algae concomitant with a decreasing biomass of about 10 mg chl. l$^{-1}$. From 30 May to 2 June about 540 mg C m$^{-2}$ of the bacterial production is not readily explainable. Using a phytoplankton biomass decrease of 10 mg chl. l$^{-1}$, a chl./carbon ratio of 30-50 (Riemann, pers. comm.), and loss of 60% of the phytoplankton carbon via lysis (Hansen et al. 1986) would supply 756-1,260 mg C m$^{-2}$ to cover a bacterial gross production of about 900 (growth yield assumed to be 0.6). Thus, a decrease in algal biomass due to nutrient limitation resulted in a change in the control of bacterial production from EOC release to algal lysis.

In the control enclosures only 13-17% of the bacterial production could be attributed to EOC. Inorganic nutrients were available in high concentrations, so physiological stress on the phytoplankton was probably minor. Algal death and lysis are thus expected to be of minor importance. The main contribution of DOC to bacterial production was probably due to zooplankton activity. The loss of primary production of about 3 g C m$^{-2}$ (Table 4.10) and the decrease in biomass of about 1 to 2 g C m$^{-2}$ were fully explained by an independent estimation of zooplankton grazing at 5.4 g C m$^{-2}$ (S. Bosselmann, pers. comm.). If the organic carbon to support a bacterial gross production of about 2.6 g C m$^{-2}$ (EOC contribution subtracted) was supplied by zooplankton activity, the DOC release has to be 50 to 60% of the ingested carbon. Bacterial utilization of the ambient DOC pool and some algal lysis would decrease this rather high estimate. However, bacterial production basically seemed to be controlled by the herbivores providing substrates as a function of their feeding activity.

That macrozooplankton indeed has a capacity to release DOC has been shown in several studies (Conover 1967, Johannes and Satomi 1967, Lampert 1978, Riemann et al. 1986). Dissolved organics released in a phytoplankton-herbivore system may enter the environment in (at least) three different ways (Lampert 1978): 1. Direct excretion (Webb and Johannes 1967, Johannes and Webb 1970). Lampert (1978) considered this contribution to be of minor importance. For *Daphnia* species, Gardner and Miller (1981) reported excrection of about 0.13 nmol amino acids mg$^{-1}$ dw h$^{-1}$ under laboratory conditions. The other two processes are related to the grazing activity of the animals, 2. Leaching from fecal materials and 3. "Sloppy feeding".

Knowledge concerning leaching from fecal materials is scarce, but studies on total DOC release from grazing zooplankton species have been carried out. Release of 15 to 30% of the particulate carbon removed from suspension was reported by Conover (1967) and Johannes and Webb (1970).

The term "sloppy feeding" is used to describe the damage to and leaching from algae not swallowed whole during feeding. This process was specifically investigated by Lampert (1978) for the freshwater species *Daphnia pulex* grazing different algae. With diatoms like *Stephanodiscus* and *Asterionella* the release of DOC was 10 to 17% of the invested carbon, but for a small *Stichoccoccus* which is swallowed, the total release was about 4%.

The contribution of specific substances as dissolved free amino acids (DFAA) to the released DOC has been investigated by Webb and Johannes (1967), Epply et al. (1981) and Riemann et al. (1986). It was found that grazing zooplankton could increase the ambient DFAA concentrations, and that the DFAA were of algal origin.

All these results show that zooplankton - especially due to "sloppy feeding" - create an input of dissolved compounds available for bacterial uptake. Zooplankton can then initiate higher bacterial production (Copping and Lorenzen 1980, Eppley et al. 1981, Riemann et al. 1986) followed by an increase in bacterial number (Crisman et al. 1981, Eppley et al. 1981).

The effects of macrozooplankton om ambient DFAA concentrations and bacterial cell production in a eutrophic lake are shown in Fig. 4.19. An artificial increase of the zooplankton density to three times the natural density, both increased DFAA concentrations and bacterial production. However, other reaction patterns were also reported, supporting the idea that only sometimes released DFAA caused an increase in bacterial production and that the factors regulating the amount of DFAA concentrations are not fully understood (Riemann et al. 1986, see also 4.5).

In theory, the grazing activity of macrozooplankton influences bacterial production by removing bacterial cells (Riemann 1985) and by increasing the substrate pool. The quantitative importance for the level of bacterial production is unpredictable, as the grazing activity varies on diel and seasonal scales according to the migration, succession, density and size of the zooplankton. Furthermore, the species composition of the phytoplankton influences the level of EOC release.

From this short summary on DOC release from zooplankton activity it is clear that the calculated 50-60% release of the amount of carbon ingested by the zooplankton in the control enclosure presented earlier, is high, and is very uncertain. Nevertheless, zooplankton presumably provided the quantitatively most important DOC contribution in that system.

Table 4.11. Estimations of the quantitative importance of different organic carbon pathways from phytoplankton to bacterial net production in enclosure experiments in Lake Hylke, 20 May-2 June, 1983.

| Enclosure | $B_n + EOC_n$ | Lysis | Zooplankton | DOC | $B_p$ |
|---|---|---|---|---|---|
| | | g C m-2 | | | |
| Control | 0.3 | 0 (?) | 1.5 | ? | 1.8 |
| Fish | 1.6 | 1.3 | 0 | ? | 2.9 |
| Fish + nutrients | 2.2 | 0.7 | 0 | ? | 2.9 |

From the different situations exemplified by the enclosures experiments in Lake Hylke we can distinguish three major pathways of organic carbon transport from algae to bacteria and define the conditions, which creates quantitative dominance of each specific process (Table 4.11):

1. EOC seems to dominate when algal populations are growing without physiological stress, and zooplankton grazing is low.

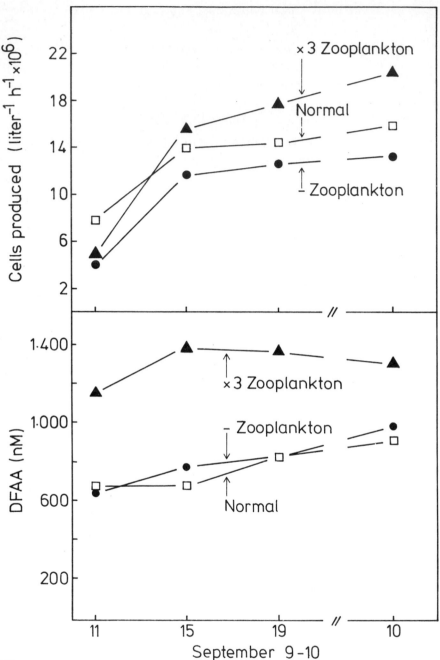

Fig. 4.19.
*The effect of removing zooplankton and increasing their density on the concentration of dissolved free amino acids (above) and bacterial production (below) in Lake Esrom, Denmark. From Riemann et al. 1986.*

2. With low grazing by macrozooplankton and nutrient deficien cies (or other physiological stress factors), phytoplankton lysis controls the dynamic and level of bacterial produc tion.

3. In situations where the phytoplankton is controlled by grazing e.g. "clear water phase" the zooplankton mediated release of DOC is most important.

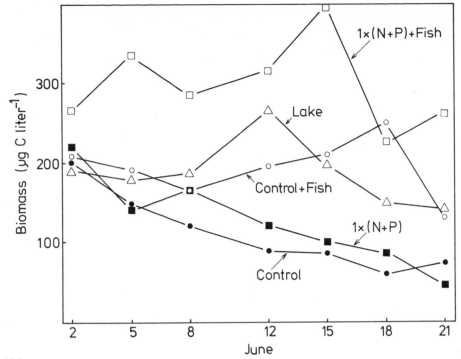

*Fig. 4.20.*
*Daily changes in the bacterial biomass in eutrophic Frederiksborg Slotssø as well as in experimental enclosures manipulated with nutrients (N and P) as well as with pelagic planktivorous fish. Redrawn from Riemann and Søndergaard (1986).*

Finally, it should be stated that all three pathways are, of course, in function at all times in natural situations, but change in relative importance over the seasonal cycle.

## 4.7. Bacterial biomass in eutrophic lakes

As described in section 4.2, determinations of the biomass of natural populations of bacteria is made from a) number of cells, b) mean cell volume, and c) a factor to convert cell volume to cell carbon. All three factors can be determined, however, in most routine studies, only the cell number and sometimes also the mean cell volume is determined.

In this section, two examples of the size-changes of the bacterial biomass are presented from freshwater environments.

In the first example (Fig. 4.20), changes in the bacterial biomass were followed in eutrophic Frederiksborg Slotssø during a period of 21 days. The results were compared with those from experimental enclosures in which the lake water was manipulated with nutrients (N and P) and pelagic, planktivorous fish (Riemann and Søndergaard 1986). During the experiment, the bacterial biomass was gradually reduced in the enclosures without fish and reached at the end an average value of 58 µg C liter[-1]. In contrast, the biomass was always higher in the enclosures containing fish as well as in the open lake. Values jumped somewhat irregularly between 128 and 390 µg C liter$^{-1}$ and the average value at the end of the period was 174 µg C liter$^{-1}$, three times higher than the biomass in the enclosures without fish. Thus, the size of natural populations of freshwater bacteria may change within few days. It is interesting that there were two types of changes in the bacterial biomass. The first one, in the enclosure without fish, was a gradually decrease caused by grazing of large *Daphnia* sp. *(D. galeata* and *D. cucullata), (*Riemann 1985). The other pattern was characterized by more dramatic changes in the fish containing enclosures and in the lake. Such changes are only possible by means of a close coupling between the growth of the prey and the predators, and in fact Riemann (1985) demonstrated that in the fish-containing enclosures and in the open lake, small microflagellates (4-8 µm in size) were responsible for most of the grazing on the bacteria. These flagellates have generation times in the same range as the bacteria (Fenchel 1982), so rapid changes in the growth pattern of the flagellates were probably responsible for the marked day to day changes in the bacterial biomass in the fish-containing environments.

Table 4.12. Examples of bacterial biomasses from eutrophic lakes

| Locality | Bacterial biomass (µg C liter$^{-1}$) | Reference |
| --- | --- | --- |
| Lake Mendota | 4-45 | Pedrós-Alió and Brock (1982) |
| Lake Erken | 7-47 | Bell et al. (1983) |
| Frederiksborg Slotssø | 50-390 | Riemann and Søndergaard (1986) |
| Lake Tystrup | 32-285 | Riemann (1983) |
| Lake Mossø | 86-140 | Riemann et al. (1982b) |
| Lake Hylke | 64-121 | Riemann and Søndergaard (1986) |

In the second example (Table 4.12), published values of the bacterial biomass are presented from various lakes. A range of nearly two orders of magnitude seems to cover the size of the bacterial carbon biomass in eutrophic lakes. The bacterial biomass constitutes most often a minor part of the phytoplankton biomass, whereas the zooplankton biomass during spring and late summer periods may be significantly lower than the biomass of the bacteria (Riemann et al. 1982b; Pedrós-Alió and Brock 1982).

Summarizing these evidences, rapid changes in the size of the bacterial biomass in natural populations may occur. These changes can be caused by both macrozooplankton *(Daphnia)* and microheterotrophic flagellates. In eutrophic lakes, the bacterial biomass varies two orders of magnitude (ê4-400 mg C liter$^{-1}$).

## Acknowledgements

The authors (BR and MS) want to thank Steven Y. Newell, Jed A. Fuhrman, Jonathan Cole and W. Vincent for comments and linguistic improvements of parts of the manus, and the author (NOGJ) wants to thank Cindy Lee for her careful and critical revision of section 4.5.

## 4.8. References

Admiraal, W., Laane, R. W. P, & Peletier, H. 1984. Participation of diatoms in the amino acid cycle of coastal waters: Uptake and excretion in cultures. Mar. Ecol. Prog. Ser. 15: 303-306.

Albright, L. J. 1983. Heterotrophic bacterial biomasses, activities, and productivities within the Fraser River plume. Can. J. Fish. Aquat. Sci. 40 Suppl. 1: 216-220.

Al-Hasan, R. H., Coughlan, S. J. & Aditipant, G. E. Fogg 1975. Seasonal variations in phytoplankton and glycollate concentrations in the Menai Straits, Anglesay. J. Mar. Biol. Ass. U.K. 55: 557-565.

Allen, H. L. 1971. Dissolved organic carbon utilization in size-fractioned algal and bacterial communities. Int. Revue ges. Hydrobiol. 56: 731-749.

Amano, M., Hara, S. & Taga, N. 1982. Utilization of dissolved amino acids in sea water by marine bacteria. Mar. Biol. 68: 31-36.

Ammerman, J. V., Fuhrman, J. A., Hagström, Å. & Azam, F. 1984. Bacterioplankton growth in seawater: I. Growth kinetics and cellular characteristics. Mar. Ecol. Prog. Ser. 18: 31-39.

Andersen, P. & Fenchel, T. 1985. Bacterivory by microheterotrophic flagellates in seawater samples. Limnol. Oceanogr. 30: 198-202.

Andersson, A., Lee, C., Azam, F. & Hagström, Å 1985. Release of amino acids and inorganic nutrients by heterotrophic marine microflagellates. Mar. Ecol. Prog. Ser. 23: 99-106.

Andrews, P. & Williams, P. J. LeB. 1971. Heterotrophic utilization of dissolved organic compounds in the sea. III. Measurements of the oxidation rates and concentrations of glucose and amino acids in sea water. J. Mar. Biol. Ass. (UK). 51: 111-125.

Azam, F. & Ammerman, J. W. 1984. Cycling of organic matter by bacterioplankton in pelagic marine ecosystems: microenvironmental considerations, p.345-360. In: Fasham, M. J. R. (ed.), Flows of energy and nutrients in marine ecosystems. Plenum Press, New York.

Azam, F. & Fuhrman, J. A. 1984. Measurement of bacterioplankton growth in the sea and its regulation by environmental conditions., p. 179-196. In: Hobbie, J. E. & P. J. LeB. Williams, (eds.), Heterotrophic activity in the sea. Plenum Press, New York.

Azam, F. & Hodson, R. E. 1977. Size distribution and activity of marine microheterotrophs. Limnol. Oceanogr. 22: 492-501.

Azam, F. & Hodson, R. E. 1981. Multiphasic kinetics for D-glucose uptake by assemblages of natural marine bacteria. Mar. Ecol. Prog. Ser. 6: 213-222.

Bell, W. H. 1984. Bacterial adaption to low-nutrient conditions as studied with algal extracellular products. Microb. Ecol. 10: 217-230.

Bell, T. B., Ahlgren, G. M. & Ahlgren, I. 1983. Estimating bacterioplankton production by measuring ³H-thymidine incorporation in a eutrophic Swedish lake. Appl. Environ. Microbiol. 45: 1709-1721.

Bell, R. T. & Kuparinen, J. 1984. Assessing phytoplankton and bacterioplankton

186

production during early spring in Lake Erken, Sweden. Appl. Environ. Microbiol. 48: 1221-1230.

Berman, T. & Gerber, C. 1980. Differential filtration studies of carbon flux from living algae to microheterotrophs, microplankton size distribution and respiration in Lake Kinneret. Microb. Ecol. 6: 189-198.

Bern, L. 1985. Autoradiographic studies of (methyl-$^3$H) thymidine incorporation in a cyanobacterium *(Microcystis wesenbergii)*-bacterium association and in selected algae and bacteria. Appl. Environ. Microbiol. 49: 232-233.

Billen, G., Joiris, C., Wijnant, J. & Gillain, G. 1980. Concentration and microbiological utilization of small organic molecules in the Scheldt estuary, the Belgian coastal zone of the North Sea and the English Channel. Estuar. Coast. Mar. Sci. 11: 279-294.

Bjørnsen, P. K. 1986. Automatic determinations of bacterioplankton biomass by means of image analysis. Appl. Environ. Microbiol. 51: 1199-1204.

Bowden, W. B. 1977. Comparison of two direct-count techniques for enumerating aquatic bacteria. Appl. Environ. Microbiol. 33: 1229-1232.

Bratbak, G. 1985. Bacterial biovolume and biomass estimations. Appl. Environ. Microbiol. 49: 1488-1493.

Bratbak, G. & Dundas, I. 1984. Bacterial dry matter content and biomass estimations. Appl. Environ. Microbiol. 48: 755-757.

Bright, J. J. & Fletcher, M. 1983. Amino acid assimilation and electron transport system activity in attached and free-living marine bacteria. Appl. Environ. Microbiol. 45: 818-825.

Brock, T. H. & Clyne, J. 1984. Significance of algal excretory products for growth of epilimnetic bacteria. Appl. Environ. Microbiol. 47: 731-734.

Brockmann, U. H., Eberlein, K., Junge, H. D. & Maier-Reimer, E. 1979. The development of a natural plankton population in an outdoor tank with nutrient-poor sea water: II. Changes in dissolved carbohydrates and amino acids. Mar. Ecol. Prog. Ser. 1: 283-291.

Burnison, B. K. & Morita, T. Y. 1974. Heterotrophic potential for amino acid uptake in a naturally eutrophic lake. Appl. Environ. Microbiol. 27: 488-495.

Bölter, M. 1981. DOC turnover and microbial biomass production. Kieler Meeresforsch., Sonderh. 5: 304-310.

Bölter, M., Liebezeit, G., Wolter, K. & Palmgren, V. 1982. Submodels of a brackish water environment. III. Microbial biomass production and related carbon flux. P.S.Z.N.I. Mar. Ecol. 3: 243-253.

Campbell, P. G. C. & Baker, J. H. 1978. Estimation of bacterial production in fresh waters by the simultaneous measurement of ($^{35}$S) sulphate and D-($^3$H) glucose uptake in the dark. Can. J. Microbiol. 24: 939-946.

Carney, J. F. & Colwell, R. R. 1976. Heterotrophic utilization of glucose and glutamate in an estuary: Effect of season and nutrient load. Appl. Environ. Microbiol. 31: 227-233.

Casida, L. E. Jr. 1971. Microorganisms in unamended soil as observed by various forms of microscopy and staining. Appl. Microbiol. 21: 1040-1045.

Cole, J. J. 1985. Decomposition, p. 302-310. In: Likens, G. E. (ed.), An ecosystem approach to aquatic ecology. Mirror Lake and its watershed. Springer Verlag, New York.

Cole, J. J., Likens, G. E. & Strayer, D. L. 1982. Photosynthetically produced dissolved organic carbon: An important carbon source for planktonic bacteria. Limnol. Oceanogr. 27: 1080-1090.

Conover, R. J. 1966. Feeding on large particles by *Calanus hyperboreus* (Kröger), p. 187-194. In: Barnes, H. (ed.), Some contemporary studies in marine science. Allen & Unwin.

Cooney, J. J., Silver, S. A. & Beck, E. A. 1985. Factors influencing hydrocarbon degradation in three freshwater lakes. Microb. Ecol. 11: 127-137.

Copping, A. E. & Lorenzen, C. J. 1980. Carbon budget of a marine phytoplankton - herbivore system with carbon-14 as a tracer. Limnol. Oceanogr. 25: 873-882.

Crawford, C. C., Hobbie, J. E. & Webb, K. L. 1974. The utilization of dissolved free amino acids by estuarine microorganisms. Ecology 55: 551-563.

Crisman, T. L., Beaver, J. R. & Bays, J. S. 1981. Examination of the relative impact of microzooplankton and macrozooplankton on bacteria in Florida lakes. Verh. Int. Ver. Limnol. 21: 359-362.

Cuhel, R. L., Taylor, C. D. & Jannesch, H. W. 1981. Assimilatory sulphur metabolism in marine microorganisms: sulphur metabolism, protein synthesis and growth of *Pseudomonas haloclurans* and *Alteromonas luteo-vioacens* during unpertubated batch growth. Arch. Microbiol. 130: 8-13.

Daley, R. J. 1979. Direct epifluorescence enumeration of native aquatic bacteria: uses, limitations and comparative accuracy, p. 29-45. In: Costerton, J. W. & Coldwell, R. R. (Eds.), Native aquatic bacteria: enumeration, activity and ecology. - American society for testing and materials. ASTM STP 695. Philadelphia.

Daley, R. J. & Hobbie, J. E. 1975. Direct counts of aquatic bateria by a modified epifluorescence technique. Limnol. Oceanogr. 20: 875-882.

Dring, J. M. & Jewson, D. H. 1982. What does $^{14}C$ uptake by phytoplankton really measure? A theoretical modelling approach. Proc. R. Soc. Lond. (B) 214: 351-368.

Ducklow, H. W. & Hill, S. 1985a. The growth of heterotrophic bacteria in the surface waters of warm core rings. Limnol. Oceanogr. 30: 239-259.

Ducklow, H. W. & Hill, S. 1985b. Tritiated thymidine incorporation and the growth of heterotrophic bacteria in warm core rings. Limnol. Oceanogr. 30: 260-272.

Edenborn, H. M. & Litchfield, C. D. 1985. Glycolate metabolism by *Pseudomonas* sp., strain S227, isolated from a coastal marine sediment. Mar. Biol. 88: 199-205.

Eppley, R. W., Horrigan, S. G., Fuhrman, J. A., Brooks, E. R., Price, C. C. & Sellner, K. 1981. Origins of dissolved organic matter in Southern California coastal waters: experiments on the role of zooplankton. Mar. Ecol. Prog. Ser. 6: 149-159.

Faust, M. A. & Correll, D. L. 1977. Autoradiographic study to detect metabolically active phytoplankton and bacteria in the Rhode River Estuary. Mar. Biol. 41:

293-305.

Fenchel, T. 1982. Ecology of heterotrophic microflagellates. IV. Quantitative occurrence and importance as bacterial consumers. Mar. Ecol. Prog. Ser. 9: 35-42.

Ferguson, R. L. & Sunda, W. G. 1984. Utilization of amino acids by planktonic marine bacteria: Importance of clean technique and low substrate additions. Limnol. Oceanogr. 29: 258-274.

Ferguson, R. L., Buckley, E. N. & Palumbo, A. V. 1984. Response of marine bacterioplankton to differential filtration and confinement. Appl. Environ. Microbiol. 47: 49-55.

Ferguson, R. L. & Rublee, P. 1976. Contribution of bacteria tostanding crop of coastal plankton. Limnol. Oceanogr. 21: 141-145.

Forsberg, B. R. 1985. The fate of planktonic primary production. Limnol. Oceanogr. 30: 807-819.

Francisco, D. E., Mah, R. A. & Rabin, A. C. 1973. Acridine orange-epifluorescence technique for counting bacteria in natural waters. Trans Am. Microsc. Soc. 92: 416-421.

Fuhrman, J. A. 1981. Influence of method on the apparent size distribution of bacterioplankton cells: epifluorescence microscopy compared to scanning electron microscopy. Mar. Ecol. Prog. Ser. 5: 103-106.

Fuhrman, J. A. & Azam, F. 1980. Bacterioplankton secondary production estimates for coastal waters of British Columbia, Antarctica, and California. Appl. Environ. Microbiol. 39: 1085-1095.

Fuhrman, J. A. & Azam, F. 1982. Thymidine incorporation as a measure of heterotrophic bacterioplankton production in marine surface waters: evaluation and field results. Mar. Biol. 66: 109-120.

Fuhrman, J. A. & McManus, G. B. 1984. Do bacteria-sized marine eukaryotes consume significant bacterial production. Science, 224: 1257-1260.

Fuhrman, J. A., Eppley, R. W., Hagström, Å. & Azam, F. 1985. Diel variations in bacterioplankton, phytoplankton and related parameters in the Southern California Bight. Mar. Ecol. Prog. Ser. 27: 9-20.

Gardner, W. S. & Lee, G. F. 1975. The role of amino acids in the nitrogen cycle in Lake Mendota. Limnol. Oceanogr. 20: 379-388.

Gardner, W. S. & Miller, W. H. I.I.I. 1981. Intracellular composition and net release rates of free amino acids in *Daphnia magna*. Can. J. Fish. Aquat. Sci. 38: 157-162.

Garrasi, C., Degens, E. T. & Mopper, K. 1979. The free amino acid composition of seawater obtained without desalting and preconcentration. Mar. Chem. 8: 71-85.

Gocke, K., Dawson, R. & Liebezeit, G. 1981. Availability of dissolved free glucose to heterotrophic microorganisms. Mar. Biol. 62: 209-216.

Grivell, A. R. & Jackson, J. F. 1968. Thymidine kinase: evidence for its absence from *Neurospora crassa* and some other microorganisms, and the relevance of this to the specific labelling of deoxyribonucleic acid. J. Gen. Microbiol. 54: 307-317.

Haan, D. de & de Boer, T. 1979. Seasonal variations of fulvic acids, amino acids, and sugars in Tjeukemeer, The Netherlands. Arch. Hydrobiol. 85: 30-40.

Hagström, Å. & Larsson, U. 1984. Diel and seasonal variation in growth rates of pelagic bacteria, p. 249-262. In: Hobbie, J. E. & Williams, P. J. LeB. (eds.), Heterotrophic activity in the sea. Plenum Press, New York.

Hagström, Å., Larsson, U., Hörstedt. P. & Normark. S. 1979. Frequency of dividing cells, a new approach to the determination of bacterial growth rates in aquatic environments. Appl. Environ. Microbiol. 37: 805-812.

Haney, J. F. 1985. Regulation of cladoceran filtering rates in nature by body size, food concentration and diel feeding patterns. Limnol. Oceanogr. 30: 397-411.

Hansen, H. J. 1984. Assimilatory sulphate reduction as an approach to determine bacterial growth. Arch. Hydrobiol. Beih. Ergebn. Limnol. 19: 43-51.

Hansen, L., Krog, G. F. & Søndergaard, M. 1986. Decomposition of lake phytoplankton. 1. Dynamics of short-term decomposition. Oikos 46: 37-44.

Hanson, R. B., Shafer, D., Ryan, T., Pope, D. H. & Lowrey, H. K. 1983. Bacterioplankton in Antarctic ocean waters during late Austral winter: abundance, frequency of dividing cells and estimates of production. Appl. Environ. Microbiol. 45: 1622-1632.

Hellebust, J. A. 1971. Excretion of some organic compounds by marine phytoplankton. Limnol. Oceanogr. 10: 192-206.

Hobbie, J. E. & Crawford, C. D. 1969. Respiration corrections for bacterial uptake of dissolved organic compounds in natural water. Limnol. Oceanogr. 14: 528-532.

Hobbie, J. E., Crawford, C. C. & Webb, K. L. 1969. Amino acid flux in an estuary. Science 159: 1463-1464.

Hobbie, J. E., Daley, J. & Jasper, S. 1977. Use of nuclepore filters for counting bacteria by fluorescence microscopy. Appl. Environ. Microbiol. 33: 1225-1228.

Hobbie, J. E., Holm-Hansen, O., Pachard, T. T., Pomeroy, L. R.,Sheldon, R. W., Thomas, J. P. & Wiebe, W. J. 1972. A study of the distribution and activity of microorganisms in ocean water. Limnol. Oceanogr. 17: 544-555.

Hobbie, J. E. & Rublee, P. 1977. Radioisotope studies of heterotrophic bacteria in aquatic ecosystems, p. 441-471. In: Cairns, J. Jr. (ed.), Aquatic microbial communities, Garland, New York.

Hobbie, J. E. & Williams, P. J. LeB. 1984. Heterotrophic activity in the sea. Plenum Press, New York.

Hollibaugh, J. T. & Azam, F. 1983. Microbial degradation of dissolved proteins in seawater. Limnol. Oceanogr. 28: 1104-1116.

Hollibaugh, J. T., Fuhrman, J. A. & Azam, F. 1980. Radioactivivity labeling of natural assemblages of bacterioplankton for use in trophic studies. Limnol. Oceanogr. 25: 172-181.

Hoppe, H.-G. 1983. Significance of exoenzymatic activities in the ecology of brackish water: Measurements by means of methyl-umbelliferyl-substrates. Mar. Ecol. Prog. Ser. 11: 299-308.

Höfle, M. 1984. Degradation of putrescine and cadavarine in seawater cultures by marine bacteria. Appl. Environ. Microbiol. 47: 843-849.

Ingraham, J. L., Maaløe, O. & Neidhardt, F. C. 1983. Growth of the bacterial cell. Sinauer Association, Sunderland, Mass., USA.

Ittekkot, V. 1982. Variations of dissolved organic matter during a plankton bloom: Qualitative aspects, based on sugar and amino acid analysis. Mar. Chem. 11: 143-158.

Ittekkot, V., Brockmann, U., Michaelis, W. & Degens, E. T. 1981. Dissolved free and combined carbohydrates during a phytoplankton bloom in the northern North Sea. Mar. Ecol. Prog. Ser. 4: 299-305.

Iturriaga, R. 1981. Phytoplankton photoassimilated extracellular products; heterotrophic utilization in marine environments. Kieler Meeresforsch. Sonderh. 5: 318-324.

Jannasch, H. W., & Jones, G. E. 1959. Bacterial populations in seawater as determined by different methods of enumeration. Limnol. Oceanogr. 4: 128-129.

Jassby, A. D. 1975. Dark sulphate uptake and bacterial production in a subalpine lake. Ecology, 56: 627-636.

Jensen, L. M., Jørgensen, N. O. G. & Søndergaard, M. 1985. Specific activity. Significance in estimating release rates of extracellular dissolved organic carbon (EOC) by algae. Verh. Int. Verein. Limnol. 22: 2893-2897.

Johannes, R. E. & Satomi, M. 1967. Measuring organic matter retained by aquatic Invertebrates. J. Fish. Res. Bd. Can. 24: 2467-2471.

Johannes, R. E. & Webb, K. L. 1970. Release of dissolved organic compounds by marine and freshwater invertebrates. Inst. Mar. Sci. (Alaska) Occas. Publ. 1: 257-273.

Joiris, G., Billen, G., Lancelot, C., Daro, M. N., Mommaerts, J. P., Bertels, A., Bossicart, M. & Hecq, J. H. 1982. A budget of carbon cycling in the Belgian coastal zone: Relative role of zooplankton, bacterioplankton and benthos in the utilization of primary production. Neth. J. Sea. Res. 16: 260-275.

Jones, J. G. 1974. Some observations on direct counts of freshwater bacteria obained with a fluorescence microscope. Limnol. Oceanogr. 19: 540-543.

Jones, J. G. & Simon, B. M. 1975. An investigation of errors in direct counts of aquatic bacteria by epifluorescence microscopy, with reference to a new method for dyeing membrane filters. J. Appl. Bact. 39: 317-329.

Jordan, M. J., Daley, R. J. & Lee, K. 1978. Improved filtration procedures for freshwater ($^{35}$S) $SO_4$ uptake studies. Limnol. Oceanogr. 23: 154-157.

Jordan, M. J. & Likens, G. E. 1980. Measurement of planktonic bacterial production in an oligotrophic lake. Limnol. Oceanogr. 25: 719-732.

Jørgensen, N. O. G. 1982. Heterotrophic assimilation and occurrence of dissolved free amino acids in a shallow estuary. Mar. Ecol. Prog. Ser. 8: 145-159.

Jørgensen, N. O. G. 1984. Occurrence and heterotrophic turnover of dissolved free amino acids in the thermally stratified Lake Almind. Verh. Int. Verein. Limnol. 22: 785-789.

Jørgensen, N. O. G. 1986. Fluxes of free amino acids in three Danish lakes. Freshwat. Biol. 16: 255-268.

Jørgensen, N. O. G. In press. Free amino acids in lakes: Concentrations and assimilation rates in relation to phytoplankton and bacterial production. Limnol. Oceanogr.

Jørgensen, N. O. G. & Søndergaard, M. 1984. Are dissolved free amino acids free? Microb. Ecol. 10: 301-316.

Jørgensen, N. O. G., Søndergaard, M., Hansen, H. J., Bosselmann, S. & Riemann, B. 1983. Diel variation in concentration, assimilation and respiration of dissolved free amino acids in relation to planktonic primary and secondary production in two eutrophic lakes. Hydrobiologia 107: 107-122.

Karl, D. M. 1979. Measurement of microbial activity and growth in the ocean by rates of stable ribonucleic acid synthesis. Appl. Environ. Microbiol. 38: 850-860.

Karl, D. M. 1981. Simultaneous rates of RNA and DNA syntheses for estimating growth and cell division of aquatic microbial communities. Appl. Environ. Microbiol. 42: 802-810.

Karl, D. M. 1982. Selected nucleic acid precursors in studies of aquatic microbial ecology. Appl. Environ. Microbiol. 44: 891-902.

Karl, D. M., Haugsness, J. A., Campbell, L. & Holm-Hansen, O. 1978. Adenine nucleotide extraction from multicellular organisms and beach sand: ATP recovery, energy charge ratios and determination of carbon/ATP ratios. J. Exp. Mar. Biol. Ecol. 34: 163-181.

Karl, D. M. & Winn, C. D. 1984. Adenine metabolism and nucleic acid synthesis: applications to microbiological oceanography, p. 197-215. In: Hobbie, J. E. & Williams, P. J. LeB (eds.), Hetrotrophic activity in the sea. Plenum Press, New York.

Keller, M. D., Mague, T. H., Badenhausen, M. & Glover, H. E. 1982. Seasonal variations in the production and consumption of amino acids by coastal microplankton. Estuar. Coast. Shelf. Sci. 15: 301-315.

King, G. M. & Berman, T. 1984. Potential effects of isotopic dilution on apparent respiration in $^{14}C$ heterotrophic experiments. Mar. Ecol. Prog. Ser. 19: 175-180.

Kirchman, D., Ducklow, H. & Mitchell, R. 1982. Estimates of bacterial growth from changes in uptake rates and biomass. Appl. Environ. Microbiol. 44: 1296-1307.

Kirchman, D., K'Nees, E. & Hodson, R. 1985. Leucine incorporation and its potential as a measure of protein synthesis by bacteria in natural aquatic systems. Appl. Environ. Microbiol. 49: 599-607.

Krambeck, H.-J. 1974. Energiehaushalt und Stofftransport eines Sees - Beispiel einer mathematischen Analyse limnologischer Processe. Arch. Hydrobiol. 73: 137-192.

Krambeck, C., Krambeck, H.-J. & Overbeck, J. 1981. Microcomputer-associated biomass determination of plankton bacteria on scanning electron micrographs. Appl. Environ. Microbiol. 42: 142-149.

Kutnetsov, S. J. & Romanenko, V. J. 1966. Produktion der Biomasse heterotropher Bakterien und die Geschwindigkeit ihrer Vermehrung im Rybinsk-Strausee. Verh. Int. Verein. Limnol. 16: 1493-1500.

Lampert, W. 1978. Release of dissolved organic carbon by grazing zooplankton. Limnol. Oceanogr. 23: 831-834.

Lampert, W. & Taylor, B. F. 1985. In situ grazing rates and particle selection by copepods: Effects of vertical migration. Ecology 60: 68-82.

Lancelot, C. 1979. Gross excretion rates of natural phytoplankton and heterotrophic uptake of excreted products in the southern North Sea, as determined by short-term kinetics. Mar. Ecol. Prog. Ser. 1: 179-186.

Lancelot, C. & Billen, G. 1984. Activity of heterotrophic bacteria and its coupling to primary production during the spring phytoplankton bloom in the southern bight of the North Sea. Limnol. Oceanogr. 29: 721-730.

Larsson, U. & Hagström, Å. 1982. Fractionated phytoplankton primary production, exudate release and bacterial production in a Baltic eutrophication gradient. Mar. Biol. 67: 57-70.

Lazrus, A., Lorange, E. & Lodge Jr, J. P. 1968. New automated microanalyses for total inorganic fixed nitrogen and for sulphate ion in water. Adv. Chem. Ser. 73: p. 164-171.

Lee, C. & Cronin, C. 1982. The vertical flux of particulate organic nitrogen in the sea: Decomposition of amino acids in the Peru upwelling area and the equatorial Atlantic. J. Mar. Res. 40: 227-251.

Levy, J., Campbell, J. J. R. & Blackburn, T. H. 1973. Introductory microbiology. John Wiley and Sons, Inc., New York.

Li, W. K. W. 1982. Estimating heterotrophic bacterial productivity by inorganic radiocarbon uptake: Importance of establishing time courses of uptake. Mar. Ecol. Prog. Ser. 8: 167-172.

Liebezeit, G. 1980. Aminosäuren und Zucker im marinen Milieu - neuere analytische Methoden und ihre Anwendung. Thesis, University of Kiel, FRG.

Lovell, C. R. & Konopka, A. 1985a. Primary and bacterial production in two dimictic Indiana lakes. Appl. Environ. Microbiol. 49: 485-491.

Lovell, C. R. & Konopka, A. 1985b. Seasonal bacterial production in dimictic lakes as measured by increases in cell number and thymidine incorporation. Appl. Environ. Microbiol. 49: 492-500.

Marcussen, B., Nielsen, P. & Jeppesen, M. 1984. Diel changes in bacterial activity determined by means of microautoradiography. Arch. Hydrobiol. Beih. Ergebn. Limnol. 19: 141-149.

Meyer-Reil, L.-A. 1978. Autoradiography and epifluorescence microscopy combined for the determination of number and spectrum of actively metabolizing bacteria in natural waters. Appl. Environ. Microbiol. 36: 506-512.

Meyer-Reil, L.-A., Bölter, M., Liebezeit, G. & Schramm, W. 1979. Short-term variations in microbiological and chemical parameters. Mar. Ecol. Prog. Ser. 1: 1-6.

Moll, R. 1984. Heterotrophy by phytoplankton and bacteria in Lake Michigan. Verh. Int. Verein. Limnol. 22: 431-434.

Monheimer, R. H. 1972. Heterotrophy by plankton in three lakes of different productivity. Nature, 236: 463-464.

Monheimer, R. H. 1974. Sulphate uptake as a measure of planktonic microbial production in freshwater ecosystems. Can. J. Microbiol. 20: 825-831.

Monheimer, R. H. 1978. Difficulties in interpretation of microbial heterotrophy from sulphate uptake data: laboratory studies. Limnol. Oceanogr. 23: 150-154.

Monheimer, R. H. 1981. Problems in using sulphate uptake for measuring microbial heterotrophy in natural aquatic ecosystems. Verh. Int. Verein. Limnol. 21: 1393-1395.

Montesinos, E., Esteve, I. & Guerrereo, R. 1983. Comparison between direct methods for determination of microbial cell volume: Electron microscopy and electronic particle sizing. Appl. Environ. Microbiol. 45: 1651-1658.

Mopper, K. & Lindroth, P. 1982. Diel and depth variations in dissolved free amino acids and ammonium in the Baltic Sea determined by shipboard HPLC analysis. Limnol. Oceanogr. 27: 336-347.

Moriarty, D. J. W. 1984. Measurements of bacterial growth rates in some marine systems using the incorporation of tritiated thymidine into DNA, p. 217-231. In: Hobbie, J. E. & Williams P. J. LeB. (eds.), Heterotrophic activity in the sea. Plenum Press, New York.

Moriarty, D. J. W. 1986. Measurement of bacterial growth rates in aquatic systems using rates of nucleic acid synthesis. Adv. Microb. Ecol. 9: 245-292.

Moriarty, D. J. W. & Pollard, P. C. 1981. DNA synthesis as a measure of bacteria productivity in seagrass sediments. Mar. Ecol. Prog. Ser. 5: 151-156.

Morita, R. Y. 1984. Substrate capture by marine heterotrophic bacteria in low nutrient waters. In: Hobbie, J. E. & P. J. LeB. Williams (eds.), Heterotrophic activity in the sea. Plenum Publishing Corporation, New York.

Murray, R. E. & Hodson, R. E. 1985. Annual cycle of bacterial secondary production in five aquatic habitats of the Okefenokee Swamp ecosystem. Appl. Environ. Microbiol. 49: 650-655.

Münster, U. 1984. Distribution, dynamic and structure of free dissolved carbohydrates in the Plussee, a North German eutrophic lake. Verh. Int. Verein. Limnol. 22: 929-935.

Myklestad, S. 1977. Production of carbohydrates by marine planktonic diatoms. II. Influence of N/P ratio in the growth medium on the assimilation ratio, growth rate, and production of cellular and extracellular carbohydrates by *Chaetoceros affinis* var. *willei* (Gev.) Hustedt and *Skeletonema costatum* (Gev.) Cleve. J. Exp. Mar. Biol. Ecol. 29: 161-179.

Måløe, O. & Kjeldgaard, N. O. 1966. Control of macromolecular synthesis. A study of DNA, RNA, and protein synthesis in bacteria. Microbial and Molecular Biology Series. W. A. Benjamin, Inc. New York, Amsterdam.

Newell, S. Y. & Christian, R. R. 1981. Frequency of dividing cells as an estimator of bacterial productivity. Appl. Environ. Microbiol. 42: 23-31.

Newell, S. Y. & Fallon, R. D. 1982. Bacterial productivity in the water column and sediments of the Georgia (USA) coastal zone: estimates via direct counting and parallel measurements of thymidine incorporation. Microb. Ecol. 8: 33-46.

Nissen, H., Nissen, P. & Azam, F. 1984. Multiphasic uptake of D-glucose by an oligotrophic marine bacterium. Mar. Ecol. Prog. Ser. 16: 155-160.

North, B. B. & Stephens, G. C. 1971. Uptake and assimilation of amino acids by *Platymonas*. 2. Increased uptake in nitrogen-deficient cells. Biol. Bull.: 140: 242-254.

Ochiai, M., Nakajima, T. & Hanya, T. 1979. Seasonal fluctuation of dissolved organic matter in Lake Nakanuma. Jap. J. Limnol. 40: 185-190.

Ochiai, M. & Ukiya, T. 1981. Seasonal variation of dissolved organic constituents in Lake Nakanuma during March to November 1979. Verh. Int. Verein. Limnol. 21: 682-687.

Overbeck, J. 1974. Über die Kompartimentierung der stehenden Gewässer - ein Beitrag zur Struktur und Funktion des limnischen Ökosystems. Verh. Ges. f. Ökologie Saarbrücken 1973: 211-223.

Overbeck, J. 1979a. Dark $CO_2$ uptake - biochemical background and its relevance to *in situ* bacterial production. Arch. Hydrobiol. Beih. Ergebn. Limnol. 12: 38-47.

Overbeck, J. 1979b. Studies on heterotrophic functions and glucose metabolism of microplankton in Pluss-See. Arch. Hydrobiol. Beih. Ergebn. Limnol. 13: 56-76.

Parsons, T. R., Stephens, K. & Strickland, J. D. H. 1961. On the chemical composition of eleven species of marine phytoplankters. J. Fish. Res. Bd. Can., 18: 1001-1016.

Pedrós-Alió, C. & Brock, T. D. 1983. The importance of attachment to particles for plankton bacteria. Arch. Hydrobiol. 98: 354-379.

Pollard, P. C. & Moriarty, D. J. W. 1984. Validity of the tritiated thymidine method for estimating bacterial growth rates: measurement of isotope dilution during DNA synthesis. Appl. Environ. Microbiol. 48: 1076-1083.

Poltz, J. 1972. Untersuchungen über das Vorkommen und den Abbau von Fetten und Fettsäuren in Seen. Arch. Hydrobiol. Suppl. 40: 315-399.

Porter, K. G. & Feig, Y. S. 1980. The use of DAPI for identifying and counting aquatic microflora. Limnol. Oceanogr. 25: 943-948.

Rai, H. & Hill, G. 1982. Establishing the pattern of heterotrophic bacterial activity in three Central Amazonian lakes. Hydrobiologia 86: 121-126.

Ramsay, A. J. 1974. The use of autoradiography to determine the proportion of bacteria metabolizing in an aquatic habitat. J. Gen. Microbiol. 80: 363-373.

Rego, J. V., Billen, G., Fontigny, A. & Somville, M. 1985. Free and attached proteolytic activity in water environments. Mar. Ecol. Prog. Ser. 21: 245-249.

Riemann, B. 1983. Biomass and production of phyto- and bacterio-plankton in eutrophic Lake Tystrup, Denmark. Freshwat. Biol. 13: 389-398.

Riemann, B. 1984. Determining growth rates of natural assemblages of freshwater bacteria by means of [3]H-thymidine incorporation into DNA: comments on methodology. Arch. Hydrobiol. Beih. Ergebn. Limnol. 19: 67-80.

Riemann, B. 1985. Potential importance of fish predation and zooplankton grazing on natural populations of freshwater bacteria. Appl. Environ. Microbiol. 50: 187-193.

Riemann, B., Bjørnsen, P. K., Newell, S. & Fallon, B. Conversion of [3]H-thymidine incorporation into cell production of coastal marine bacteria. Limnol. Ocea nogr. In press.

Riemann, B., Fuhrman, J. A. & Azam, F. 1982a. Bacterial secondary production in freshwater measured by $^3$H-thymidine incorporation method. Microb. Ecol. 8: 101-114.

Riemann, B., Jørgensen, N. O. G., Lampert, W. & Fuhrman, J. A. (1986). Zooplankton induced changes in dissolved free amino acids and in production rates of freshwater bacteria. Microb. Ecol. 12: 247-258.

Riemann, B., Nielsen, P., Jeppesen, M., Marcussen, B. & Fuhrman, J. A. 1984. Diel changes in bacterial biomass and growth rates in coastal environments, determined by means of thymidine incorporation into DNA, frequency of dividing cells (FDC), and microautoradiography. Mar. Ecol. Prog. Ser. 17: 227-235.

Riemann, B., Søndergaard, M., Schierup, H.-H., Bosselmann, S., Christensen, G., Hansen, J. & Nielsen, B. 1982b. Carbon metabolism during a spring diatom bloom in the eutrophic Lake Mossø. Int. Revue ges. Hydrobiol. 67: 145-185.

Riemann, B. & Søndergaard, M. 1984a. Measurements of diel rates of bacterial secondary production in aquatic environments. Appl. Environ. Microbiol. 47: 632-638.

Riemann, B. & Søndergaard, M. 1984b. Bacterial growth in relation to phytoplankton primary production and extracellular release of organic matter, p. 233-248. In: Hobbie, J. E. & Williams, P. J. LeB. (eds.), Heterotrophic activity in the sea. Plenum Press, New York.

Riemann, B. & Søndergaard, M. 1986. Regulation of bacterial secondary production in two eutrophic lakes and in experimental enclosures. J. Plank. Res. 8: 519-536.

Riley, J. P. & Segar, D. A. 1970. The seasonal variation of free and combined dissolved amino acids in the Irish Sea. J. Mar. Biol. Ass. U.K. 50: 713-720.

Roberts, R. B. Cowie, D. B., Abelson, P. H., Bolton, E. J. & Britten, R. J. 1955. Studies of biosynthesis in *Eschericia coli*. Carnegie Inst. Wash. Publ. 607: 521 p.

Salonen, K. 1977. The estimation of bacterioplankton numbers and biomass by phase contrast microscopy. Ann. Bot. Fenn. 14: 25-28.

Schnitzer, M. & Khan, S. U. 1972. Humic substances in the environment. Marcel Dekker, Inc., New York.

Schoenberg, S. A. & Maccubbin, A. E. 1985. Relative feeding rates on free and particle-bound bacteria by freshwater zooplankton. Limnol. Oceanogr. 30: 1084-1089.

Sepers, A. B. J. 1981. The aerobic mineralization of amino acids in the saline Lake Grevelingen and the freshwater Haringvliet basin (The Netherlands). Arch. Hydrobiol. 92: 114-129.

Sheer, B. F., Sheer, E. B. & Berman, T. 1982. Decomposition of organic detritus: A selective role for microflagellate protozoa. Limnol. Oceanogr. 27: 765-769.

Sieburth, J. McN., Johnson, K. M., Burney, C. M. & Lavoie, D. M. 1977. Estimation of in-situ rates of heterotrophy using diurnal changes in dissolved organic matter and growth rates of picoplankton in diffusion culture. Helgoländer Wiss. Meeresunters. 30: 565-574.

Simon, M. 1985. Specific uptake rates of amino acids by attached and free-living bacteria in a mesotrophic lake. Appl. Environ. Microbiol. 49: 1254-1259.

Smith, D. F. & Horner, S. M. J. 1981. Tracer kinetic analysis applied to problems in marine biology. Bull. Fish. aquat. Sci. Can. 210: 113-127.

Sorokin, Y. I. 1969. Bacterial production, p. 128-151. In: Vollenweider, R. A. (ed.), A manual on methods for measuring primary production in aquatic environments. IBP Handbook No. 12. Blackwell, Oxford.

Sorokin, Y. I. & Pavaljeva, E. B. 1972. On the quantitative characteristics of the pelagic ecosystem of Dalnee Lake (Kamchatka). Hydrobiologia 40: 519-552.

Stanley, S. O. & Brown, C. M. 1976. Inorganic nitrogen metabolism in marine bacteria: The intracellular free amino acid pools in a marine pseudo-monad. Mar. Biol. 38: 101-109.

Starikova, N. D. & Korzhikova, R. I. 1969. Amino acids in the Black Sea. Oceanology 9: 509-518.

Stephens, G. C. & North, B. B. 1971. Extrusion of carbon accompanying uptake of amino acids by marine phytoplankters. Limnol. Oceanogr. 16: 752-757.

Stevenson, L. H. 1978. A case for bacterial dormancy in aquatic systems. Microb. Ecol. 4: 127-133.

Straskrabová, V. & Fuksa, J. 1982. Diel changes in numbers and activities of bacterioplankton in a reservoir in relation to algal production. Limnol. Oceanogr. 27: 660-672.

Søndergaard, M. 1984. Dissolved organic carbon in Danish Lakes: Concentration, composition and lability. Verh. Int. Verein. Limnol. 22: 780-784.

Søndergaard, M., Emmery, L. & Hansen, K. S. 1984. Phytoplankton development and release of extracellular organic products (EOC): Enclosure experiments, p. 35-44. In: Bosheim, S. & Nicholls, M. (eds.), Interaksjoner mellom trofiske nivoer i ferskvann. Norsk Limnologforening. ISBN 82-990973-3-9.

Søndergaard, M., Riemann, B. & Jørgensen, N. O. G. 1985. Extracellular organic carbon (EOC) released by phytoplankton and bacterial production. Oikos 45: 323-332.

Tabatabai, M. A. 1974. A rapid method for determination of sulphate in water samples. Environ. Lett. 7: 237-243.

Tamminen, T. 1982. Winter microbial activity in Lake Tuusulanjärvi. Hydrobiologia 86: 109-113.

van Es, F. B. & Meyer-Reil, L.-A. 1982. Biomass and metabolicactivity of heterotrophic marine bacteria. Adv. Microb. Ecol. 6: 111-170.

Wambeke, F. van & Bianchi, M. A. 1985. Bacterial biomass production and ammonium regeneration in Mediterranean sea water supplemented with amino acids. 2. Nitrogen flux through heterotrophic microplankton food chain. Mar. Ecol. Prog. Ser. 23: 117-128.

Wangersky, P. J. 1959. Dissolved carbohydrates in Long Island Sound, 1956-1958. Bull. Bingham Oceanogr. Soc. 17: 87-94.

Wangersky, P. J. 1977. The role of particulate matter in the productivity of surface waters. Helgol. Wiss. Meeresunters. 30: 456-564.

Watson, S. W., Novitzky, T. J., Quinby, H. L. & Valois, F. W. 1977. Determination of bacterial number and biomass in the marine environment. Appl. Environ. Microbiol. 33: 940-946.

Webb, K. L. & Johannes, G. E. 1967. Studies of the release of dissolved free amino acids by marine zooplankton. Limnol. Oceanogr. 12: 376-382.

Wiebe, W. J. & Smith, D. F. 1977. Direct measurement of dissolved organic carbon release by phytoplankton and incorporation by microheterotrophs. Mar. Biol. 42: 213-223.

Williams, P. J. LeB. 1973. The validity of the application of simple kinetic analysis to heterogenous microbial populations. Limnol. Oceanogr. 18: 159-165.

Williams, P. J. LeB. 1981. Incorporation of microheterotrophic processes into the classical paradigm of the planktonic food web. Kieler Meeresforsch. Sonderh. 5: 1-28.

Williams, P. J. LeB. & Askew, C. 1968. A method for measuring the mineralization by microorganisms of organic compounds in sea water. Deep Sea Res. 15: 365-375.

Williams, P. J. LeB., Berman, T. & Holm-Hansen, O. 1976. Amino acid uptake and respiration by marine heterotrophs. Mar. Biol. 35: 41-47.

Williams, P. J. LeB. & Yentsch, C. S. 1976. An examination of photosynthetic production, excretion of photosynthetic products, and heterotrophic utilization of dissolved organic compounds with reference to results from a coastal subtropical sea. Mar. Biol. 35: 31-40.

Witzel, K.-P., Moaledj, K., & Overbeck, H. J. 1982. A numerical taxonomic comparison of oligocarbophilic and saprophytic bacteria isolated from Lake Pluss-See. Arch. Hydrobiol. 95: 507-520.

Woldringh, C. L. 1976. Morphological analysis of nuclear separation and cell division during the life cycle of E. coli J. Bacteriol. 125: 248-257.

Wood, H. G. & Stjernholm, R. L. 1962. Assimilation of carbon dioxide by heterotrophic organisms, p. 41-117. In: Gunsalus & Stanier, The Bacteria 3. Academic Press, New York.

Wright, R. T. 1970. Glycollic acid uptake by planktonic bacteria p. 521-536. In: Hood, D. W. (ed.), Organic matter in natural waters. Institute of Marine Science, Occasional Publication No. 1, College, Alaska.

Wright, T. R. & Hobbie, J. E. 1966. Use of glucose and acetate by bacteria and algae in aquatic ecosystems. Ecology 47: 447-468.

Zimmermann, R. & Meyer-Reil, L. A. 1974. A new method for fluorescence staining of bacterial populations on membrane filters. Kieler Meeresforsch. 30: 24-27.

ZoBell, C. E. 1946. Marine Microbiology. A monograph on hydrobacteriology, Chronica Botanica Co., Waltham, Mass.

Zygmuntowa, J. 1981. Free amino acids in cultures of various algae species. Acta. Hydrobiol. 23: 283-296.

# Chapter 5. ZOOPLANKTON

by Suzanne Bosselmann and Bo Riemann

## 5.1. Introduction

Natural populations of zooplankton comprise species ranging in size from a few microns up to several centimeters. The macrozooplankton, defined as species larger than 200 μm, comprise mainly Crustacea and is the most well described part of the community with respect to classification and ecology. The microzooplankton, defined as species smaller than 200 μm, comprise rotifers, ciliates and flagellates. These have often been omitted from ecological investigations, especially the heterotrophic flagellates and smaller ciliates for which even the taxonomy is difficult.

Lack of suitable methods have rendered a close examination of the grazing on organic particulate matter difficult. Most knowledge arises from laboratory experiments whereas direct studies on natural assemblages of zooplankton are still few.

The abundance of various zooplankton species is highly variable during the year and among lakes having varying levels of primary productivity. Also the vertical and horizontal distribution of zooplankton is very heterogenous, possibly regulated by both biotic and abiotic factors. The vertical and horizontal distribution and the extent of diel migrations is, moreover, species-specific and change between lakes and seasons in a still unpredictable way. The ecological importance of the vertical migrations are further complicated by the diel variations in individual ingestion rates which have been demonstrated especially in large daphnids.

The spatial heterogeneity and diel changes in grazing make the estimation of grazing of the zooplankton community extremely difficult.

A large literature deals with the spatial and temporal distribution of the zooplankton, the population dynamic aspects, and the environmental factors influencing them. These subjects are outside the scope of this chapter and will only be treated in relation to zooplankton grazing.

This chapter has been designed to evaluate most importantly the grazing of zooplankton on natural populations of phytoplankton and bacteria. Special attention is paid to various techniques to determine grazing, and both species-specific and population grazing rates are discussed.

## 5.2. The structure of zooplankton communities in eutrophic lakes

A list of the most abundant zooplankton species in eutrophic lakes will comprise rather few species of Crustacea and Rotifera. These taxonomic groups constitute most of the zooplankton biomass, except for some periods in winter, when Ciliata can be dominating (Nauwerck 1963, Godeanu 1978, Riemann et al. 1982a). The herbivorous Crustacea are represented primarily by the filter-feeding Cladocera

*Daphnia, Bosmina, Diaphanosoma* and *Ceriodaphnia,* the filter-feeding Copepoda *Eudiaptomus,* and *Cyclops* and *Mesocyclops,* raptorial as adults. Among the herbivorous Rotifera species of the genera *Keratella, Brachionus, Polyarthra, Synchaete,* and *Kellicottia* seem to be by far the most numerous.

The species distribution of crustaceans and rotifers does not exhibit any general relationships to the primary productivity in different lakes. In contrast, the Cladocera *Holopedium* occurs only in softwater lakes (Larsson 1968), the Cladocera *Chydorus* is found in eutrophic lakes during blooms of cyanobacteria, and the Copepoda *Eurytemora* is found only in the deeper layers of large mesotrophic lakes.

The most diverse zooplankton community regarding species richness and evenness seems to exist in some intermediate trophic state (Patalas and Patalas 1966, Brooks 1969, Godeanu 1978).

More pronounced differences appear, however, when dominance of the various species or genera is considered. In a large number of Norwegian lakes, Rognerud and Kjellberg (1985) (Fig. 5.1) found that Cladocera constituted 40-50% of the total zooplankton biomass in oligotrophic-mesotrophic lakes and 60-70% in the most eutrophic conditions. The calanoid Copepoda were the dominant Crustaceans in the mesotrophic state, but were replaced by Cyclopoida in more productive lakes. Similar trends were demonstrated by Patalas and Patalas (1966), Patalas (1972), and Pederson et al. (1975).

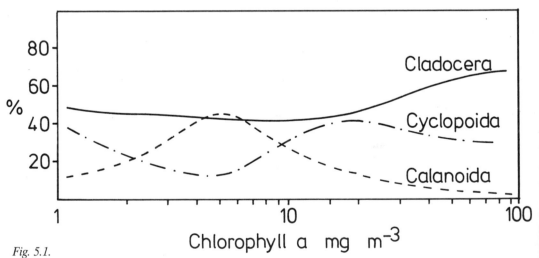

Fig. 5.1.
*The relative occurrence of the biomass of Cladocera, Calanoida, and Cyclopoida (% of total) in relation to phytoplankton biomass (from Rognerud and Kjellberg 1984).*

The dominance of Daphniidae in eutrophic lakes at the expense of Calanoida has been ascribed to the different abilities to utilize particles of different concentrations as exemplified by *Eudiaptomus gracilis* and *Daphnia longispina* (Muck and Lampert 1984). The incipient limiting food level (see below) of *E. gracilis* was lower than that of *D. longispina,* i.e. the copepod expended less energy related to food intake at the low concentrations. At the higher food concentrations *D. longispina* increased its ingestion rate further compared to *E. gracilis.* At fluctuating food concentrations at low levels *Eudiaptomus* was favoured and only little affected, whereas *Daphnia* lost its weight and reduced its filtering rate correspondingly.

Among the Cladocera a shift of species appears in the transition from oligotrophy to eutrophy. Generally, large *Daphnia* species are replaced by smaller ones and smaller genera like *Chydorus* increase in numbers (Patalas and Patalas 1966, Petrowitch 1966, Bays and Crisman 1983, Rognerud and Kjellberg 1984). The absence of large daphnids from some eutrophic lakes is caused by the size selective predation by fish (Hrbacek 1962, Brooks and Dodson 1965).

The importance of fish predation will increase with increasing eutrophication of a lake (Larkin and Northcote 1969). The size efficiency hypothesis by Brooks and Dodson (1965) also explains the dominance of large species in the absence of fish predation: All planktonic herbivores should compete for the smallest particles, < 15 µm. The larger species are most efficient and can take larger particles also; therefore, the small species will be competitively eliminated by large daphnids and calanoid copepods. Dodson (1974), however, demonstrated that the absence of small daphnids might be caused by predation by invertebrate predators, *Chaoborus* larvae and carnivorous cladocerans, preferring smaller zooplankters as prey. However, the relative importance of the various predators on zooplankton in lakes of different productivity is still a question.

A third hypothesis was proposed by Gliwicz (1977), who found that small species were too insufficient filterers to survive in low phytoplankton concentrations in contrast to the larger ones. The feeding of these, however, were reduced by interference by large algae. This interference was avoided by the small daphnids, as they were able to reduce the gape between the carapace valves (see below). This ability was also demonstrated by Porter and McDonough (1984) and Gliwicz and Siedlar (1985). Also calanoid copepods may be able to avoid interference by the large algae and filaments (Richman and Dodson 1983) and due to this ability they may be dominant in very eutrophic lakes (Haney 1973, Edmondson and Litt 1982, Dodson and Richman 1983).

The species composition of the crustacean zooplankton in lakes changes distinctly during the year. In Lake Mikolajskie (Fig. 5.2) (Gliwicz 1977), a small winter population of copepods is replaced during spring by maxima of *Bosmina coregoni*, *Daphnia cucullata, D. galeata* and *Eudiaptomus graciloides*. During mid-summer these species decline, *D. cucullata* and *Eudiaptomus* more slowly, remaining in low numbers. In late summer maxima of *Diaphanosoma brachyurum, Chydorus sphaericus* and *Ceriodaphnia quadrangula* develop. In the autumn the disappearance of these summer species is followed by second maxima of the spring species. This scheme is in agreement with the succession found in various eutrophic lakes ( Nauwerck 1963, Hakkari 1969, Bosselmann 1974, Nowak 1975, Gliwicz and Prejs 1977, Andersen and Jacobsen 1979). Gliwicz (1977) explained this succession of species by their different response to high concentrations of large algae. The carapace gapes of *Daphnia* and *Bosmina* were only little reduced in the presence of large algae; thus the ingestion is reduced by the interference of inedible particles and

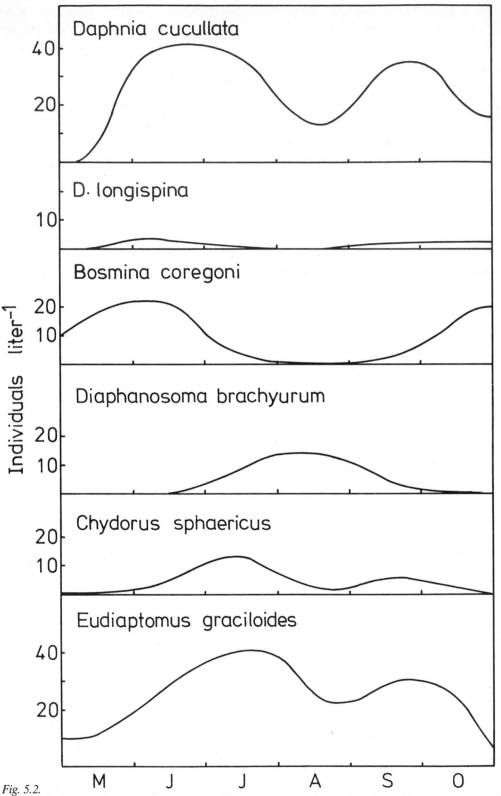

Fig. 5.2.
*The seasonal variations in the occurrence of filter-feeding Crustacea species during May-October in Lake Mikolajskie (Redrawn from Gliwicz 1977).*

the fecundity declines correspondingly. The summer species reduce the gape between valves to a higher degree, reducing the size range of particles retained. Asmaller reduction of the ingestion is counteracted by the high concentration of small particles found in late summer, small algae, bacteria, and detrital particles resulting from the excess production of large algae (see below). *D. cucullata* was found to be intermediate to *D. longispina* and the summer species, *Chydorus* and *Diaphanosoma,* explaining the superiority of this species in summer as well as in eutrophic lakes in general. Correspondingly, Webster and Peters (1978) found that larger Cladocera were more sensitive than smaller ones to the adverse influence of large filamentous algae.

The feeding behaviour of the various species and possibly different responses to toxic algae deserve attention as toxic effects may also be avoided by the feeding strategy.

An alternate explanation may be proposed to explain seasonal succession of species, i.e. a size specific predation from fish with greatest impact in June to August. *Daphnia cucullata, Chydorus* and *Ceriodaphnia* are small Cladocera species escaping predation. *Diaphanosoma* may similarly be favoured by its high transparency, and fast escape response.

Furthermore, in the case of copepods the development of separate generations through the year triggered by the temperature is an important factor for the development of population maxima during the year.

These alternative explanations may be complimentary, whereby similar factors regulate the zooplankton structure during the year and in the course of a eutrophication.

## 5.3. Filtering, grazing, and ingestion rates

The feeding of zooplankton depends on various endogenous and environmental factors. The ecologically most important factors will be discussed below. Comparisons and evaluation of results found in the literature are difficult due to inconsistent use of terms and the limitations of various techniques. The feeding activities of zooplankton are conventionally described by filtering rates, grazing rates, and ingestion (or feeding) rates; therefore, a short examination will be made of their definitions and of the limitations in their application.

The filtering rate is defined as the volume of water containing the number of particles removed by an animal per unit of time (Erman 1956, cited in Starkweather 1980, Rigler 1961), i.e. the filtering rate is synonymous with the clearance rate, a concept widely used in physiology. This definition overcomes the trivial discussion of the difference between effective and actual filtering rates. In the present context the grazing rate will be defined as the percentage of food of the total present removed by the zooplankton community or population per unit of time.

The ingestion rate is the amount of food eaten per animal per unit of time as number or volume of food particles, calories, wet weight, dry weight, or carbon.

The food concentration must be regarded when filtering rates from different experiments are compared. Filtering rates are comparable only with food concentrations below the incipient limiting level ILL (see below), whereas ingestion rates can be compared when food concentrations are above ILL.

The filtering rate is size selective in Diaptomids (Gliwicz 1969, McQueen 1970, Bogdan and McNaught 1975, Vanderploeg 1981), in Daphniidae (Burns 1969a, 1969b, Gliwicz 1969, Porter 1973, Berman and Richman 1974), as well as in Rotifera (Gliwicz 1969, Gilbert and Starkweather 1980).

The size selectivity implies that the filtering rate depends on the particles present in the medium. Furthermore, the estimation of a filtering rate will depend on the recording of food particles (number, weight, activity) in the experiment in question, if particles of various sizes are present. Thus, a filtering rate must be defined for a specific particle or particle size, and rates are not comparable unless determined on similar particles. From this it appears that a filtering rate should not be given for the feeding on a natural phytoplankton assemblage.

These considerations explain the often found disagreements between results discussed in the literature. The filtering rates obtained from feeding on particles of optimum size (McMahon and Rigler 1968, Petersen et al. 1978, Porter et al. 1982) exceed those obtained from natural populations of nannoplankton (Nauwerck 1959, Richman 1964, Gulati 1978, Zankai 1983), presumably because a number of the nannoplanton-sized particles were filtered with lower efficiencies.

In conclusion, filtering rates obtained from experiments in which zooplankton are fed one-sized food particles are valid for these particles only. Ingestion rates cannot be assessed from filtering rates determined on other particles than the ambient, i.e. ingestion of natural phytoplankton must be determined from experiments involving all edible algae. The grazing rate should be used for a defined spectrum of particles or the total community.

Thus, the feeding in a natural plankton community should be expressed by the ingestion rate regarding the zooplankton and by the grazing rate regarding the phytoplankton.

## 5.4. Methods

The feeding of zooplankton has been studied extensively in the laboratory, whereas experiments in natural environments are few. Most experiments have dealt with the influence of environmental factors on the filtering rates of especially Cladocera and Copepoda on algal cultures.

Methods to determine zooplankton grazing include counting procedures in which the number or biomass of particles is determined before and after grazing, various radiotracer procedures in which the food source is labeled with a radioactive compound, procedures involving the use of fluorescent and/or labeled artificial beads, and rough estimates from changes in the biomass of the zooplankton.

Recordings of particles (algae and bacteria) before and after grazing yield filtering rates and ingestion rates with the proviso that recirculation of particles does not take place, i.e. that particles do not pass the gut of the animals without being digested. Particles have been counted with the microscope (Ryther 1958, Gliwicz 1969, Buikema 1973, Holm et al. 1983, Børsheim and Olsen 1984) or by Coulter

Counter (McQueen 1970, Berman and Richman 1974, Kersting and van der Leuw 1976, Vanderploeg 1981). In order to find significant difference between particle numbers in these experiments, the zooplankton must be concentrated above natural densities, or the experiments must run for several hours with the risk of sedimentation of particles, recirculation of particles, interference of fecal material, or other adverse bottle effects.

### 5.4.1. Zooplankton feeding on algae

The above mentioned difficulties are partly overcome in radiotracer experiments. In most experiments using algae as food, $C^{14}$ has been used. The filtering or grazing rate can be estimated from the radioactivity of the animals after grazing in relation to the radioactivity of the medium. The ingestion can be calculated from the specific activity of the food. Low concentrations of zooplankton or algae can be used if the particles have been sufficiently labeled, and moreover the feeding period can be kept short. The feeding period must be shorter than the time of gut passage; otherwise a feeding rate between the ingestion and assimilation will be obtained. The time of gut passage has been found to vary between 2-5 min at 23°C (Zankai 1983) and 30 min. at 22°C (Burns 1969c) for Daphnia species, 1 hr for Diaptomus species at 20°C (Richman 1966, Kibby 1971), and 30 min. for a mixed zooplankton population at 14°C (Bosselmann et al., in prep.a). Differences in food concentration may account for some of this variation; Geller (1975) found that the time of gut filling declined with increasing concentration of food particles. A review of gut passage times is found in Peters (1984).

Labeled algae experiments have been used widely during the last decades, see review by Peters and Downing (1984). In these filtering rates and ingestion rates of various zooplankton species on monocultures of algae have been determined.

Also grazing rates and ingestion rates on natural phytoplankton communities have been estimated from labeling experiments by various authors (Nauwerck 1959, Richman 1964, Bell and Ward 1970, Gulati 1975, 1978, Bogdan and McNaught 1979, Gulati et al. 1982, Zankai 1983, Janicki and DaCosta 1984). Filtering rates obtained in such experiments, however, depend on the amount of unavailable algae in the labeled fraction of phytoplankton, and thereby on the size range of this fraction. Therefore, they are of limited value for comparative purposes. Furthermore, it is emphasized that filtering rates as well as grazing rates refer to the labeled fraction only.

The manipulation of animals in these experiments can influence the feeding, with the most important effect in short time experiments.

Handling of the zooplankton is reduced to a minimum using the *in situ* procedure of Haney (1971): A small volume of uniformly labeled algae are resuspended in a grazing chamber containing the natural plankton community. Thus the filtering rate can be estimated in situ as well as in the laboratory without significant changes in the ambient medium and the density of the zooplankton community. Although this method has been used by several investigators over the last ten years, it must be emphasized that the filtering rates obtained only refer to particles

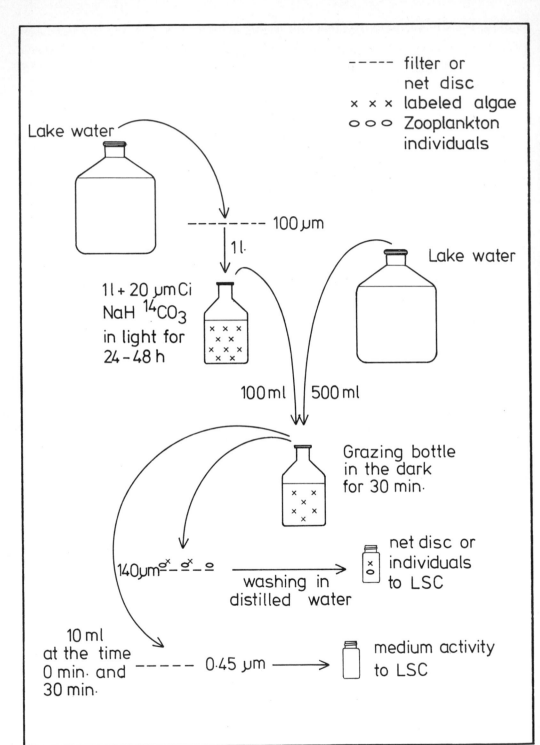

Fig. 5.3.
*A diagram of grazing experiments using phytoplankton <100 μm and zooplankton >140 μm (from Bosselmann et al. In prep.).*

of the same size and quality as the particles added, and generalizations to grazing or ingestion rates on the phytoplankton as a whole cannot be made. This limitation is overcome in the modification of the method introduced recently (Haney and Trout 1985), in which natural phytoplankton < 151 mm was labeled, concentrated and added to the grazing chamber.

In the experiment using natural assemblages of phytoplankton, the algae (or in some cases the zooplankton) are concentrated by filtering or centrifuging. Both processes may influence the natural conditions adversely, but they are nearly eliminated in the approach used by Bosselmann et al. (in prep.) (Fig. 5.3). Here, natural populations of phytoplankton < 100 μm (or 50 μm) were incubated with $C^{14}$ (20 μCi l$^{-1}$) for 24-48 hrs in the light at the ambient temperature. One proportion of labeled algae was added to a grazing bottle containing five portions of lake water with unlabeled algae. The grazing was carried out either *in situ* or in darkness in the laboratory at the ambient temperature. After a feeding period (30 min at 15°C), the zooplankton was filtered through a 140 μm mesh net (25 mm), transferred to vials and dissolved in a tissue solubilizer for counting in a liquid scintillation counter. The activity of the medium was measured before and after the grazing.

The most important problem in experiments using labeled phytoplankton assemblages is the different specific labeling of various phytoplankton species, as demonstrated by Desortova (1976). The very large differences in specific labeling of various phytoplankters, however, were obtained after very short (1-3 h) incubations. In Frederiksborg Slotssø smaller variations (about 15% deviation from the average) were found between various size fractions when natural phytoplankton was incubated for 48 hours in the lake (Bosselmann et al. in prep. a). Still species specific variations were not considered.

Gulati (1975) obtained ingestion rates from grazing rates of nannoplankton < 15 mm and the amount of seston of similar size. It was assumed that the zooplankton ingested detritus and algae with the same efficiency. However, especially during periods with large amounts of detritus, the ingestion rates were overestimated. At least *Bosmina* and *Diaptomus* have been demonstrated to filter detrital particles less efficiently than algae (Starkweather and Bogdan 1980). On the other hand, Gulati (1975), Gulati et al. (1982), and various other authors underestimate the ingestion by excluding phytoplankton larger than e.g. 15 or 33 μm. Various Cladocera and Copepoda species filtered algae between 50 and 100 μm less efficiently than algae < 50 μm and the grazing on algae > 100 μm is thought to be negligible (Bosselmann et al. in prep.a) If algae > 100 μm are also filtered, ingestion rates will also represent understimates using the Bosselmann et al. (in prep. a) approach.

A method which estimates the *in situ* grazing of algae and detritus as well as bacteria deserves to be mentioned. Gliwicz (1969) counted all categories of particles in chambers with active and narcotized zooplankton, respectively, and the ingestion was calculated from the difference in particles between the chambers after a period of 5 hours. Due to the length of the experimental period the inhibitory effect of the anaesthetics on the primary production (Wium-Andersen 1975) should not influence the results importantly. However, the method must be considered too time consuming for routine investigations.

The zooplankton grazing may be estimated from population biomass, applying turnover rates and various energetic efficiencies. Energy budgets of single species found in the laboratory may not be applied to populations in the lake without difficulty, but some general trends can be outlined.

The production per biomass unit of zooplankton depends on various factors as the age structure of the population, food conditions, temperature, predation and other mortality. However, from production estimates based on population dynamic analyses (Bosselmann 1975, 1979, Petersen 1983, Hillbricht-Ilkowska 1967, Hillbricht-Ilkowska and Weglenska 1970), the following daily turnover rates seem applicable:

Rotifera 0.20 $d^{-1}$
Cladocera 0.15 $d^{-1}$
Copepoda 0.08 $d^{-1}$

Values of assimilation efficiencies given in the literature range from 1 to 100%, but the extreme values have in most cases been obtained from experiments involving monocultures. Most values concerning natural phytoplankton or a mixed diet are in the range from 60% to 80%.

Net growth efficiencies (production/assimilation) of zooplankton species are less numerous in the literature. Winberg (1971) suggested values from 25% to 40% for a mixed zooplankton community. This results in a gross growth efficiency (production/ingestion) between 15% and 32%.

The estimated ingestion rates based on the mentioned efficiencies include algae as well as detritus and bacteria and should, therefore, exceed the experimental values, if other than the counted or radiolabeled particles were important as food. The importance of algae and bacteria as food for zooplankton will be discussed below.

Thus, from a zooplankton biomass, a range of ingestion may be estimated, the maximum value, $I_{c\,max}$ being about twice that of the minimum value $I_{c\,min}$. A population of copepods will graze 16% to 33% of its biomass per day, for Cladocera the figures will be 31% to 67%. A mixed community will exhibit intermediate values. These estimates are in good agreement with ingestion rates found for various zooplankton species or populations (Horn 1981).

In the Danish eutrophic Lake Hylke (Bosselmann et al., in prep.b) a close agreement was found between the experimentally determined ingestion rates of algae, $I_e$, and $I_{c\,max}$ in May (Fig. 5.4) and between $I_e$ and $I_{c\,min}$ in September (Fig. 5.5). The high value of ingestion in May may be caused by a high turnover of biomass of a young increasing population. The low value of $I_e$ in September is probably caused by an inhibition from the cyanobacteria in this period (discussed below), and with this a low ingestion and low turnover of biomass. The effects of assim

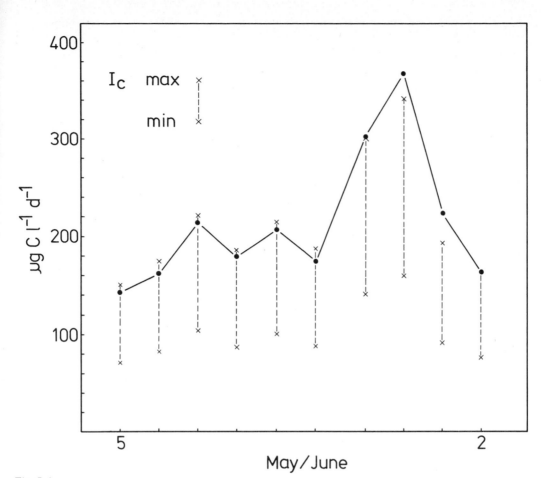

*Fig. 5.4.*
*Ingestion rates of the zooplankton community > 140 mm fom Lake Hylke calculated from grazing on*
*[14]C-labeled algae (· —— ·) or from population estimates (x-----x). (From Bosselmann et al. In prep.a).*

ilation efficiency, growth efficiency and population structure, however, are linked together in a complex way, and more investigations are necessary in this field. However, the method seems to yield a fair estimate of the grazing of a community.

### 5.4.2. Zooplankton grazing on bacteria

Cultures or natural populations of bacteria may be labeled by means of almost any compound which is taken up by the bacteria. Most often glucose, acetate or thymidine are used. After incubation with the radiotracer, the labeled bacteria are given to the animals with or without removal of the excess radiotracer that had not been assimilated by the bacteria. Hollibaugh et al. (1980) removed excess [3]H-thymidine by filtration and then subsequently resuspension of the labeled bacteria from the filter. However, Bjørnsen et al. (1986), Geertz-Hansen et al. (1986), and Riemann (1985) all found it unnecessary to remove dissolved labeled compounds using natural bacterial assemblages and [3]H-thymidine as the tracer. As an alternative to the labeling procedure of the bacteria prior to grazing experiments, Roman

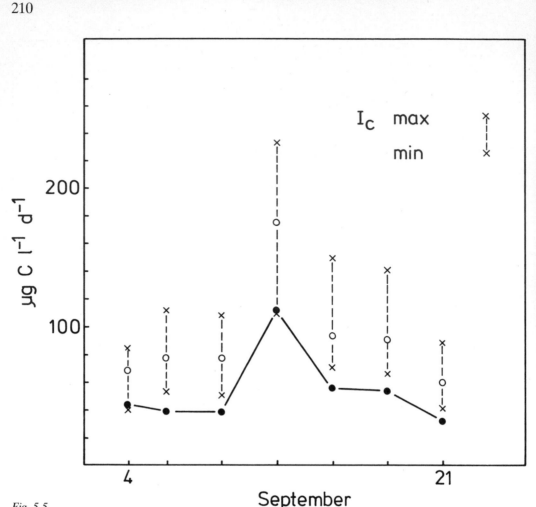

Fig. 5.5.
*Ingestion rates of the zooplankton community from Lake Hylke calculated from grazing on $^{14}$C labeled algae (· —— ·) or from population estimates (x-----x). (From Bosselmann et al. In prep.a).*

and Rublee (1981) added $^{3}$H-thymidine directly to the grazing champer, containing zooplankton and bacteria ($> 3$ μm). However, only 0.02-2% of the total radioactivity added was found in their particulate fraction, and this limits the use of this procedure when grazing is low. When pre-labeled bacteria are used, 2-20% of the radioactivity is found in the particles (Bjørnsen et al. 1986). If prelabeled bacteria together with the dissolved radiotracer are added to the grazing chamber containing natural populations of unlabeled bacteria and natural populations of zooplankton, then two precautions have to be mentioned. Firstly, additional labeling of unlabeled bacteria may occur during the grazing period. This will cause an overestimate of the filtering rate. When $^{3}$H-thymidine is used, 50 ml bacteria are labeled with most often 4-10 nM $^{3}$H-thymidine. This volume is added to 1 liter of natural lake water diluting the thymidine concentration to 0.2-0.5 nM. This thymidine concentration is below the level of 10 nM, which is most often required to maximally label freshwater bacteria (Riemann et al. 1982b), so additional labeling of unlabeled bacteria is not expected (Riemann 1985, Bjørnsen et al. 1986). If glucose or acetate are used as substrates for the bacteria, some additional labeling may occur. As a second point, heterotrophic microflagellates (4-8 μm in size) are present in lake water (Fenchel 1982, Riemann 1985). If the flagellates graze the labeled bacteria, radioactivity will be present in the flagellates as well. This will cause an error in the calculated grazing rate of macrozooplankton grazing on bacteria.

It is possible to further increase the specific activity of the bacteria by additions of more substrate, increasing the temperature and the incubation period etc. These manipulations may work well, using cultures, but often they accelerate succession and growth of large bacteria, when natural populations are used (Newell and Christian 1981, Bjørnsen et al. 1986).

The specific activity of the bacteria is high enough to detect even low grazing rates, using natural populations of bacteria and zooplankton. The grazing period should be within the period of linear uptake. Most often 20-60 min are sufficient to get enough activity in the animals and still be within the period before defecation decreases the activity in the animals. The choice of incubation period depends roughly on the temperature. Still, however, it is important to run time series, since the gut clearance rate is to some extent species-specific and also depends on the concentration of food (Gulati 1974, Gulati et al. 1982).

Some authors have used concentrated zooplankton (Pedrós-Alió and Brock 1982), whereas others have used natural zooplankton communities at their natural densities (Riemann 1985, Geertz-Hansen et al. 1986).

As an alternative to the radiolabeled bacteria, fluorescent and/or labeled beads may be used as artificial food particles for zooplankton. With respect to fluorescent monodisperse latex beads, these are easily distinguished from detritus and living particles. They may be purchased with diameters from 0.25 μm to 5.5 μm, and in experiments they are offered in small quantities to transparent microzooplankton and counted under an epifluorescence microscope. Although very little is known about the effects of shape and surface properties of the beads on the grazing rates of certain groups of microzooplankton, e.g. many ciliates do not discriminate between small particles by properties other than size (Fenchel 1982). Børsheim (1984) tested two sizes of latex beads on *Epistylis rotans* and *Strombidium* sp. and found linear uptake rate during 6 min incubations. Clearance rates were, however, markedly lower compared with the rates presented by Fenchel (1982). When experiments were carried out, using *Daphnia longispina* and *Bosmina longispina,* no quantitative determinations could be made from counting (Børsheim 1984). Probably, the use of labeled beads could extend the use of this technique to include members of the zooplankton community which eat bacteria-sized particles, but are not transparent for a microscopic examination.

## 5.5. Individual feeding rates

Ingestion rates (Table 5.1) of Copepoda range from 0.11 mg C ind$^{-1}$ d$^{-1}$ of *E. gracilis* feeding on phytoplankton < 320 μm at 3-5°C (Zankai 1978) and 0.7 μg C ind$^{-1}$ d$^{-1}$ of the same species on Scenedesmus at 19°C (Muck and Lampert 1984),

Table 5.1. Maximum ingestion rates in mg C ind$^{-1}$ d$^{-1}$ of various species of Cladocera and Copepoda.

| Species | Animal length (mm) or biomass (µg C) | Ingestion rate (µg C ind$^{-1}$d$^{-1}$) | Ingestion (% of biomass) | Temp (°C) | Food | Author |
|---|---|---|---|---|---|---|
| **Daphnia magna** | 2,5-3 mm 90 µg | 12,0 | 13 | 20 | **Chlorella vulgaris** | Kersting (1978) |
| **D. magna** | 2,8-3,3 mm 90-140 µg | 4,8 | 4 | 20 | **Escherichia coli** | McMahon and Rigler (1963) |
| do. | do. | 16,3 | 15 | 20 | **Chlorella vulgaris** | do. |
| do. | do. | 15,4 | 14 | 20 | **Saccharomyces cerevisiae** | do. |
| do. | do. | 49,0 | 44 | 20 | **Tetrachymena nyriformis** | do. |
| **D. magna** | 2,7 mm 90 µg | 31,1 | 35 | 20 | **Chlamydomonas rheinhardi** | Porter et al. (1982) |
| **D. magna** | 40 µg | 79,2 | 198 | 18-20 | **Chlorella vulgaris** | Ryther (1954) |
| **D. hyalina** | | | 20 | 20 | **Asterionella formosa** | Horn (1981) |
| | | | 56 | 20 | **Scenedesmus quadricauda** | Horn (1981) |
| **D. pulex** | 16,5 µg | 2,4-5,4 | 14-33 | 20 | **Phytoplankton + detritus** 25-153 µm | Bell and Ward (1970) |
| do. | 16,5 µg | 4,4 | 27 | 9 | **Phytoplankton** 25-150 µm | do. |
| do. | 16,5 µg | 1,8 | 11 | 4 | **Detritus** <153 µm | do. |
| do. | 16,5 µg | 3,2 | 19 | 8,8 | do. | do. |
| **D. pulex** | 1,9 mm | 6,6 | 18 | 18 | **Aphanizomenon flos aque** (filaments and small flakes) | Holm et al. (1983) |
| **D. pulicaria** | 2 mm | 20,2 | | 20 | **Scenedesmus acutus** | Lampert (1981) |
| | | 22,8 | | 20 | **S. acutus and Aphanizomenon gracile** | do. |
| **D. longispina** | 10 µg | 7,7 | 77 | 19 | **Scenedesmus acutus** | Muck and Lampert (1984) |
| **D. spp** | | | | 16-20 | Phytoplankton <33 µm | Gulati (1978) |
| **D. galeata** | 2 µg | 2,0 | 100 | 12 | Phytoplankton <50 µm | Bosselmann et al. (in prep.a) |
| do. | 2 µg | 2,2 | 110 | 12 | Phytoplankton <70 µm | do. |
| do. | 1,7 µg | 1,0 | 58 | 14 | Phytoplankton <50 µm | do. |

|  |  |  |  |  |  |  |
|---|---|---|---|---|---|---|
| **Bosmina coregoni** | 1,0 µg | 0,7 | 70 | 12 | Phytoplankton <50 µm | do. |
| **Chydorus sphaericus** | 0,6 µg | 1,0 | 167 | 12 | do. | do. |
| **Diaphanomoma brachyurum** | 0,5 µg | 0,2 | 40 | 14 | do. | do. |
|  | 1,3 µg | 0,3 | 25 | 14 | do. | do. |
| **Bosmina longirostris** |  | 0,7-1,5 |  | 16-20 | Phytoplankton <33 µm | Gulati (1978) |
| **Diaphanosoma brachyurum** |  | 0,6-0,7 |  | 16-20 | do. | do. |
| Cladocera |  |  | 11-143 | 14-15 | Chrysomonads | Cushing (1976) |
| **Eudiaptomus gracilis** |  | 2,4-2,8 |  | 16-20 | Phytoplankton <33 µm | Gulati (1978) |
| **E. gracilis** | 3,4 µg | 1,8 | 53 | 17 | Phytoplankton <320 µm | Kibby (1971) |
| **E. gracilis** | 3 µg | 0,7 | 23 | 19 | **Scenedesmus acutus** | Muck and Lampert (1984) |
| **E. gracilis** | 2,8 µg | 1,22 | 44 | 20 | Phytoplankton <320 µm | Zankai (1978) |
| do. | 4,4 µg | 0,11 | 2,5 | 3-5 | do. | do. |
| **E. gracilis** |  |  | 16 | 20 | **Asterionella formosa** | Horn (1981) |
| **E. gracilis** |  |  | 24 | 20 | **Scenedesmus quadricauda** |  |
| **E. gracilis** | 3,7 µg | 2,2 | 59 | 12 | Phytoplankton <50 µm | Bosselmann (in prep.) |
|  | 3,7 µg | 3,1 | 89 | 12 | Phytoplankton <70 µm | do. |
|  | 3,9 µg | 1,4 | 36 | 14 | do. | do. |
|  | 3,9 µg | 1,7 | 44 | 14 | Phytoplankton <140 µm | do. |
| **E. graciloides adults** |  |  | 60 | 20 | **Chlorella vulgaris** | Kryuchkova and Rybak (1975) |
| do. |  |  | 40 | 20 | **Chlamydomonas eugametos** | do. |
|  |  |  | 20-38 | 20 | **Scenedesmus sp.** | do. |
| nauplii and copepodites |  |  | 100-270 | 20 | **Chlorella vulgaris** | do. |
| **Diaptomus oregonensis** | 4,4 µg | 1,3 | 30 | 22 | Nannoplankton | Richmann (1964) |
| **Cyclops vicinus** |  |  | 15 | 20 | **Asterionella formosa** | Horn (1981) |
| do. |  |  | 9 | 20 | **Scenedesmus quadricauda** | do. |
| **Mesocyclops** | 0,7 µg | 0,6 | 86 | 12 | Phytoplankton <50 µm | Bosselmann et al. (in prep.a) |
| do. | 0,7 µg | 0,3 | 43 | 14 | do. | do. |
| do. | 0,7 µg | 0,4 | 57 | 14 | Phytoplankton <140 µm | do. |
| Copepoda |  |  | 0,4-20 | 14 | Green algae | Cushing (1976) |

whereas *E. gracilis* ingested 2.8 µg C ind$^{-1}$ d$^{-1}$ of phytoplankton <33 µm at 16-20°C (Gulati 1978). All results concern adult individuals. In Cladocera an even larger variation is found. *Chydorus sphaericus* ingested 0.2 µg C ind$^{-1}$ d$^{-1}$ and *D. galeata* 1.0 µg C ind$^{-1}$ d$^{-1}$ of natural phytoplankton <50 µm at 14°C (Bosselmann et al. In prep.a), whereas *D. magna* ingested 79.2 µg C ind$^{-1}$ d$^{-1}$ of *Chlorella vulgaris* at 18-20°C (Ryther 1954) and 49 µg C ind$^{-1}$ d$^{-1}$ of *Tetrahymena* at 20°C (McMahon and Rigler 1963) (Table 5.1).

When related to the individual biomass, ingestion rates vary from a minimum of 0.4% d$^{-1}$ of Copepoda ingesting green algae at 14°C (Cushing 1976) and 1.5% d$^{-1}$ in *Eudiaptomus gracilis* ingesting phytoplankton <320 µm at 3-5°C d$^{-1}$ (Zankai 1978). Maximum values are 270% d$^{-1}$ in nauplii of *Eudiaptomus graciloides* ingesting *Chlorella vulgaris* at 20°C (Kryuchkova and Rybak 1975) and 84% d$^{-1}$ of adult *E gracilis* ingesting phytoplankton <70 µm at 12°C (Bosselmann et al., in prep.a).

In Cladocera minimum ingestion rate is 4% d$^{-1}$ in *Daphnia magna* on *Escherishia coli* at 21°C, and maximum is 167% d$^{-1}$ in *Chydorus sphaericus* feeding phytoplankton at 12°C (Bosselmann et al., in prep.a).

It appears from Table 5.1 that generally, ingestion rates obtained with natural phytoplankton as food exceed those on algal cultures. Reasons for the large variation in ingestion rates should be found in food quality and quantity as well as in other environmental factors. The most important of these will be discussed.

The ingestion increases with increasing food concentration until a level, the incipient limiting level of feeding, ILL, is reached. Above this ingestion is constant. This pattern has been demonstrated for Diaptomids (Richman 1964, Horn 1981, Muck and Lampert 1984), for Cladocera (Ryther 1956, Rigler 1961, Burns 1968, Berman and Richman 1974, Porter et al. 1982, Holm et al. 1983, Muck and Lampert 1984) and for Rotifera (Starkweather 1979). The relationship between ingestion rate and the concentration of food (Fig. 5.6) can be described by a curvilinear function (Porter et al. 1982) or a rectiliniear function (Lampert 1977). The question which is the best fit, is still not answered.

The incipient limiting level results from a saturation of the filtering apparatus and thus supports the idea of a mainly mechanical filtering mode (Vanderploeg 1981). Porter et al. (1982) studied the filtering of *Daphnia magna* in detail. Below ILL the filtering appendages beat with a constant rate, above this level the beat rate declined. The mandibular rate increased with increasing food concentrations below ILL, but was constant above ILL. Rejection rate increased at concentrations above the limiting level. The limiting level could thus be interpreted from the mechanical movements of the various appendages involved. Burns (1968) made similar observations on the filtering appendages and in part of the mandibular rate, which furthermore was found to decrease rapidly at very low concentrations of food, indicating a decrease of the filtering rate. A diminished filtering activity below a treshold concentration would be an energetic advantage for the animals.

The incipient limiting level of *Eudiaptomus gracilis* feeding on *Scenedesmus* at 19°C was 76 µg C l$^{-1}$ (Table 5.2) (Muck and Lampert 1984), whereas it was >860 µg C l$^{-1}$, when the same species ingested *Stichococcus.* Among Cladocera a similar

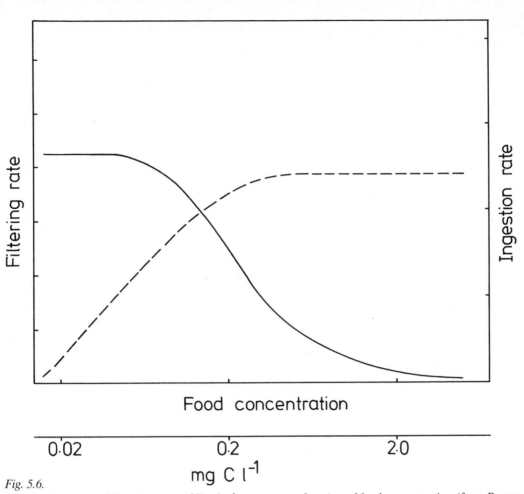

*Fig. 5.6.*
*The filtering rate and ingestion rate of Daphnia magna as a function of food concentration (from Porter et al. 1982).*

variation was found. The ILL of assimilation of *Daphnia magna* was 160 μg C l$^{-1}$, when *Nitschia* was ingested at 15°C, whereas the level was 510 μg C l$^{-1}$ for this species feeding on *Asterionella* (Lampert 1977).

   The incipient limiting level depends on the size of particles (Lampert 1977, Starkweather 1980, Muck and Lampert 1984), which explains the variations mentioned above. The influence of the temperature on ILL is not unambiguous. Lampert (1977) and Muck and Lampert (1984) found increasing as well as decreasing levels with increase of the temperature. A similar disagreement is observed regarding the influence of the individual size of the animals. Lampert (1977) found that the ILL of assimilation increased with the length of *Daphnia pulex* when it ingested *Scenedesmus,* but decreased when it was fed *Stichococcus.*

   Thus, the limiting level for a zooplankton population in natural phytoplankton communities is difficult to determine, as various particle sizes and qualities are

Table 5.2. The incipient limiting level of feeding (mg C 1⁻¹) of various species of Cladocera and Copepoda on different types of food.

| Species | Length (mm) | Incipient limiting level ($\mu$g C 1⁻¹) | Temperature (°C) | Food | Author |
|---|---|---|---|---|---|
| **Daphnia magna** | 2,7 | 200 | 20 | **Chlamydomonas rheinhardi** | Porter et al. (1982) |
| **D. magna** | | 290 | 10 | **Chlorella vulgaris** | Kersting and van der Leuw (1976) |
| do. | | 370 | 18 | do. | do. |
| **D. pulex** | 3 | 250 | 15 | **Scenedesmus acutus** | Lampert (1977*) |
| do. | 3 | 210 | 20 | do. | do. |
| do. | 3 | 350 | 25 | do. | do. |
| do. | 1 | 240 | 25 | do. | do. |
| do. | 3 | 160 | 15 | **Nitschia** | do. |
| do. | 1 | 200 | 15 | do. | do. |
| do. | 3 | 300 | 15 | **Asterionella** | do. |
| do. | 1 | 510 | 15 | do. | do. |
| do. | 3 | 440 | 15 | **Staurastrum** | do. |
| **D. pulex** | 2,3 | 340 | 18 | **Aphanizomenon flos-aquae** | Holm et al. (1983) |
| **D. pulex** | 2 | 180 | 10 | **Scenedesmus** | Geller (1975) |
| do. | 2 | 300 | 20 | do. | do. |
| **D. catawba** | 1,5 | 210 | 22 day exp. | Seston < 31 $\mu$m | Haney (1985) |
| do. | 1,5 | 430 | night exp. | do. | do. |
| **D. rosea** | 1,55 | 400 | | | DeMott (1982) |
| **D. rosea** | 1,5 | 120 | 20 | **Rhodotoria glutinis** | Burns and Rigler (1967) |
| **D. longispina** | 1,75 | 260 | 19 | **Scenedesmus acutus** | Muck and Lampert (1984) |
| do. | 1,75 | 190 | 7 | do. | do. |
| do. | 1,75 | 360 | 19 | **Stichococcus minutissimus** | do. |
| do. | 1,75 | 520 | 7 | do. | do. |
| **Eudiaptomus gracilis** | | 76 | 19 | **Scenedesmus acutus** | do. |
| do. | | 110 | 7 | do. | do. |
| **do.** | | > 860 | 19 | **Stichococcus minutissimus** | do. |
| do. | | 100 | 7 | do. | do. |
| **Diaptomus oregonensis** | | 300 | 20 | **Chlamymodomonas rheinhardi** | Richmann (1966) |

*ILL of assimilation

involved, and as unavailable algae and detritus particles may interfere. However, a concentration of about 200-300 µg C l$^{-1}$ is suggested.

The filtering of crustacean zooplankton as well as of many rotifers depends strongly on the size of food particles. Regarding size limits for the filtering, results may seem conflicting. Adult *Eudiaptomus* have been found to filter most efficiently on particles between 25 and 40 mm (Gliwicz 1977), between 6 and 40 mm (Vanderploeg 1981), and between 6 and 16 mm (McQueen 1970). In all cases the food was natural phytoplankton assemblages, and particles larger than 40 µm were not regarded. Bosselmann et al. (in prep.a) demonstrated that *Eudiaptomus* ingested phytoplankton between 50 and 140 µm, though less efficiently than the algae <50 µm. Bogdan and McNaught (1975) found that *Eudiaptomus* filtered nannoplankton (<22 µm) more efficiently than net plankton (22-100 µm).

Small individuals of *Daphnia* preferred bacteria for larger particles (Gophen et al. 1974, Lampert 1975), while larger individuals ingested bacteria and large algae (a pennate diatom of 28 µm length) equally well (Gliwicz and Siedlar 1980). A selectivity for smaller algae in *Daphnia* was demonstrated by Burns (1969a), Gliwicz (1969), Berman and Richman (1971) and Porter (1973), whereas Bogdan and McNaught (1975) and Janicki and DaCosta (1984) found that net plankton (22-100 µm) and nannoplankton (<22 µm) were filtered by the same efficiency. Bosselmann et al. (In prep. a) found that *Daphnia* ingested algae between 50 µm and 140 µm, but less efficiently than algae <50 µm. Also Gliwicz (1969) and Muck and Lampert (1984) found that *Daphnia* species utilized a broad spectrum of particle size. The variable size of individuals in a *Daphnia* population may contribute to this conception.

Algae too large to be filtered can be ingested by some zooplankton species which break the filaments or colonies (Infante 1971, Holm et al. 1983). This type of feeding may be important, especially in the case of low nannoplankton concentrations, or in the case of high concentrations of large algae which, on the other hand, may interfere in the filtering of the smaller algae.

It may be concluded that crustacean zooplankters feed by a passive-mechanical filtering, which in Copepoda may be supplemented by a raptorial mode of feeding that selects large particles of high quality (Vanderploeg 1981). This mechanical filtering can be disturbed by high concentrations of particles, whether these are of accessible size or too large to be filtered. Generally, *Daphnia* species utilize a broad size spectrum, whereas *Eudiaptomus* species are more strongly selective with respect to size and quality.

Apart from particle size, food quality has consequences for the feeding. In this respect the role of large cyanobacteria is important because of their dominance in many eutrophic lakes.

Some cyanobacteria may be readily ingested in small colonies or filaments (Lampert 1977, DeBernardi et al. 1981); larger colonies may be broken and ingested (Lampert 1977, Holm et al. 1983). Their food value, however, is low because they are assimilated less efficiently than most other algae (Arnold 1971, Lampert 1977, DeBernardi et al. 1981, Holm et al. 1983). Furthermore, some cyanobacteria seem to influence the zooplankton adversely by reducing their feeding rate (Lampert 1981, Richman and Dodson 1983), their fecundity and growth (Lampert 1977, 1981).

The value of detritus as food for zooplankton is poorly known. Bell and Ward (1980) found that *Daphnia* ingested detritus particles almost as efficiently as live algae, but that the assimilation efficiency on this food was lower. Starkweather and Bogdan (1980) also found *Daphnia* indifferent with respect to algae and detritus, whereas *Eudiaptomus* preferred algae and some Rotifera species preferred the detrital material. The nutritional value of detritus is lower than that of algae, as they are assimilated less efficiently (Bell and Ward 1980, Gulati et al. 1982).

Generally, the filtering rate increases with the size of the individuals. The filtering was found to have a linear relationship to the square or cube of the length of various *Daphnia* species (McMahon and Rigler 1963, Burns 1968, Holm et al. 1983).

The temperature may be the most important environmental factor influencing the feeding, apart from food quality and quantity. The filtering rate increases about linearly to the temperature in the naturally occurring temperature regime (Geller 1975). At a higher temperature a decline may appear (Lampert 1977).

Diurnal variations in the filtering rate of zooplankton was demonstrated by Haney (1973), Haney and Hall (1975), Starkweather (1975, 1983) and Haney (1985). Haney and Hall (1975) demonstrated that the diurnal variations in filtering were related to the vertical migrations of the zooplankton only in some seasons, and not always in a predictable pattern. This diurnal rhythm was ascribed to an endogenous component triggered by changes in light densities. Some rhythmicity was maintained when *Daphnia* were kept in darkness after exposure to normal light-darkness cycle (Starkweather 1975), and also temperature was found to influence the feeding rhythm. Haney (1985) found differences between day and night feeding of *Daphnia* corresponding to those of prefed and hungry animals (Geller 1975) respectively, regarding individual ingestion rates as well as changes in the correlations between filtering rates and body length. Thus the high nocturnal feeding compensated for the costs of migration and of the stay during day at lower temperature with reduced filtering activity.

Lampert and Taylor (1984) found large diel variations in the epilimnetic grazing due to vertical migrations of the zooplankton, but grazing per unit biomass did not fluctuate. In any case, the possible diel variation in individual filtering and the possible vertical migrations imply that an extrapolation of an all day and night grazing from individual filtering rates is a complicated task, especially in deep, stratified lakes.

## 5.6. Zooplankton community grazing rates on phytoplankton

Most investigations on quantitative relationships between phytoplankton and zooplankton in lakes cover only shorter periods of the year, or one zooplankton species. Nevertheless a summary of major seasonal changes is made below in an attempt to design a more general ecological frame for our understanding the growth and grazing of natural zooplankton populations in eutrophic lakes.

The zooplankton can reduce the phytoplankton biomass and influence the species composition of algae by grazing. On the other hand, the quality and quantity of the phytoplankton can influence the production of zooplankton. The population dynamics and production of *Eudiaptomus graciloides* in Lake Esrom fluctuated in accordance with the structure and production of the phytoplankton (Bosselmann 1975a, b). The nauplii, dominating the population from April to early June, presumably utilized smaller particles like bacteria resulting from excess production of spring diatoms. The larger copepodites grazed the dominating nannoplanktonic algae in June, and a low algae biomass was found during this period. The production and turnover of *E. graciloides* declined rapidly in July, when the phytoplankton was dominated by *Ceratium hirundinella* and cyanobacteria. This minimum production was due to a high mortality of the larger copepodites, possibly caused by the low level of accessible food. During the subsequent period dominated by cyanobacteria and diatoms, the population consisted of all stages utilizing the diverse food. The mortality in August was ascribed to predation by carnivorous Cladocera, *Leptodora* and *Bythotrephes, Chaoborus* larvae, or fish.

Also the population of *Daphnia galeata* in Lake Esrom was regulated by food during early summer and by predation in late summer and autumn (Petersen 1983).

The relative importance of the various predators was not known in Lake Esrom. In Lake Mikolajskie Gliwicz et al. (1978) found that the main predators *Leptodora* and *Mesocyclops* were more important than the fish. In contrast to this, it has been demonstrated from numerous enclosure experiments and from removal of natural fish populations from lakes that fish are important for the regulation of zooplankton biomass (Hrbacek et al. 1965, Andersson et al. 1978, Stenson et al. 1978, De Bernardi et al. 1978, Lynch and Shapiro 1981, Vojtech 1983, Goad 1984). More information on the relative importance of the various predators and interrelations is strongly needed.

In Lake Mossø a very low zooplankton grazing was found during the spring maximum of diatoms in April (Riemann et al. 1982a). The total zooplankton grazing averaged 0.2% $d^{-1}$ of the phytoplankton biomass, corresponding to 0.4% $d^{-1}$ of the phytoplankton production. Ciliates and rotifers constituted the main part of the zooplankton biomass, indicating that bacteria were important as food in this period.

In eutrophic Lake Hylke zooplankton and phytoplankton relations were studied in May and September (Bosselmann et al. in prep.b). During May (Fig. 5.7), the total phytoplankton biomass declined from initially 990 $\mu$mg C $l^{-1}$ to a level of 200-300 $\mu$g C $l^{-1}$, and nannoplankton <50 $\mu$m constituted 80-90% of the biomass. In the same period, the zooplankton biomass increased from 250 $\mu$g C $l^{-1}$ to a maximum of 450 $\mu$g C $l^{-1}$, after which a rapid decline appeared. During the entire month the grazing exceeded the phytoplankton production. The grazing rate increased from 15% $d^{-1}$ to 60% $d^{-1}$, corresponding to 200% $d^{-1}$ to 600% $d^{-1}$ of the

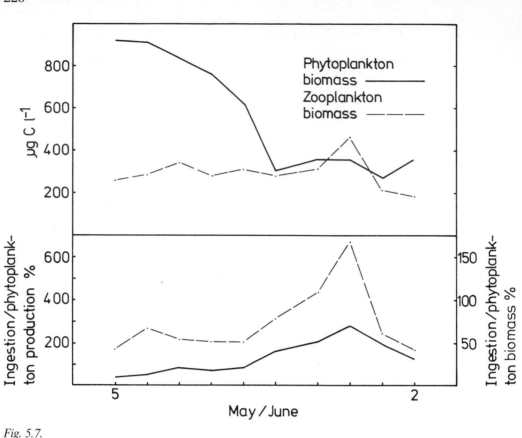

Fig. 5.7.

*Phytoplankton (————)and zooplankton biomass > 140 µm (——·——·——)Lake Hylke in May-June and ingestion of the zooplankton as a percentage of the phytoplankton biomass and production, respectively (from Bosselmann et al. In prep.b)*

primary production. At the end of the period the grazing suddenly declined to 30% d⁻¹, corresponding to 40% of the phytoplankton production. This decline was due to a reduction in the zooplankton biomass, especially *Daphnia galeata.* So in May the zooplankton grazing appeared to be the main factor regulating the biomass of algae. In the second half of the month the phytoplankton biomass was below 300 µg C l⁻¹. This is thought to be the incipient limiting level of at least some zooplankton species and may explain their declining biomass. In *Daphnia* the production of resting eggs was provoked in this period.

During September (Fig. 5.8) the phytoplankton biomass reached a higher level, fluctuating between 1.5 and 2 mg C l⁻¹, out of which the nannoplankton <50 µm constituted 30 to 40%. The average zooplankton biomass was about 200 µg C l⁻¹. Correspondingly, the grazing rate was low, only exceeding 5% d⁻¹ on one single date. So the zooplankton grazing did not have any influence on the development of the phytoplankton biomass in this month.

It was demonstrated from enclosure experiments that the zooplankton was to a high degree controlled by predation from fish in September. When fish were

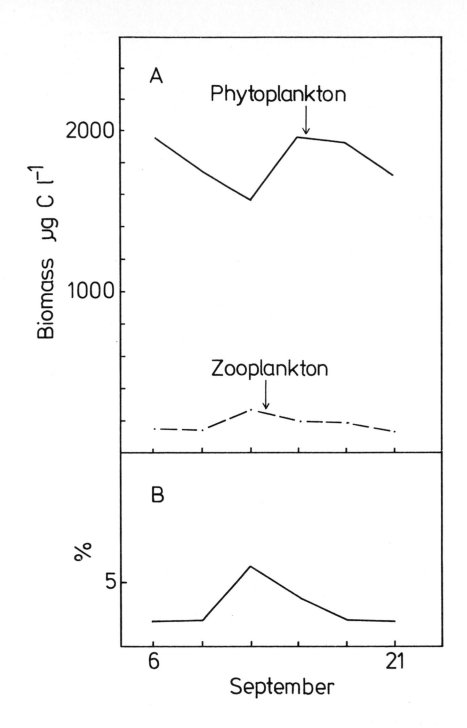

*Fig. 5.8.*
*As Fig. 7 for Lake Hylke in September.*

removed, a 3-fold increase in biomass was found within 15 days. But predation alone could not explain the low grazing rate. The individual ingestion rates of the dominating zooplankton species (Table 5.1) were significantly smaller in Septem-

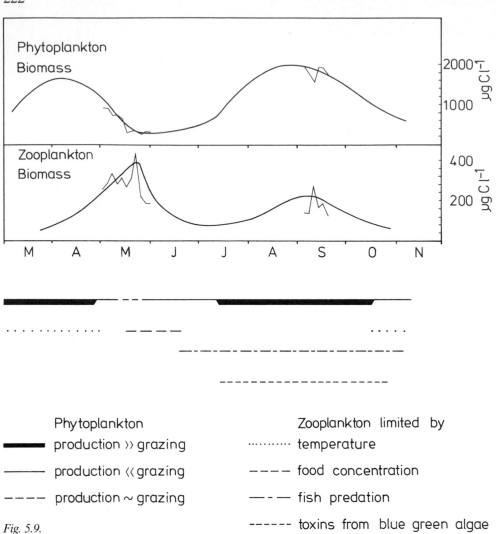

Fig. 5.9.
*Generalized picture of the seasonal variations in phytoplankton - zooplankton relationships on the example of Lake Hylke. Biomass of phytoplankton and zooplankton in Lake Hylke in May-June and September are indicated.*

ber than in May in spite of higher temperature and non-limiting food concentration. The food level was also high when only the nannoplankton was considered. So the lower ingestion rates are ascribed to either toxicity of the cyanobacteria or interference from the large algae.

Predation by fish was not important in Lake Hylke in May. This is in accordance with the results from Norwegian Lake Haugatjern (Reinertsen and Olsen 1984), where the elimination of the fish stock had a significant effect on the phytoplankton in summer, but not in spring.

From these investigations and from the results from Lake Vechten (Gulati et al. 1982), Lake Mikolajskie (Gliwicz and Hillbricht-Ilkowska 1975), Lake Plussee (Krambeck 1981), and Lake Schierensee (Nowak 1975) some general trends appear in the seasonal relations between zooplankton and phytoplankton in eutrophic lakes (Fig. 5.9). When the phytoplankton production increases during early spring, the zooplankton grazing is low, limited mainly by the temperature, and the phytoplankton biomass increases. After a period of some weeks the growth of Copepoda and of Cladocera hatching from resting eggs starts.

The resultant increased grazing and possible nutrient limitations and increased sedimentation of algae reduce the biomass and production of algae. A shortlived balance may be established, followed by an overgrazing of the phytoplankton, which may result in a clear water phase. The food level in this phase may limit the zooplankton, i.e. the concentration is below ILL, and this may in turn reduce the zooplankton biomass and grazing.

The zooplankton apparently overexploits its food resources, and the resulting reduced grazing pressure allows the summer algae to develop, but due to the higher temperature in this period and the dominance of Daphniidae, which have a higher growth potential than Copepoda, the zooplankton is able to respond more quickly to this second maximum of primary production. So the time lag between primary and secondary producers is smaller than in spring.

The ILL of grazing is far exceeded in these months. In spite of this the zooplankton production does not reach the level possible judged from the available food amount. The grazing is low and not able to limit the biomass of algae. This results in a high phytoplankton biomass, and low transparency.

In the autumn, when the algae production is limited by the light, the lower temperature may cause a shift in the species composition of the algae, and another period of significant grazing may appear.

The maxima of microzooplankton species in spring and late summer - autumn presumably reflect the amount of detritus and bacteria resulting from the periods of excess production of phytoplankton.

The main reasons for the poor development of the zooplankton, the low grazing and resulting high phytoplankton biomass during late summer has recently been extensively discussed. Major questions include: Are the cyanobacteria toxic, or are they merely too large to be eaten? Conversely, is the zooplankton exclusively controlled by predators? From the examples above it appears that these various regulating factors may operate simultaneously, and together they result in a less efficient planktonic system, as the direct grazing chain becomes less important.

A balance between phytoplankton production and zooplankton grazing is only found during short periods in a eutrophic lake, and these will be reduced during continued eutrophication. If the phytoplankton production is never limited by lack of nutrients, the zooplankton grazing may not reach the level balancing the supply of algae biomass. An early summer period of clear water may not appear. Also the production of cyanobacteria will increase and extend in time, and the impact of predation by fish and *Chaoborus* may be larger. The trends described for the late summer in moderately eutrophic lakes will correspondingly extend. As a result the zooplankton grazing will decrease relative to the phytoplankton production, and the excess of algae biomass will extend in quantity as well as in time.

## 5.7. Zooplankton community grazing rates on natural populations of bacteria

The lack of suitable methods to determine bacterial production also influences the interpretation of results obtained from zooplankton grazing on natural bacterial assemblages. Since the biomass of bacteria may be turned over up to about 3 times per 24 hours, the bacterial production is necessary, when data on zooplankton grazing is evaluated in an ecological context. Also the grazing methodology is new and still under discussion, so only few data on zooplankton grazing on bacteria are published.

Peterson et al. (1978), Porter et al. (1983), Pace et al. (1983) and Børsheim and Olsen (1984) demonstrated that *Daphnia* grazed on natural bacterial assemblages. Compared with the filtering rates on yeast, those obtained from natural bacterial assemblages were much lower (Peterson et al. 1978). Haney (1973), and Bogdan and Gilberts (1982) examined *in situ* grazing rates, however, cultured bacteria were used as food source. These data suggest that the concentration of natural bacteria should be high before the bacteria could contribute as food for the animals.

Among the few published reports on natural zooplankton grazing on natural populations of bacteria, Pedrós-Alió and Brock (1983) evaluated the role of macrozooplankton as bacteriovores. Zooplankton ($>64$ mm) feeding rates were determined from cell count procedures and from assimilation of $^{14}$C-labeled bacterioplankton. During a period of two years in Lake Mendota, the zooplankton ingested 1-10% of the bacterial production. These rather low values indicate that other processes might be important in the regulation of aquatic bacteria, e.g. grazing from organisms $<64$ µm (Sorokin and Pavelja 1972, Fenchel 1982), sedimentation of bacteria attached to particles (Pedrós-Alió and Brock 1983b) or lysis.

As an example of the potential importance of various size fractions of zooplankton as bacteriovores, grazing of macrozooplankton ($>140$ µm) and microzooplankton were measured by means of $^{3}$H-thymidine labeled bacteria in eutrophic Frederiksborg Slotssø as well as in experimental enclosures manipulated with planktivorous fish and nutrients. The natural density of zooplankton were allowed to graze *in situ* on the natural bacterial assemblages. In Fig. 5.10 is presented the average chlorophyll a content (A) and the average zooplankton biomass ($>140$ µm) (B) in the lake and in the experimental enclosures during a period of 19 days. The phytoplankton chlorophyll a content in the experimental enclosures without fish were about half of those in the lake. Additions of fish increased the chlorophyll a to the same value as found in the lake, whereas additions of both fish and nutrients increased the chlorophyll a content to about 1.8 times those found in the lake.

Marked changes were also found in the zooplankton biomass. Nearly 4 times higher values were found in the enclosures without fish, compared with the values found in the enclosures with added fish. *Daphnia cucullata,* and *D. galeata* dominated, but *Eudiaptomus graciloides* also occurred frequently.

The measured ingestion rates of zooplankton $>140$ µm (Fig. 5.11) increased in the absence of fish to about 3 µg bacterial carbon liter$^{-1}$hour$^{-1}$. In contrast, the

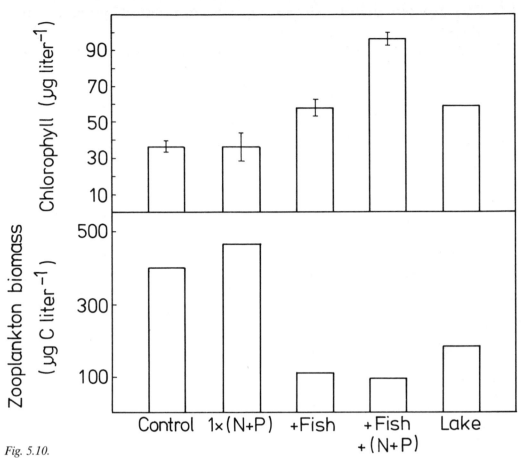

Fig. 5.10.
*Average daily chlorophyll a content (A) and average daily biomass of zooplankton sized > 140 μm (B) in Frederiksborg Slotssø and in experimental enclosures. (Redrawn from Riemann 1985).*

ingestion rates dropped below 0.3 μg bacterial carbon liter⁻¹hour⁻¹in the fish containing enclosures as well as in the lake. Grazing by zooplankton <140 μm >50 μm dropped below 0.1 μg bacterial carbon liter⁻¹hour⁻¹in all enclosures and in the lake. The entire grazing of the zooplankton >140 μm + <140 μm >50 μm made out 48-51% of the bacterial net production in the enclosures without fish, compared with 4% in the enclosures with added fish and 21% in the open lake. These values are mean average values for the entire period. In fact, at the end of the experimental period, zooplankton >140 μm were responsible for removing up to 90% of the bacterial net production. So, in the natural zooplankton communities, grazing by cladocerans can control the biomass of the bacteria. However, fish predation may eliminate their importance in controling bacteria.

Who was then responsible for the control of the bacterial biomass in the enclosures with added fish and in the open lake? During the last two sampling days in Frederiksborg Slotssø, the number of flagellates (4-8 μm in size) were counted, and their grazing was calculated from a literature-based mean clearance rate of 8 x 10⁻⁶ml individual⁻¹hour⁻¹(Fenchel 1982, Andersen and Fenchel 1985). The number of flagellates were around 5 x 10⁻³ml⁻¹and their ingestion accounted for 94-99% of

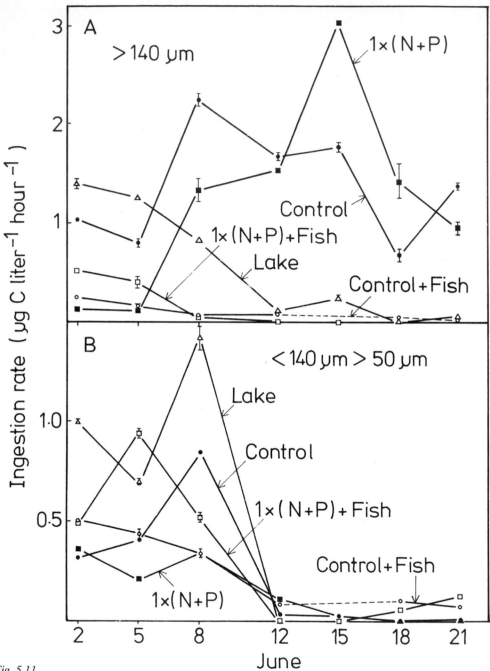

*Fig. 5.11.*
*Daily changes in rates of zooplankton ingestion of bacteria in Frederiksborg Slotssø and in experimental enclosures. (A) Zooplankton sized >140 mm; (B) Zooplankton sized 50 to 140 mm. (From Riemann 1985).*

the total ingestion in the enclosures containing fish and in the lake. However, in the enclosures without fish, zooplankton $>140$ μm made out 80-90% of total ingestion (Table 5.3). Flagellates occurred in numbers from 50-300 ml[-1], and their ingestion made out between 2 and 6% of the total ingestion of bacteria.

Table 5.3. Ingestion rates for various bacterivores in Frederiksborg Slotssø and from experimental enclosures (Data from Riemann 1985).

| Enclosures:<br>Ingestion rates<br>($\mu g$ C $1^{-1}$ $d^{-1}$) | Control | + (N + P) | + Fish | + Fish + (N + P) | Lake |
|---|---|---|---|---|---|
| 1. Zooplankton $>140$ $\mu m$ | 24.3 | 28.1 | 0.8 | 0.5 | 0.7 |
| 2. Zooplankton $<140$ $\mu m$ $>50$ $\mu m$ | 0.1 | 0.1 | 2.2 | 2.2 | 0.7 |
| 3. Flagellates (4-8 $\mu m$) | 6.0 | 2.8 | 153.1 | 233.0 | 120.0 |
| (1-3) | 30.4 | 31.3 | 156.2 | 235.7 | 121.4 |

These experiments demonstrate that large crustacean zooplankton in lakes are important in the control of the bacteria. However, fish predation may reduce and even eliminate their role as bacteriovores. Flagellates can occur in lakes in large numbers, but a reduced fish predation can change the trophic structure by development of a crustacean dominated zooplankton community which takes over the dominant role as bacteriovores and also cause a reduction in the number of flagellates.

This scenario for the control of the bacterial biomass is far from complete. A lot of simple questions regarding seasonal changes in zooplankton communities in relation to species succession of both zooplankton and phytoplankton communities have still not been answered. If large crustaceans are inhibited by cyanobacteria, marked changes will occur in the trophic structure of the bacteriovores. Such changes are important for the size of the bacterial biomass as well as for an efficient transport of dissolved organic carbon to higher trophic levels (macrozooplankton and fish).

## 5.8. Phytoplankton and bacteria as food sources for zooplankton

Considering the knowledge about the flux of organic matter from both phytoplankton and bacteria to zooplankton, an important question is the extent to which either of these potential food sources dominate the food for zooplankton in natural environments.

Several members of freshwater zooplankton communities have physical capabilities to filter even small bacteria (Peterson et al. 1978, Riemann and Bosselmann 1984), and *Daphnia* species may survive for months, exclusively fed with bacteria. However, a number of copepods change their filtering rates when the size of the food particles varies, and bacterial-sized particles are most often not retained to any measurable extent. As copepods are common members of natural freshwater assemblages, they would probably not survive a prolonged period in which the food particles were of bacterial sizes.

The potential importance of bacteria as food for natural poulations of zooplankton was evaluated in experiments in which zooplankton grazing on both phytoplankton and bacteria was measured (Table 5.4). The experiments were carried

Table 5.4. Biomass of zooplankton $> 140$ µm, phytoplankton and bacteria, grazing rates of zooplankton on phytoplankton and bacteria and the percentage of bacteria in the zooplankton ingestion in Frederiksborg Slotssø and experimental enclosures during the period 2-21 June.

| | Zooplankton biomass (µg C l$^{-1}$) (% d$^{-1}$) | Phytoplankton biomass (µg C l$^{-1}$) | Bacterial biomass (µg C l$^{-1}$) | $I_{bact.}/I_{total}$ (%) | Grazing rate on phytoplankton (% d$^{-1}$) | Grazing rate on bacteria (% d$^{-1}$) |
|---|---|---|---|---|---|---|
| **Enclosures:** | | | | | | |
| – Fish | 404 | 890 | 106 | 17 | 8 | 32 |
| – Fish + NP | 470 | 1005 | 125 | 13 | 23 | 23 |
| + Fish | 104 | 2101 | 197 | 1 | 1 | 1 |
| + Fish + NP | 84 | 3210 | 303 | 1 | 2 | 1 |
| Lake | 198 | 1668 | 195 | 15 | 4 | 6 |

out in large plastic enclosures using natural populations of zooplankton from the lake. During the initial part of the experiment, the phytoplankton biomass dominated over the bacterial biomass, however, at the end of the experiment, zooplankton ingestion rates on phytoplankton were much higher than the grazing rates on the bacteria, and the bacterial carbon biomass accounted for 75-84% of the total carbon grazed by the zooplankton. As an average value for the entire period, *Daphnia pulex* accounted for 88% of the zooplankton biomass, and the selectivity coefficient (defined as the ratio between clearance rate of algae and clearance rate of bacteria) was in average 1.12, indicating that bacteria and algae were filtered with the same efficiency.

In eutrophic Frederiksborg Slotssø the ingestion of phytoplankton and bacteria by zooplankton $> 140$ µm was examined during three weeks in June, in the lake, in enclosures without fish, and enclosures with fish present, and in parallel enclosures with N and P added (Andersen et al., in prep.). Table 5.4.

Maximum average percentage of bacteria in the total zooplankton ingestion, 17%, appeared in the enclosure without fish, in which the zooplankton biomass was high and concentrations of phytoplankton and bacteria low. Grazing rates were 32% d$^{-1}$ on bacteria and 18% d$^{-1}$ on phytoplankton, i.e. bacteria were prefered by the zooplankters. In enclosures with added fish, bacteria constituted only 1% of the zooplankton diet and grazing rates were similar, about 1%, on algae and bacteria. Zooplankton biomass was low, whereas the biomasses of algae and bacteria were high.

A high percentage of bacteria in the zooplankton ingestion appeared when zooplankton biomass was high. In these cases *Daphnia* species were dominating. They filtered bacteria more efficiently than phytoplankton (or with the same rate), and they were able to reduce the concentrations of both algae and bacteria.

During the period of investigation similar trends appeared in all enclosures and in the lake: In the first half of the period bacteria were filtered more efficiently than phytoplankton, whereas the algae were prefered in the second half. These trends were ascribed not only to changes in the age structure of the *Daphnia* populations and the relative numbers of *Daphnia* in the total zooplankton, but also to differences in the quality and quantity of the phytoplankton.

In view of the knowledge on the efficiency of *Daphnia* species to graze even small particles, the data of Børsheim and Olsen (1984) and Andersen et al. (in prep) confirm the theory that in *Daphnia* dominated communities of zooplankton bacteria may be filtered with the same efficiency as algae or even more efficiently, and that bacteria can constitute the major part of the food for the animals, if concentrations of algae are low. It should be mentioned that the role of particulate detritus was not included in these evaluations of various food sources for the zooplankton.

## Acknowledgements

We wish to thank W. Lampert and J. Haney for valuable criticism, suggestions and linguistic improvements.

## 5.9. References

Andersen, F., Riemann, B., & Bosselmann, S. . The potential importance of phyto- and bacterioplankton as food for natural populations of freshwater zooplankton. Submitted.

Andersen, G., Berggreen, H., Cronberg, G., & Gelin, C. 1978. Effects of planktivorous and benthivorous fish on organisms and water chemistry in eutrophic lakes. Hydrobiologia 59: 9-15.

Andersen, J. M. & Jacobsen, O. S. 1979. Production and decomposition of organic matter in eutrophic Frederiksborg Slotssø, Denmark. Arch. Hydrobiol. 85: 511-542.

Andersen, P. & Fenchel, T. 1985. Bacterivory by microheterotrophic flagellates in seawater samples. Limnol. Oceanogr. 30: 198-202.

Arnold, D. E. 1971. Ingestion, assimilation, survival, and reproduction by *Daphnia pulex* fed seven species of blue-green algae. Limnol. Oceanogr. 16: 906-920.

Bays, J. S. & Crisman, T. L. 1983. Zooplankton and trophic state relationships in Florida lakes. Can. J. Fish Aquat. Sci. 40: 1813-1819.

Bell, R. K. & Ward, F. J. 1970. Incorporation of organic carbon by *Daphnia pulex*. Limnol. Oceanogr. 15: 713-726.

Berman, M. S. & Richman, S. 1974. The feeding behaviour of *Daphnia pulex* from Lake Winnebago, Wisconsin. Limnol. Oceanogr. 19: 105-109.

Bjørnsen, P. K., Larsen, J. B., Geertz-Hansen, O. & Olesen, M. 1986. A field technique to the determination of zooplankton grazing on natural bacterioplankton. Freshwater Biol.16: 245-253.

Bogdan, K. G. & Gilbert, J. J. 1982. Seasonal patterns of feeding by natural populations of *Keratella, Polyarthra,* and *Bosmina:* Clearance rates, selectivities, and contributions to community grazing. Limnol. Oceanogr. 27: 918-934.

Bogdan, K. G. & McNaught, D. C. 1975. Selective feeding by *Diaptomus* and *Daphnia.* Verh. Int. Ver. Limnol. 19: 2935-2942.

Bogdan, K. G., Gilbert, J. J., & Starkweather, P. L. 1980. In situ clearance rates of planktonic rotifers. Hydrobiologia 73: 73-77.

Bogdan, K. G. & Gilbert, J. J. 1982. Seasonal pattern of feeding by natural populations of *Keratella, Polyarthra* and *Bosmina:* clearance rates, selectivities and contribution to community grazing. Limnol. Oceanogr. 27: 918-934.

Bosselmann, S. 1974. The Crustacean plankton of Lake Esrom. Arch. Hydrobiol. 74: 18-31.

Bosselmann, S. 1975a. Population dynamics of Eudiaptomus graciloides in Lake Esrom. Arch. Hydrobiol. 75: 329-346.

Bosselmann, S. 1975b. Production of *Eudiaptomus graciloides* in Lake Esrom, 1970. Arch. Hydrobiol. 76: 43-64.

Bosselmann, S. 1979. Production of *Keratella cochlearis* in Lake Esrom. Arch. Hydrobiol. 87: 304-313.

Bosselmann, S. 1981. Population dynamics and production of *Keratella hiemalis* and *K. quadrata* in Lake Esrom. Arch. Hydrobiol. 427-447.

Bosselmann, S., Hansen, J. & Andersen, F. The ingestion of phytoplankton by the zooplankton in a eutrophic lake. A comparison of methods. Submitted.

Bosselmann, S., Andersen, F., & Hansen, J. Relations between phytoplankton and zooplankton in eutrophic Lake Hylke. Submitted.

Brooks, J. L. & Dodson, S. I. 1965. Predation, body size, and composition of plankton. Science 150: 28-35.

Brooks, J. L. 1969. Eutrophication and changes in the composition of the zooplankton p. 236-255. In Eutrophication: Causes, consequences, correctives. National Academy of Sciences. Washington.

Buikema, A. L. 1973. Filtering rate of the Cladocera, *Daphnia pulex,* as a function of body size, light and acclimation. Hydrobiologia 41: 515-527.

Burns, C. W. 1968. Direct observations of mechanisms regulating feeding behaviour of *Daphnia,* in lake water. Int. Revue ges. Hydrobiol. 53: 83-100.

Burns, C. W. 1969a. Particle size and sedimentation in the feeding of two species of *Daphnia.* Limnol. Oceanogr. 14: 392-402.

Burns, C. W. 1969b. The relationship between body size of filter-feeding Cladocera and the maximum size of particle ingested. Limnol. Oceanogr. 14: 675-678.

Burns, C. W. 1969c. Relation between filtering rate, temperature, and body size in four species of *Daphnia.* Limnol. Oceanogr. 14: 693-700.

Burns, C. W. and Rigler, F. H. 1967. Comparison of filtering rates of *Daphnia rosea* in lake water and suspensions of yeast. Limnol. Oceanogr. 12: 492-502.

Børsheim, Y. B. 1984. Clearance rates of bacteria-sized particles by freshwater ciliates measured with monodisperse fluorescent latex beads. Oecologia 63: 286-288.

Børsheim, K. Y. & Olsen, Y. 1984. Grazing activities by *Daphnia pulex* on natural populations of bacteria and algae. Verh. Int. Ver. Limnol. 22: 644-648.

Cushing, D. H. 1976. Grazing in Lake Erken. Limnol. Oceanogr. 21: 349-356.

De Bernardi, R. & Guissano, G. 1978. Effect of fish mortality on zooplankton structure and dynamics in a small Italian lake (Lago di Annone). Verh. Int. Ver. Limnol. 20: 1045-1048.

De Bernardi, R., Giussano, G., & Pedretti, E. L. 1981. The significance of blue-green algae as food for filterfeeding zooplankton. Experimental studies on *Daphnia* spp. fed *Microcystis aeruginosa.* Verh. Int. Ver. Limnol. 21: 477-483.

DeMott, W. R. 1982. Feeding selectivities and relative ingestion rates of *Daphnia* and *Bosmina.* Limnol. Oceanogr. 27: 518-527.

Desortova, B. 1976. Productivity of individual algal species in natural phytoplankton assemblages determined by means of autoradiography. Arch. Hydrobiol. Supp. 49: 415-449.

Erman, L. A. 1956. On the quantitative aspects of the feeding of Rotifers *(Rotifera Phylum)* (in Russian) Zool. Zh. 35: 965-971.

Fenchel, T. 1982. Ecology of heterotrophic microflagellates. IV. Quantitative occurrence and importance as bacterial consumers. Mar. Ecol. Prog. Ser. 9: 35-42.

Geertz-Hansen, O., Olesen, M., Bjørnsen, P. K., Larsen, J. B. & Riemann, B. Zooplankton consumption of bacteria in a eutrophic lake and in experimental enclosures. Arch. Hydrobiol.

Geller, W. 1975. Die Nahrungsaufnahme von *Daphnia pulex* in Abhängigkeit von der Futterkonzentration, der Temperatur, der Körpergrösse und dem Hungerzustand der Tiere. Arch. Hydrobiol. Suppl. 48: 47-107.

Gliwicz, Z. M. 1969. Studies on the feeding of pelagic zooplankton in lakes with varying trophy. Ekol. pol. A 17: 663-708.

Gliwicz, Z. M. 1977. Food size selection and seasonal succession of filter feeding zooplankton in a eutrophic lake. Ekol. pol. 25: 179-225.

Gliwicz, Z. M. & Hillbricht-Ilkowska, A. 1975. Ecosystem of the Mikolajskie Lake. Elimination of phytoplankton biomass and its subsequent fate in lake through the year. Pol. Arch. Hydrobiol. 22: 39-52.

Gliwicz, Z. M. & Prejs, A. 1977. Can planktivorous fish keep in check planktonic Crustacean populations? A test of size-efficiency hypothesis in typical Polish lakes. Ekol. Pol. 25: 567-591.

Gliwicz, Z. M. & Siedlar, E. 1980. Food size limitation and algae interfering with food collection in *Daphnia.* Arch. Hydrobiol. 88: 155-177.

Gliwicz, Z. M., Hillbricht-Ilkowska, A., & Weglenska, T. 1978. Contribution of fish and invertebrate predation to the elimination of zooplankton biomass in two Polish lakes. Verh. Int. Ver. Limnol. 20: 1007-1011.

Goad, J. 1984. A biomanipulation experiment in Green Lake,Seattle, Washington, Arch. Hydrobiol. 102: 137-152.

Godeanu, S. P. 1978. Specificity of the zooplankton in several lakes from northern Germany with different degree of eutrophication. Verh. Int. Ver. Limnol. 20: 963-968.

Gophen, M., Cavari, B. Z. & Berman, T. 1974. Zooplanktonfeeding on differentially labelled algae and bacteria. Nature 247: 393-394.

Gulati, R. D. 1975. A study on the role of herbivorous zooplankton community as primary consumers of phytoplankton in Dutch lakes. Verh. Int. Ver. Limnol. 19: 1202-1210.

Gulati, R. D. 1978. Vertical changes in the filtering, feeding and assimilation rates of dominant zooplankters in a stratified lake. Verh. Int. Ver. Limnol. 20: 950-956.

Gulati, R. D., Siewertsen, K. & Postema, G. 1982. The zooplankton: its community structure, food and feeding, and role in the ecosystem of Lake Vechten. Hydrobiologia 96: 127-163.

Hakkari, L. 1969. Zooplankton studies in the lake Längelmävesi, South Finland. Ann. Zool. Fennici 6: 313-326.

Haney, J. F. 1971. An in situ method for the measurement of zooplankton grazing rates. Limnol. Oceanogr. 16: 970-977.

Haney, J. F. 1973. An in situ examination of the grazing activities of natural zooplankton communities. Arch. Hydrobiol. 72: 87-132.

Haney, J. F. 1985. Regulations of cladoceran filtering rates in nature by body size, food concentration, and diel feeding patterns. Limnol. Oceanogr. 30: 397-411.

Haney, J. F. & Hall, D. J. 1975. Diel vertical migration and filter-feeding activities of *Daphnia*. Arch. Hydrobiol. 75: 413-441.

Haney, J. F. & Trout, M. A. 1985. Size selective grazing by zooplankton in Lake Titicaca. - Arch. Hydrobiol. Beih. 21: 147-160.

Hillbricht-Ilkowska, A. 1977. Attempt at evaluation of the production and turnover of plankton rotifers on the example of *Keratella cochlearis* (Gosse). Bull. Polon. Acad. Sci. Ser., Sci. Biol. 15: 35-40.

Hillbricht-Ilkowska, A. & Weglenska, T. 1970. The effect ofsampling frequency and the method of assessment on the production values obtained for several zooplankton species. Ekol. Pol. 18: 539-557.

Hollibaugh, J. T., Fuhrman, J. A. & Azam, F. 1980. Radioactivity labeling of natural assemblages of bacterioplankton for use in trophic studies. Limnol. Oceanogr. 25: 172-181.

Holm, N. P., Ganf, G. G. & Shapiro, J. 1983. Feeding and assimilation rates of *Daphnia pulex* fed *Aphanizomenon flos-aquae*. Limnol. Oceanogr. 28: 677-678.

Horn, W. 1981. Phytoplankton losses due to zooplankton grazing in a drinking water reservoir. Int. Revue ges. Hydrobiol. 66: 787-810.

Hrbacek, J., Dvorakova, M., Korinek, V., & Prochazkova, L. 1961. Demonstration of the effect of the fish stock on the species composition of zooplankton and the intensity of metabolism of the whole plankton association. Verh. Int. Verein. Limnol. 14: 192-195.

Infante, A. 1973. Untersuchungen zur Ausnutzbarkeit verschie dener Algen durch das Zooplankton. Arch. Hydrobiol. Suppl. 42: 340-405.

Janicki, A. & DeCosta, J. 1984. The filtering rates of four herbivorous crustaceans on nannoplankton and net plankton in an acid lake in West Virginia. Int. Rev. ges. Hydrobiol. 69: 643-652.

Kersting, K. 1978. Some features of feeding, respiration and energy conversion of *Daphnia magna*. Hydrobiologia 59: 113-120.

Kersting, K. & Leuw, W. van der. 1976. The use of Coulter Counter for measuring the feeding rates of *Daphnia magna*. Hydrobiologia 49: 233-237.

Kibby, H. V. 1971. Energetics and population dynamics of *Diaptomus gracilis*. Ecol. Monogr. 41: 313-327.

Krambeck, H.-J. 1981. Longterm monitoring and modelling of the nutrient cycle in a Baltic lake. Verh. Int. Verein. Limnol. 21: 460-465.

Kryuchkova, N. M. & Rybak, V. K. 1975: Growth of *Eudiaptomus graciloides* (Lill.) under different feeding conditions. Hydrobiol. J. 10: 30-36.

Lampert, W. 1975. A tracer study on the carbon turnover of *Daphnia pulex*. Verh. Int. Verein. Limnol. 19: 2913-2921.

Lampert, W. 1977. Studies on the carbon balance of *Daphnia pulex* de Geer as related to environmental conditions. II. The dependence of carbon assimilation on animal size, temperature, food concentration and diet species. Arch. Hydrobiol. Suppl. 48: 310-335.

Lampert, W. 1978. Climatic conditions and planktonic interrelations as factors controlling the regular succession of spring algal bloom and extremely clear water phase in Lake Constance. Verh. Int. Ver. Limnol. 20: 969-974.

Lampert, W. 1981. Inhibitory and toxic effects of blue green algae on *Daphnia*. Int. Revue ges. Hydrobiol. 66: 285-298.

Lampert, W. & Taylor, B. 1985. Zooplankton grazing in a eutrophic lake: Implications of diel vertical migration. Ecology 66: 68-82.

Larkin, P. A. & Northcote, T. L. 1969. Fish as indicators of eutrophication. In Eutrophication: causes, consequences, correctives. Washington D. C.

Larsson, P. 1968: *Holopedium gibberum* - en indikator på kalkfattig vann - Fauna, Oslo 21: 130-132.

Lynch, M. & Shapiro, J. 1981. Predation, enrichment, and phytoplankton community structure. Limnol. Oceanogr. 26: 86-102.

McMahon, J. W. & Rigler, F. H. 1965. Feeding rate of *Daphnia magna* Straus in different foods labeled with radioactive phosphorus. Limnol. Oceanogr. 10: 105-113.

McNaught, D. C. 1975. A hypothesis to explain the succession from calanoids to cladocerans during eutrophication. Verh. Int. Verein. Limnol. 19: 724-731.

McQueen, D. J. 1970. Grazing rate and food selection in *Diaptomus oregonensis* (Copepoda) from Marion Lake, B. C. J. Fish. Res. Bd. Can. 27: 13-20.

Muck, P. & Lampert, W. 1984. An experimental study on the importance of food conditions for the relative abundance of calanoid copepods and cladocerans. I. Comparative feeding studies with *Eudiaptomus gracilis* and *Daphnia longispina*. Arch. Hydrobiol. Suppl. 66: 157-169.

Nauwerck, A. 1959. Zur Bestimmung der Filtrierrate limnischer Planktontiere. Arch. Hydrobiol. Suppl. 25: 83-101.

Nauwerck, A. 1962. Nicht-algische Ernährung bei *Eudiaptomus gracilis* (Sars). Arch. Hydrobiol. Suppl. 25: 393-400.

234

Nauwerck, A. 1963. Die Beziehungen zwischen Phytoplankton und Zooplankton im See Erken. Symp. Bot. Uppsaliensis 17(5): 1-163.

Newell, S. Y. & Christian, R. R. 1981. Frequency of dividing cells as an estimator of bacterial productivity. Appl. Environ. Microbiol. 42: 23-31.

Nowak, K.-E. 1975. Die Bedeutung des Zooplanktons für den Stoffhaushalt des Schierensees. Arch. Hydrobiol. 75: 140-224.

Pace, M. L. & Orcutt, J. D. Jr. 1981. The relative importance of protozoans, rotifers, and crustaceans in a fresh water zooplankton community. Limnol. Oceanogr. 26: 822-830.

Pace, M. L., Porter, K. G. & Feig, Y. S. 1983. Species- and age-specific differences in bacterial resource utilization by two co-occurring cladocerans. Ecology 64: 1145-1156.

Pastorok, R. A. 1981. Prey vulnerability and size selection by *Chaoborus* larvae. Ecology 62: 1311-1324.

Patalas, K. 1972. Crustacean plankton and the eutrophication of St. Lawrence Great Lakes. J. Fish. Res. Bd. Can., 29: 1451-1462.

Patalas, J. & Patalas, K. 1966. The crustacean plankton communities in Polish lakes. Verh. Int. Ver. Limnol. 16: 204-215.

Pederson, G. L., Welch, E. B., & Litt, T. 1975. Plankton secondary productivity and biomass: Their relations to lake trophic state. Hydrobiologia 50: 129-144.

Pedrós-Alió, C. & Brock, T. D. 1982. Assessing biomass and production of bacteria in eutrophic Lake Mendota, Wisconsin. Appl. Environ. Microbiol. 44: 203-218.

Pedrós-Alió, C. & Brock, T. D. 1983a. The impact of zooplankton feeding on the epilimnetic bacteria of a eutrophic lake. Freshwat. Biol. 13: 227-239.

Pedrós-Alió, C. & Brock, T. D. 1983b. The importance of attachment to particles for plankton bacteria. Arch. Hydrobiol. 98: 354-379.

Peters, R. H. 1984. Mehods for the study of feeding, grazing and assimilation by zooplankton, p.336-412. In: J. A. Downing & Rigler, F.H. (eds.): Secondary productivity in fresh waters, 2. ed. Blackwell.

Petersen, F. 1983. Population dynamics and production of *Daphnia galeata* (Crustacea, Cladocera) in Lake Esrom. Holarct. Ecol. 6: 285-294.

Peterson, B. J., Hobbie, J. E. & Haney, J. F. 1978. *Daphnia* grazing on natural bacteria. Limnol. Oceanogr. 23: 1039-1044.

Petrowitsch, P. G. 1966. Relative Bedeutung der Hauptformen des See-Zooplanktons im Produktionsprozess. Verh. Int. Verein. Limnol. 16: 425-431.

Porter, K. G. 1973. Selective grazing and differential digestion of algae by zooplankton. Nature 244: 179-180.

Porter, K. G. & McDonough, R. 1984. The energetic cost of response to blue green algal filaments by cladocerans. Limnol. Oceanogr. 29: 365-369.

Porter, K. G., Gerritzen, J. & Orcutt, J. D. Jr. 1982. The effect of food concentration on swimming patterns, feeding behaviour, ingestion, assimilation, and respiration by *Daphnia.* Limnol. Oceanogr. 27: 935-949.

Porter, K. G., Feig, Y. S. & Vetter, E. F. 1983. Morphology, flow regimes and filtering rates of *Daphnia, Ceriodaphnia,* and *Bosmina* fed natural bacteria. Oecologia 58: 156-163.

Reinertsen, H. & Olsen, Y. 1984. Effects of fish elimination on the phytoplankton community of a eutrophic lake. Verh. Int. Verein. Limnol. 22: 649-657.

Richman, S. 1964. Energy transformation studies on *Daphnia oregonensis.* Verh. Int. Verein. Limnol. 15: 654-659.

Richman, S. 1966. The effect of phytoplankton concentration on the feeding rate of *Diaptomus oregonensis.* Verh. Int. Ver. Limnol. 16: 392-398.

Richman, S. & Dodson, S. I. 1983. The effect of food quality on feeding and respiration by *Daphnia* and *Diaptomus.* Limnol. Oceanogr. 28: 948-956.

Riemann, B. 1985. Potential importance of fish predation and zooplankton grazing on natural populations of freshwater bacteria. Appl. Environ. Microbiol. 50: 187-193.

Riemann, B. & Bosselmann, S. 1984. *Daphnia* grazing on natural populations of bacteria. Verh. Int. Verein. Limnol. 22: 795-799.

Riemann, B., Søndergaard, M., Schierup, H.-H., Bosselmann, S., Christensen, G., Hansen, J., & Nielsen, B. 1982a. Carbon metabolism during a spring diatom bloom in the eutrophic Lake Mossø. Int. Revue ges. Hydrobiol. 67: 145-185.

Riemann, B., Fuhrman, J. A. & Azam, F. 1982b. Bacterial secondary production in freshwater measured by $^3$H-thymidine incorporation method. Microb. Ecol. 8: 101-114.

Rigler, F. H. 1961. The relation between concentration of food and feeding rate of *Daphnia magna* Straus. Can. J. Zool. 39: 857-868.

Rognerud, S. & Kjellberg, G. 1984. Relationships between phytoplankton and zooplankton biomass in large lakes. Verh. Int. Verein. Limnol. 22: 666-671.

Roman, M. R. & Rublee, P. A. 1981. A method to determine in situ zooplankton grazing rates on natural particle assemblages. Mar. Biol. 65: 303-309.

Ryther, J. H. 1954. Inhibitory effects of phytoplankton upon the feeding of *Daphnia magna* with reference to growth, reproduction and survival. Ecology 35: 522-533.

Schindler, D. W. 1968. Feeding, assimilation, and respiration rates of *Daphnia magna* under various environmental conditions and their relation to production estimates. J. Anim. Ecol. 37: 369-385.

Sorokin, J. I. & Paveljeva, E. B. 1972. On the quantitative characteristics of the pelagic ecosystem of Dalnee Lake (Kamchatka). Hydrobiologia 40: 519-552.

Starkweaterher, P. L. 1975. Diel patterns of grazing in *Daphnia pulex* Leydig. Ver. Int. Ver. Limnol. 19: 2951-2857.

Starkweather, P. L. 1980. Aspects of the feeding behaviour and trophic ecology of suspension feeding rotifers. Hydrobiologia 73: 63-72.

Starkweather, P. L. & Bogdan, K. G. 1980. Detrital feeding in natural zooplankton communities. Discrimination between live and dead algal foods. Hydrobiologia 73: 83-85.

Starkweather & Gilbert, J. J. 1977. Feeding in the rotifer *Brachionus calyciflorus* 2. Effect of food density on feeding rates using *Euglena gracilis* and *Rhodotorula glutinis.* Oecologia 28: 133-139.

236

Stenson, J., Bohlin, T., Henrikson, B. I., Oscarson, H. G., & Larson, P. 1978. Effects of fish removal from a small lake. Verh. Int. Ver. Limnol. 20: 794-801.

Vanderploeg, H. A. 1981. Seasonal particle-size selection by *Diaptomus sicilis* in offshore Lake Michigan. Can. J. Fish. Aquat. Sci. 38: 504-517.

Vyhnalek, V. 1983. Effect of filter-feeding zooplankton on phytoplankton in fish ponds. Int. Revue ges. Hydrobiol. 68: 397-410.

Webster, K. E. & Peters, R. H. 1978. Some size dependent inhibitions of large Cladoceran filterers in filamentous suspensions. Limnol. Oceanogr. 23: 1238-1245.

Welsh, K. E. & Peters, R. H. 1978. Some size dependent inhibitions of large Cladoceran filterers in filamentous suspensions. Limnol. Oceanogr. 23: 1238-1245.

Winberg, C. G. 1972. Some interim results of Sovjet IBP investigations on lakes. In: Kajak, Z. & Hillbricht-Ilkowska, A. (Eds.), Productivity Problems of Freshwaters. Eds. (Warszawa-Krakow).

Wium-Andersen, S. 1975. The influence of the zooplankton anaesthetising substance *Physostigminum salicylicum* on photosynthesis. - Arch. Hydrobiol. 76: 379-383.

Zankai, N. 1978. Jahreszeitliche Änderung der Filtrierrate des Copepoden *Eudiaptomus gracilis* (G. O. Sars) im Balaton-See. Verh. Int. Verein. Limnol. 20: 2551-2555.

Zankai, N. 1983. Ingestion rates of some *Daphnia* species in a shallow lake (Lake Balaton, Hungary). Int. Revue ges. Hydrobiol. 68: 227-237.

# Chapter 6. THE FISH COMMUNITY OF TEMPERATE EUTROPHIC LAKES

By Lars Johansson and Lennart Persson

## 6.1. Introduction

The approach of this chapter differs somewhat from those of the preceding chapters as only few quantitative comparisons of rates and biomasses can be presented. This is related to the fact that the number of studies on biomass, production and ingestion rates carried out on fish in eutrophic lakes are very low compared to studies on bacteria, phytoplankton and zooplankton. The main reason for this is probably the huge effort that has to be put into capturing and marking fish. Estimating food consumption rates at the fish level also involves an immense effort. Actually only one study from a European eutrophic lake covering both population estimates and estimates of ingestion rates has so far been published (Persson 1983 a, b, 1986). As this study involved the development of models and methods for the estimation of food consumption rates in the field, these techniques will be reviewed in this chapter. Accurate estimates of fish population size based on multiple capture recapture methods, on the other hand, have been fully reviewed by Robson & Sangler (1978) and Youngs & Robson (1978) so there is no reason to consider them here. The same is true for the estimation of biomasses and production fully reviewed by Chapham (1978 a, b). The assumptions inherent in these methods are also discussed in the above papers.

The paucity of studies on biomass, production and consumption rate make it difficult to discuss general patterns at the fish level in eutrophic lakes in terms of quantitative rates. The qualitative aspects of the fish community in temperate eutrophic lakes including resource utilization by fish species have, however, been more generally studied (Svärdson 1976, Leach et al. 1977, Niederholzer & Hofer 1980, Persson 1983 a, b, c, 1986). In many respects our knowledge within this area at the fish level exceeds that at other trophic levels. With regard to species interactions, the changes in fish communities with eutrophication have been the focus of several studies (Svärdson 1976, Hartmann et al. 1977, Leach et al. 1977, Persson 1983 c). One main purpose of this chapter is to discuss how species interactions are affected by the eutrophication-induced change in the resource spectra.

The fish community does not merely respond passively to changes in the environmental conditions. Predation by fish on zooplankton affects both size structure and species composition of the zooplankton community (Hrbácek 1962, Brooks & Dodson 1965, Stenson 1972, Anderson 1984). We will in this chapter review this topic and discuss the mechanisms behind it. Of special interest will be to discuss whether cyprinid species, numerically dominant in eutrophic lakes, differ from other fish species with respect to their effects on lower trophic levels.

## 6.2. Methods for estimating food consumption rates

Although several methods have been suggested to estimate the rate of food consumption in fishes (see Windell 1978), only two main approaches have been used in recent years. The first approach is based on a direct estimation of food consumption in the field and is historically a development of the method proposed by Bajkov (1934). The second approach is based on a combination of laboratory estimated metabolic demands and field data on growth, stomach contents and temperature.

### 6.2.1. Field methods for estimating food consumption rates

The most commonly used model for the direct estimation of food consumption in the field is the model developed by Elliott & Persson (1978):

$$DR = \sum_{i=1}^{n} C_i = \sum_{i=1}^{n} F_i t = \sum_{i=1}^{n} \frac{(S_i - S_{i-1} e^{-Rt}) \, Rt}{1 - e^{-Rt}} \qquad (1)$$

where $DR$ is the daily ration, $C_i$ is the consumption rate over $t$ hrs, $F$ the consumption rate per hr, $S_i$ and $S_{i-1}$ are the weights of the stomach contents at sampling events $i$ and $i-1$ respectively, $t$ is the time between sampling events, n number of sampling events over 24 hrs ($= 24/t$) and $R$ is the instantaneous rate of food evacuation. $R$ can be estimated from field data (Thorpe 1977, Craig 1978), but as laboratory estimates will give both lower variance and are evidently applicable to the field situation (Persson 1982, 1983 a), it is preferable to use laboratory estimates of $R$. The time between sampling events ($t$) should, according to Elliott & Persson (1978) not exceed 3 hrs.

If the amount of food in the stomach is the same at the beginning and end of the 24 hr sampling period, equation 1 reduces to the model suggested by Eggers (1978):

$$DR = 24 \, \bar{S} R \qquad (2)$$

where $\bar{S}$ is the mean content of food in the stomach over 24 hours.

The Elliott & Persson model has been tested for total stomach contents (Persson 1982, 1983 a). As the evacuation rates of different prey items in the stomach are not independent (Persson 1984), the model cannot be used to estimate the food consumption of different prey items. Persson (1984) suggested that the daily consumption $(DR_j)$ of each prey item $j$ should be calculated as the fraction the mean weight of it over 24 hrs constitutes of the total mean stomach content times the total daily ration

$$([\Sigma S_{ij}/\Sigma S_i]DR) \qquad (3)$$

In equation 1 $F$ is assumed to be constant between sampling events. If the consumption rate is not constant but decreases with time due to satiation effects, equation 1 will underestimate the rate of food consumption (Elliott & Persson

1978). In this case the amount of food consumed will increase to an aymptote $(C_{max})$ and $C_i$ is now given by

$$C_i = C_{max} - ae^{-bt} \tag{4}$$

where $a = C_{max} - S_{i-1} \cdot C_{max}$ and $b$ are given from the equation:

$$S_i = S_{i-1} e^{-Rt} + e^{-Rt} \int_0^t abe^{(R-b)t}dt \tag{5}$$

This equation is more complex than equation 1 and requires information on both $C_{max}$ and $b$. As fish in the field max will rarely feed to satiation, the simpler equation 1 is adequate in most cases.

The rate of food evacuation $(R)$ in the above model is assumed to be exponential, an assumption which has been shown to be true for a large spectrum of food items and fish species (Persson 1986). For fish feeding on large and nutritious prey items the assumption of an exponential decay may not be warranted (Persson 1986), hence the assumption of an exponential decay should be tested in every study. To circumvent the problem with this assumption, Olson & Mullen (1986) proposed a general model to estimate the rate of food consumption. The model is a development of the model originally derived by Robson (1970), and assumes that feeding occurs in discrete steps. The model predicts the feeding rate $(r_j)$ of a prey item $j$ by dividing the mean stomach contents of the prey item $\Sigma S_{ij} / n$ by the integral $(A_j)$ of the evacuation function for this prey item. The evacuation function, in contrast to the Elliott & Persson model, can take many forms. For a predator consuming a variety of prey types which are evacuated at different rates the rate of feeding $(r)$ is:

$$r = \sum_{j=1}^{m} \frac{\Sigma S_{i,j}/n}{A_j} \tag{6}$$

where $j$ refers to the prey type. Daily meal is calculated by multiplying $r$ by the number of hours the fish is feeding per day.

Though Robson's model makes no assumption concerning the pattern of food evacuation, one precarious assumption made is that every prey item is evacuated independently of other prey items. Since this assumption is in disagreement with the results of Persson (1984), the usefulness of Robson's model remains to be investigated.

### 6.2.2. The bioenergetic method for estimating food consumption rates

The second approach, the bioenergetic, is basically a development of Winberg's (1956) balance equation. The consumption is estimated from the following equation (Kitchell et al. 1977 a, Rice et al. 1983, Stewart et al. 1983, Rice & Cochran 1984):

$$dB/dt = C - (R_{S+A} + R_{SDA} + F + U) \tag{7}$$

where $dB/dt$ is the specific growth rate, $C$ is the specific consumption rate, $R_{S+A}$ is the specific rate of activity dependent metabolism, $R_{SDA}$ is the specific dynamic action, $F$ is the specific egestion rate and $U$ is the specific excretion rate.

Data on stomach contents, growth and temperature are derived from field data, while the information on the other parameters are usually taken from the literature. $R_{S+A}$ is dependent on temperature and size and $R_{SDA}$, $F$ and $U$ are dependent on temperature and ration. The specific values given to the constants to estimate different parameters are species-specific (Rice et al. 1983, Stewart et al. 1983). The indirectly estimated food consumption rate $C$ is partitioned into different food categories based on information on stomach contents derived from the field.

### 6.2.3. Comparison between the field and the bioenergetic approach

The estimates of the two approaches discussed above have been compared by Rice & Cochran (1984), who showed that the bioenergetic model and the Elliott & Persson model gave the same results within an error of no more than 8%. One advantage of the Elliott & Persson model is that the daily ration can be partitioned into different time intervals (Persson 1983 a, b). This is of vital importance for studies on interactions between predator and prey on a diel scale. One disadvantage of the field method is that the rate of food consumption may vary from day to day and thus an estimate on a specific day may be unrepresentative for the period. The field method demands frequent samplings ($t \leq 3$ hrs) over 24 hrs, and hence the field effort is much higher than that demanded when using the bioenergetic approach. For studies running over several years, the bioenergetic approach is therefore the only practicable option in many cases. It should, however, be pointed out that the bioenergetic approach also requires numerous stomach samples from the field in order to give accurate estimates of food consumption of different prey types. The field method involves fewer assumptions concerning the metabolism of fishes which is a significant advantage because, as mentioned above, the specific values of different metabolic demands are species-specific. The information on species-specific values is, however, increasing and so the estimates obtained using the bioenergetic method should also increase in accuracy and precision in the future.

## 6.3. Community structure and succession
### 6.3.1. Community changes with eutrophication

Eutrophication of lakes will affect the fish community both with respect to changes in total biomass and species composition. While the increase in the total biomass of the fish community is obvious and expected in view of the increased production by lower trophic levels (Fig. 6.2), the change in species composition demands a more thorough analysis, which is the goal of this section.

The main change in species composition with a change from oligotrophic to mesotrophic conditions is the replacement of salmonids (i.e. the families *Salmonidae* and *Coregonidae)* with percids as the most common group in the fish community (Nümann 1972, Hartmann & Nümann 1977, Kitchell et al. 1977 b). This

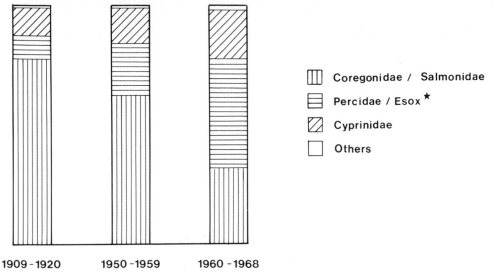

Coregonidae / Salmonidae

Percidae / Esox [*]

Cyprinidae

Others

1909 - 1920          1950 - 1959          1960 - 1968

*Fig. 6.1.*
*The change in the fish community of Lake Bodensee from 1900 to 1970 (data from Nümann 1973).*
*\*mainly perch.*

change was observed in Lake Bodensee, a lake that changed from an oligotrophic to a mesotrophic state between 1900 and 1970 (Nümann 1972) (Fig. 6.1) and in Lake Constance (Hartmann & Nümann 1977) both of which are situated in West Germany. The decrease in salmonids is related to the disappearance of a competitive refuge for salmonids versus other species, i.e the hypolimnion with its low water temperature. In Lake Constance oxygen saturation in the hypolimnion decreased from 82% in the period 1925-1936 to 29% in the period 1952-1975 (Hartmann & Nümann 1977), turning the hypolimnion into an unsuitable habitat for salmonids.

Cyprinids will increase in abundance in both absolute and relative terms with a change from oligotrophic to mesotrophic conditions (Fig. 6.1). It is, however, with the progression to eutrophic conditions that cyprinids become the numerically dominating group at the expense of percids (Svärdson 1976, Kitchell et al. 1977 b, Leach et al. 1977, Persson 1983 c). This decrease in abundance of percids with eutrophication in European lakes is, however, only brought about by reductions in the abundance of perch *(Perca fluviatilis)* and not for the other two common species in Europe, ruffe *(Gymnocephalus cernua)* and zander *(Stizostedion lucioperca)* (Rundberg 1977, Leach et al. 1977) (Fig. 6.2). Both these species increase in absolute abundance with eutrophication though their percentage contribution to the fish community may not change significantly. When viewed as a single group, the relative importance of percids will, however, decrease with eutrophication (Fig. 6.2).

The reasons behind the observed changes in the fish community are related to a number of environmental and biological factors, of which the interactions between cyprinids and percids in relation to eutrophication will be our focal point. The discussion concerning cyprinid-percid interactions suffers from two weaknesses that should be pointed out. Firstly, our knowledge concerning competitive interactions

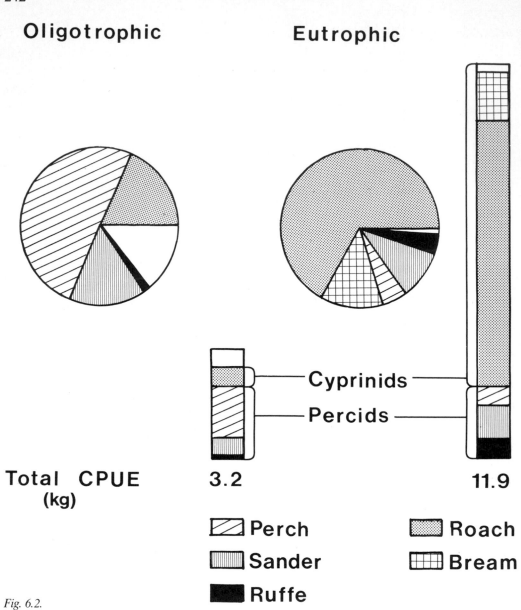

**Total CPUE (kg)**     3.2             11.9

*Fig. 6.2.*
*Catch per unit effort (CPUE, catch per survey link per 24 hrs) of different fish species in an oligotrophic (Lake Bolmen) and a eutrophic (Lake Vombsjön) lake. Circles show the percentage composition and histograms show the actual biomasses (data from Hamrin 1984).*

is much higher than that concerning predatory interactions, hence there will be a strong bias in the discussion towards the former. Predatory interactions between fish species are present in eutrophic systems but their relative importance in relation to competitive interactions is smaller than that in oligotrophic systems (see chapter 7). Secondly, studies on the interactions between percids and cyprinids have mainly dealt with the interactions between perch and roach *(Rutilus rutilus)*, while knowledge concerning other percid and cyprinid species is poor. As a consequence, the discussion will largely be restricted to perch-roach interactions, while other species will be dealt with only briefly.

### 6.3.2. Resource utilization of percids

From what is known about the food resource utilization of the ruffe it appears that it largely reflects the resource availability in the habitat which it occupies. As a species mainly restricted to the bottom, ruffe consumes chironomids and benthic cladocerans (Aleksandrov 1972, Federova & Verkasov 1972, Brabrand 1983). Considering the resource use of ruffe, competition between ruffe and perch and cyprinids in eutrophic lakes is likely to occur, although this has not yet been documented. The difference between perch and ruffe shown in response to eutrophication is probably related to the adaptation of ruffe to feed under poor light conditions. Around the mouth and along the lateral line the ruffe has well developed sensory organs (Disler & Smirnow 1977) that will allow it to detect and capture prey under the poor light conditions found in eutrophic lakes.

The zander also has adaptations for feeding in turbid waters. A pigmented epithelium, the *tapedum lucidum,* increases the light sensitivity of the eyes (Ali et al. 1977). The adaptation to low light levels is reflected in the vertical distribution of zander in transparent lakes, where it is found in the deep waters (Hamrin 1974). Though zander is common in many oligotrophic lakes, a fact that in many cases is a result of introduction by man (Svärdson 1976, Svärdson & Molin 1973), it is more abundant in eutrophic lakes (Fig. 6.2). Except when young, the zander is piscivorous and will prey upon smelt *(Osmerus eperlanus),* cyprinids, ruffe and perch depending on the availability of these species (Willemsen 1969, 1977). It has been suggested that predation by zander may regulate the population size of perch (Svärdson 1976), a fact that might contribute to the low abundance of perch in eutrophic lakes. However, the evidence for this hypothesis is poor and other studies (Willemsen 1977) have suggested that zander predation on perch is negligible. Presumably the effects of zander predation on perch is dependent on the presence or absence of other prey fishes, for example smelt (Svärdson 1976).

Compared to the situation with ruffe and zander, studies on the feeding habitats of perch in different environments are numerous. During its ontogenetic development perch will go through potentially 3 different stages with respect to food resource utilization; a zooplanktivorous stage, a macroinvertebrate feeding stage and a piscivorous stage. The length of the zooplankton feeding period is dependent on the zooplankton abundance and size structure of the perch population. In lakes with very low densities or total absence of roach, yearlings of perch will feed almost exclusively on zooplankton (Gumáa 1977, Craig 1978, Nyberg 1979) and zooplankton may form a significant part of the diet for several years (Thorpe 1977, Craig 1978, Nyberg 1979) (Table 6.1). In contrast, in eutrophic lakes with dense population of roach, perch may switch to feeding on macroinvertebrates when very young (Persson 1983 a, 1986).

When the duration of a specific stage in the life history of individuals in a size structured population has marked effects on population abundance and structure, this stage is termed a bottleneck (Werner & Gilliam 1984). It has been shown both in field studies (Persson 1983 a, 1986) and in theoretical models (Werner & Gilliam 1984) that intense competition during one stage of an individual's life period

Table 6.1. Food utilization of perch ( > 1 year) in the absence of roach (a) and coexisting with a dense population of roach (b).

| Prey | Proportion a | (%) b |
|---|---|---|
| Zooplankton | 40 | 2 |
| Chironomids | 21 | 64 |
| Non chironomid macroinvertebrates | 34 | 25 |
| Others | 5 | 8 |

Sources: (a) Nyberg 1979, (b) Persson 1983 a.

in size structured populations may lead to a bottleneck. Bottlenecks are well known from studies on stunted perch populations in Scandinavian forest lakes (Alm 1946). Persson (1983 a) has shown that a similar situation may be present in eutrophic lakes. In perch populations the macroinvertebrate feeding stage is usually the bottleneck in the recruitment to larger size classes (Keast 1977, Persson 1983 a, 1986). An early switch of 0+ perch to feeding on macroinvertebrates, which was shown for a perch population in the shallow and eutrophic Lake Sövdeborg (Persson 1983 a, 1986), will increase the effects of this bottleneck. In this study the food consumption rate of perch was very low compared to that of other perch populations in oligotrophic and mesotrophic lakes (Thorpe 1977, Craig 1978) and comparable to that of slow growing perch populations in forest lakes (Nyberg 1979). The consumption was also far below the maximum food intake of perch as measured in the laboratory (Lessmark 1983), suggesting a severe resource limitation. This suggestion was also supported by extremely low growth rates (Persson 1983 a) comparable to those of stunted perch populations in forest lakes. As a consequence of the early switch of yearling perch to feeding on macroinvertebrates, the food spectrum of the perch population was totally dominated by the most limiting resource, i.e. macroinvertebrates, in contrast to the food spectrum of perch populations in lakes where roach was absent (Table 6.1). Different age classes of perch also fed largely on the same type of macroinvertebrates (Table 6.1, Fig. 6.3), and very few perch entered the piscivorous stage. The low consumption rates, a result of intense competition, resulted in a population totally dominated by small individuals (Persson 1983 a).

To which extent the situation in Lake Sövdeborg can be applied to eutrophic lakes in general is not known due to the scarcity of data from other eutrophic lakes. Studies on the perch population in Lake Ijssel, Netherlands (Willemsen 1977), for example, suggest that the individual growth rate of perch in eutrophic lakes may be high in some cases. There is, however, clear evidence that the size class diversity of perch populations in eutrophic lakes is low with the populations dominated by small individuals (Persson 1983 a, c). This suggests that the situation present in Lake Sövdeborg may be common in many eutrophic lakes. In Lake Ijssel, the availability of a profitable prey fish, smelt, might have had a significant

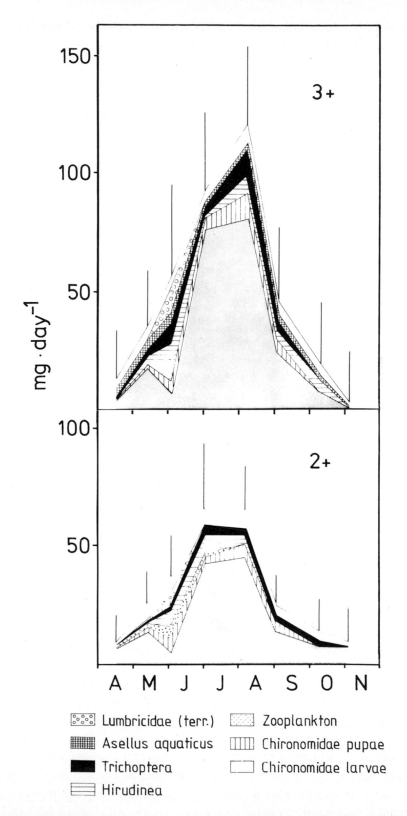

Fig. 6.3.
*Daily food consumption of an individual 2+ perch and 3+ perch respectively in Lake Sövdeborg in 1980 (from Persson 1983a).*

impact. The possibility of perch individuals switching from feeding on macroinvertebrates to piscivory at a relatively early stage in this lake may have diminished the importance of the macroinvertebrate feeding bottleneck.

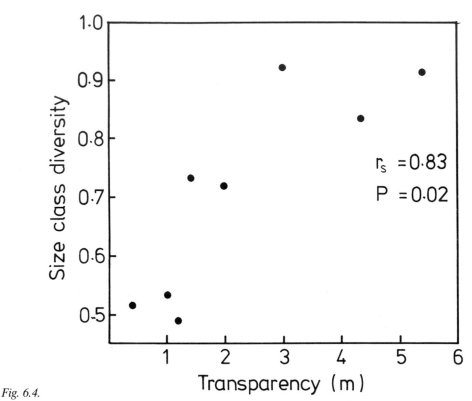

*Fig. 6.4.*
*Size class diversity of perch populations versus transparency of lakes (from Persson 1983a).*

It has been suggested that habitat may be important for segregating different age classes of the same species (Keast 1977, 1978, Laughlin & Werner 1978; Brandt 1980, Persson 1983 a). For a visual predator such as perch it is reasonable to assume that light is an important factor limiting the degree of habitat subdivision. This is supported by the positive correlation between the size class diversity of perch populations and the transparency of lakes (Fig. 6.4). The importance of habitat heterogeneity for segregating age classes of perch has also been indicated by other studies (e.g. Sandheinrich & Hubert 1984). The turbid waters of eutrophic lakes will prevent habitat segregation between age (size) classes which may result in intense competition. The high size class diversity of perch populations in transparent lakes is, however, not solely explained by reduced between age class competition as a result of increased habitat heterogeneity. A high within age class competition will also result in low individual growth rates and consequently low size class diversity. The hypothesis of the importance of habitat heterogeneity in segregating age classes must therefore also include an assumption of a significant predation pressure from larger perch on smaller individuals reducing within age class competition in the youngest age classes (Persson 1983 a).

### 6.3.3. Resource utilization of cyprinids
### 6.3.3.1. Digestion of non-animal food items

In contrast to percids which are obligate carnivores, cyprinid species can utilize both animnal and non-animal food items. It has been shown that for the roach bluegreen algae are generally more nutritious than other vegetable food types (Sorokin 1968, Lessmark 1983) (Table 6.2). This has been related to the higher protein content of bluegreen algae compared to other algae and higher plants (Lessmark 1983). Though bluegreen algae are more nutritious than other vegetable food types it should be pointed out that they are still considerably less nutritious than animal food items (Sorokin 1968, Lessmark 1983). Sorokin (1968) suggested that cyprinid species differ in their ability to assimilate bluegreen algae. As his experiments were carried out with young cyprinids which are essentially zooplanktivorous in the field it is not known whether the same holds for older fish.

Table 6.2. Assimilation efficiencies of roach and bream respectively and specific growth rate of roach fed different vegetable food items.

| Food item | | Assimilation defficienty (%)[a] | | Specific growth rate (%)[b] |
|---|---|---|---|---|
| | | Bream | Roach | Roach |
| Bluegreen algae | **Aphanizomenon** | 0.9 | 62 | $0.90 \pm 0.33$[***] |
| Green algae | **Scenedesmus** | 0.3 | 14 | - |
| Filamentous algae | **Spirogyra** | - | - | $-0.64 \pm 0,65$[ns] |
| Starving fish | | | | $-0.72 \pm 0.33$ |

[a] Data from Sorokin (1968)      [***] $P < 0.001$
[b] Data from Lessmark (1983)    [ns] non significant

The mechanism behind the capacity of cyprinids to assimilate bluegreen algae is poorly known, though several hypotheses have been put forward. For African cichlids it has been shown that bluegreen algae are digested by acid lysis at a very low pH (pH $<2$) (Moriarty 1973, Bowen 1976). This mechanism cannot, however, explain the high assimilation efficiency for cyprinids feeding on vegetable food, as cyprinids lack a distinct stomach with a low pH (Kapoor et al. 1975). Both acid and pepsin secreting cells are missing, and the pH of the alimentary canal does not drop below 5. It is obvious that the pharyngeal teeth, which are well developed in cyprinids, play a significant role by macerating the food particles before they enter the intestine as intact food particles are not digested at all (Hickling 1966). Lobel & Ogden (1981) suggested that the pharyngeal teeth function as a mill breaking up the cells and making them accessible to enzymes. Different cyprinid species have different morphologies of their pharyngeal teeth which are related to their main diet (Prejs 1984). Cyprinids lack cellulase production to break down cell walls but as cellulase activity has been found in the alimentary canal (Stickney & Shumway 1974, Niederholzer & Hofer 1979, Prejs 1984), it has been suggested that cellulase originating from consumed prey will make the digestion of cellulose possible.

### 6.3.3.2. Food utilization of cyprinids

Alongside the roach, bream is another cyprinid which increases with eutrophication (Fig. 6.2). In some shallow and very eutrophic lakes this species seems to replace roach as the most abundant species. The food of bream is dominated by zooplankton, chironomids and sediment (Andersson et al. 1975, Lammens 1982). Whether the detritus found in the intestines of bream is an accidental uptake or not and whether bream can utilize detritus as an energy source remain to be investigated. Studies on carp (Kevern 1966) have shown that this species can assimilate detritus to some extent. Sorokin (1968) found that young bream did not have the same capacity as young roach to assimilate cyanobacteria (Table 6.2), but as mentioned above this may be a result of the young age of the fish studied. In contrast to roach, bream (>20 cm) has the capacity to filter feed on zooplankton (Lammens 1985), which may be one explanation for its success in eutrophic lakes.

In contrast to perch in eutrophic lakes, roach may not be resource limited by the quantity of food available. It was shown by Persson (1983 b) that roach in the field consumed food at a rate close to its maximum consumption rate as measured in the laboratory. The food consisted to a large extent of cyanobacteria and detritus (80% of the diet) (Fig. 6.5). Due to the lower assimilation efficiency of this food compared to animal food items, individual growth rates were still very low, which suggested a resource limitation with respect to the quality of food (Persson 1983 b). The lower protein content of detritus/algae compared to animal food is probably the main limiting factor (Menzel 1959, Bowen 1979 a, b).

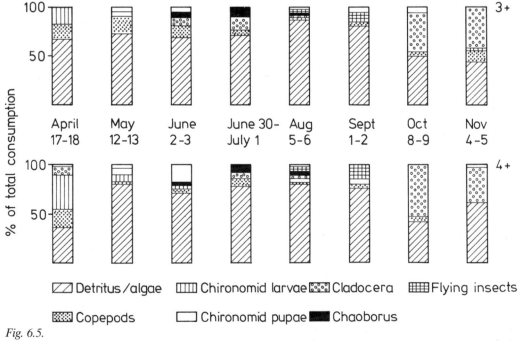

Fig. 6.5.

*Percentage contribution of different food items to the diet of 3+ and 4+ roach in Lake Sövdeborg in 1980 (from Persson 1983b).*

Most cyprinids show ontogenetic shifts in diet with age (size). This generally involves a shift from zooplanktivory during young stages to omnivory or herbivory during older stages (Brabrand 1983, Prejs 1984). The ontogenetic shift in diet may also include changes in histology and morphology of the alimentary canal (Sinha & Moitra 1975). Some cyprinid species such as bream do, however, continue to feed to a large extent on zooplankton with increasing size (Andersson et al. 1975, Lammens 1982, 1985). The continued feeding on zooplankton with an increase in size of bream is related to a shift from particulate feeding to filter feeding (Lammens 1985). Seasonal changes in diet of young cyprinids remain essentially unstudied and so it is not known whether they show the same seasonal changes in diet as older individuals.

The high availability of detritus and algae in eutrophic lakes will increase the carrying capacity of roach in these water bodies when compared to oligotrophic lakes, especially when compared to carnivorous percids (Fig. 6.2). This, in conjunction with the high feeding effieiency of roach on zooplankton (Table 6.3), will strongly favour roach in relation to perch in eutrophic lakes. Though roach can utilize cyanobacteria as a resource, the nutritional value of cyanobacteria when compared to animal food items is low. As a consequence, the individual growth rate will be low with consequent effects on the size structure of the population. Roach populations feeding mainly on detritus and algae will thus mainly consist of small individuals (Persson 1983 b).

Table 6.3. Capture rate, handling time and swimming speed for roach and perch feeding on zooplankton of various densities

| Prey item | density (no/1) | capture rate (no/s) | | handling time (s) | | swimming speed (cm/s) | |
|---|---|---|---|---|---|---|---|
| | | roach x̄ ± 1SD | perch x̄ ± 1SD | roach x̄ ± 1SD | perch x̄ ± 1SD | roach x̄ ± 1SD | perch x̄ ± 1SD |
| **Daphnia** | 0.5 | 1.0±0.4 | 0.3±0.1 | | | | |
| | | | | 0.6±0.1 | 1.1±0.5 | 46.9±8.0 | 5.2±0.3 |
| | 4.0 | 1.6±0.3 | 1.0±0.2 | | | | |
| **Cyclops** | 2.5 | 0.5±0.1 | 0.1±0.02 | | | | |
| | | | | 1.1±0.1 | 2.9±0.2 | | 3.0±0.1 |
| | 10.0 | 0.7±0.1 | 0.2±0.03 | | | | |

### 6.3.4. Laboratory studies on feeding efficiency of roach and perch on zooplankton and macroinvertebrates

The trophic interactions between fish species can be analyzed by studying the feeding behaviours of the fish when feeding on different prey species and densities in the laboratory. Such studies give us information on the utilization efficiencies of the studied species for different types of resources. One of the first attempts to connect laboratory studies of feeding behaviours to the species distribution in the field was made by Werner (1977). Werner determined the prey utilization efficiency on different prey types for three different species of centrarchids, bluegill sunfish *(Lepomis macrochirus),* green sunfish *(Lepomis cyanellus)* and largemouth bass *(Micropterus salmoides)* in laboratory studies. With the knowledge of the resource distribution in different habitats in the field, he was able to successfully predict the distributions of the three species in the field. This approach, involving the relation of results from feeding efficiency studies in the laboratory to the situation in the field, has also proved useful in other studies (Mittelbach 1981 a, Giller & McNeill 1981, Werner et al. 1981, 1983).

With the same intention we studied the utilization efficiency of roach and perch in the laboratory on three different prey types; *Daphnia magna* (1.3 mm), *Cyclops* sp. (1.5 mm) and *Chironomus plumosus* larvae (8 mm). Roach was the most efficient of the two species when feeding on *Daphnia* (Table 6.3). This reflects the differences in search strategies between the species. The perch swims slowly while searching and attacks every single prey sighted, whereas the roach swims continuously at a high speed (Table 6.3) and attacks only those prey which appear in front of it. A parallel to the differences in feeding behaviour of perch and roach is found in the study of Janssen (1982). He studied the search behaviour of an obligate planktivore, the alewife *(Alosa aestivalis),* and a facultative planktivore, bluegill sunfish, and found a striking difference between them. The bluegill searched while stationary like perch, while the alewife was a swim-searcher like the roach. Because it is easier to detect moving prey against a stationary background, Janssen suggested that the search strategy of the bluegill was an adaptation to heterogenous backgrounds near the shore. The swim-search behaviour is more adaptive in the open water, giving the predator a higher feeding rate in this habitat. This hypothesis is in agreement with the distinction between roach and perch since perch is more often found in the littoral and the roach in the pelagial areas of a lake (Stenson 1979, Persson 1983 c). Handling time, given by the reciprocal of the maximum feeding rate, expresses the time the predator is occupied with a single prey item (Ware 1972, Werner 1977). The superiority of roach over perch when feeding on *Daphnia* is also reflected by the differences in handling time between the two species (Table 6.3). While cladocerans have very little ability to escape from a fish predator, copepods with their greater capacity to make evasive jumps are more difficult to capture (Confer & Blades 1975, Drenner et al. 1978, Winfield et al. 1983). This is clearly reflected in the lower efficiency of both roach and perch when feeding on copepods (Table 6.3). Also for copepods roach was more efficient than perch.

Chironomid larvae was the most important food item for perch in Lake Sövdeborgssjön (Fig. 6.3). As expected, the perch in the laboratory experiment were by far the most efficient utilizers of the chironomids (Table 6.4). Roach only found the larvae by accident, and their feeding rate was independent of prey density while perch showed an increasing feeding rate with increasing chironomid density (Table 6.4). In the field, roach feed mainly on small, pelagic prey, whereas perch feed more on larger, littoral prey (Stenson 1979, Persson 1983 c). In Lake Sövdeborgssjön there is a marked difference in the prey size distribution between the lit-

Table 6.4. Capture rate and handling time for roach and perch feeding on **Chironomid** larvae of two densities

| Density (no. m$^{-2}$) | capture rate (no. s$^{-1}$) | | handling time (s) | |
|---|---|---|---|---|
| | roach x̄ ± 1SD | perch x̄ ± 1SD | roach x̄ ± 1SD | perch x̄ ± 1SD |
| 300 | 0.001 ± 0.001 | 0.004 ± 0.0001 | | |
| | | | 6.9 ± 1.9 | 5.5 ± 0.2 |
| 1500 | 0.002 ± 0.002 | 0.13 ± 0.006 | | |

toral and the pelagial, the latter dominated by small zooplankton and the littoral by macroinvertebrates (Persson 1985). These differences seem likely to have general applicability since the same principal patterns have also been shown by Keast and Harker (1977) and Mittelbach (1981 b). These facts reflect the importance of the littoral zone for the feeding of perch. Under eutrophication the littoral zone becomes narrow as submerged macrophytes decrease, while phytoplankton and emergent macrophytes increase (Wetzel 1979, Sand-Jensen 1980). This will in turn result in that the macroinvertebrates on which the perch is more efficient than roach become relatively scarcer.

### 6.3.5. Interactions between roach and perch

We have mentioned above that roach may influence the resource utilization of perch in eutrophic lakes. A number of studies (Sumari 1971, Svärdson 1976, Holcik 1977, Burroughs et al. 1979, Persson 1983 c) have suggested that the abundance relationships between roach and perch and between percids and cyprinids in general are dependent on the competitive interactions between them. The advantage enjoyed by roach in eutrophic lakes can be related to (1) its ability to utilize cyanobacteria abundant in eutrophic lakes (Table 6.2), (2) its high feeding efficiency on both cladoceran and copepod zooplankton (Table 6.3), (3) its ability to feed on smaller zooplankton than can perch (Stenson 1979, Lessmark 1983), zooplankton forms which dominate in lakes with heavy fish predation and (4) the decreased abundance of submersed macrophytes with eutrophication (Wetzel 1979, Sand-Jensen 1980), which reduces the shelter and food source for many macroinvertebrates on which perch is a more efficient feeder than roach (Table 6.4). Taken together these factors suggest that roach should be the more abundant species in eutrophic systems, while the opposite should be the case in oligotrophic systems, a fact which has also been observed (Fig. 6.2, Fig. 6.6).

The food overlap between roach and perch in eutrophic lakes may be very low, as roach feed mainly on detritus and algae (Fig. 6.5) (Persson 1983 c), but the competitive effect of roach on perch can still be high due to the differences in population size of the two species in eutrophic lakes. This is supported by data on the amount of food removed by the perch and the roach populations respectively in Lake Sövdeborg (Fig. 6.7).

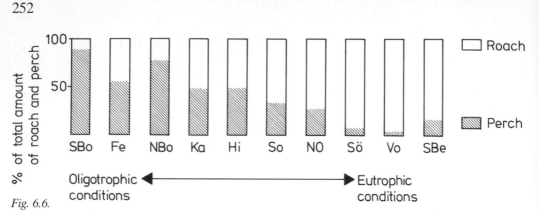

*Fig. 6.6.*
*Quantitative relationships of roach and perch in lakes of different trophic conditions (from Persson 1983c).*

An effect of roach on the size structure of perch populations is suggested by a number of factors. Firstly, heavy predation by roach on zooplankton will reduce the overall density of zooplankton and change the zooplankton community towards an assemblage dominated by small forms (Henriksson et al. 1978, Andersson 1984). This change results in an early switch by 0+ perch from a diet of zooplankton to a diet of macroinvertebrates with the result of an increase in intraspecific competition in the perch population as shown earlier. Roach will also have a eutrophicating effect on the water system (Henriksson et al. 1978, Andersson 1984) reducing transparency and hence habitat heterogeneity for perch. This results in an increased intraspecific competition within the perch population and a low size class diversity (Fig. 6.4).

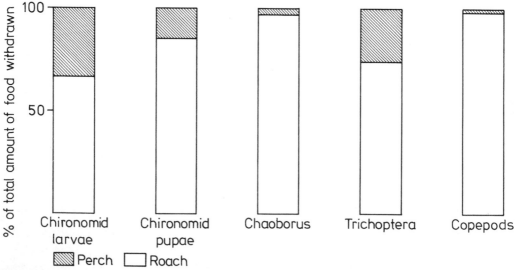

*Fig. 6.7.*
*Amount of food consumed by the roach and perch populations respectively as a percentage of the total consumption in Lake Sövdeborg in 1980 (from Persson 1983c).*

### 6.3.6. Experimental evidence for the presence of competitive interations between roach and perch

The sort of evidence so far presented for the presence of competition between perch and roach is either based on negative correlations between abundance of perch and roach in different lakes (Sumari 1971, Svärdson 1976, Person 1983 c) or on correlations between changes in population size of perch and roach respectively in the same lake (Holcik 1977, Persson 1983 c). As correlations may be noncausal, other explanations for the observed patterns cannot be dismissed. To test whether competition is actually present and how it may affect the resource utilization of the species in natural systems, experimental perturbations are necessary (Connell 1980, Schoener 1983).

Such an experiment was performed in the shallow and eutrophic Lake Sövdeborg, the study site for previous work on resource utilization of roach and perch (Persson 1983 a, b, c). By a large-scale removal of individuals, the population size of roach was reduced by 70% and its biomass by 30%. Several hypotheses regarding the response of the perch population to the reduction in population size of roach were formulated (Fig. 6.8) (Persson 1986). An increase in zooplankton abundance was expected due to the high feeding efficiency of roach on these prey (Table 6.3). As an effect of this, the switch from zooplankton feeding to macroinvertebrate feeding in 0+ perch was predicted to be delayed compared to the situation before the reduction. The prolonged zooplankton utilization of 0+ perch was expected to decrease the duration of the macroinvertebrate feeding stage, and thus the intraspecific induced macroinvertebrate bottleneck should become less important. This was predicted to in turn result in an increased food consumption rate, increased individual growth rate of adult perch and a consequent increase in the population size of perch.

The effects of the roach reduction were in complete agreement with the above hypotheses (Fig. 6.8). Before the reduction, 0+ perch switched to a macroinvertebrate diet during their first summer, while after the reduction of the roach population they never switched to feed on macroinvertebrates (Fig. 6.9). In contrast to the situation before the reduction 0+ perch stayed in the pelagic zone of the lake (Persson 1986). Individual growth rate of perch also increased, and the consumption rate of adult perch increased twofold compared to the situation before the reduction (Table 6.5). An increase in the population size of perch was also observed (by 140%), which was mainly a result of an increase in abundance of 0+ perch. Other fish species such ad rudd *(Scardinius erythrophthalmus)* and belica *(Leucaspius delineatus)* also increased in abundance, which was probably a result of competitive release from the roach. Since these species are potential prey for perch, this resulted in an earlier switch of adult perch to piscivory, which itself further reduced the effects of the macroinvertebrate bottleneck stage (Persson 1986) (Table 6.5).

As the study was a whole lake experiment, a control was inevitably lacking. The fact that all events were consistent with the hypothesis that roach competition greatly influenced population size and resource utilization of the perch population

254

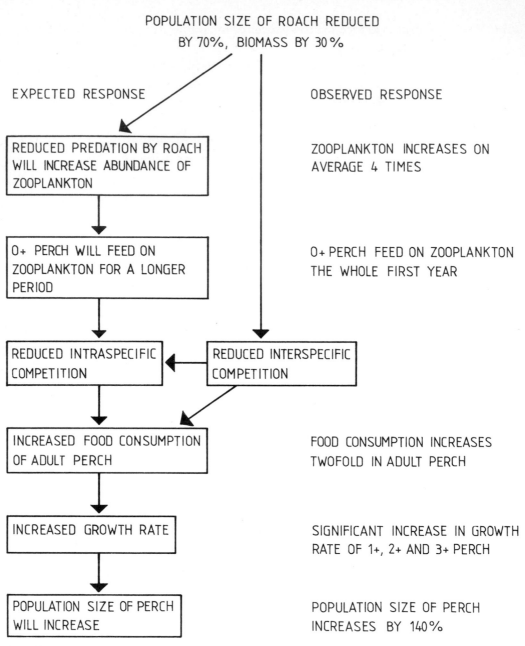

POPULATION SIZE OF ROACH REDUCED
BY 70%, BIOMASS BY 30%

EXPECTED RESPONSE

OBSERVED RESPONSE

REDUCED PREDATION BY ROACH WILL INCREASE ABUNDANCE OF ZOOPLANKTON

ZOOPLANKTON INCREASES ON AVERAGE 4 TIMES

0+ PERCH WILL FEED ON ZOOPLANKTON FOR A LONGER PERIOD

0+ PERCH FEED ON ZOOPLANKTON THE WHOLE FIRST YEAR

REDUCED INTRASPECIFIC COMPETITION

REDUCED INTERSPECIFIC COMPETITION

INCREASED FOOD CONSUMPTION OF ADULT PERCH

FOOD CONSUMPTION INCREASES TWOFOLD IN ADULT PERCH

INCREASED GROWTH RATE

SIGNIFICANT INCREASE IN GROWTH RATE OF 1+, 2+ AND 3+ PERCH

POPULATION SIZE OF PERCH WILL INCREASE

POPULATION SIZE OF PERCH INCREASES BY 140%

*Fig. 6.8.*
*Expected and observed response of the perch population to the reduction in the population size of roach in Lake Sövdeborg.*

is, however, strong although not conclusive evidence that competition was causing the observed patterns.

## 6.4. Effects of fish predation on the zooplankton community
### 6.4.1. Fish predation as a factor regulating zooplankton in general

The traditional view that fish populations merely harvest the surplus of a lake's production without affecting lower trophic levels was first challenged by Hrbacek

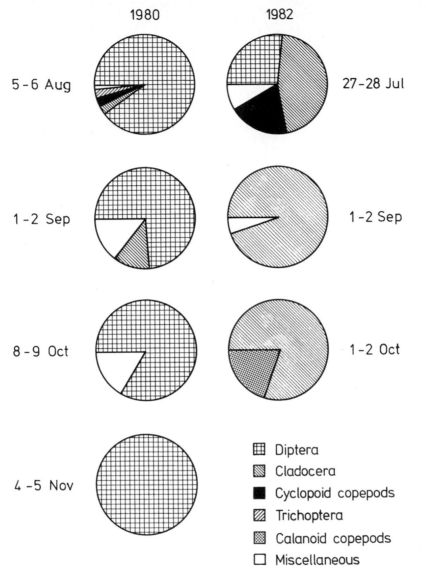

Fig. 6.9.
*Changes in the food spectra of 0+ perch from 1980 to 1982 in Lake Sövdeborg (from Persson 1986).*

(1962) who studied the zooplankton fauna in relation to the fishstock. The idea that fish affect the zooplankton community has subsequently received much attention and numerous studies have demonstrated the impact of fish on the size structure and species composition of zooplankton communities (Brooks & Dodson 1965, Galbraith 1967, Reif & Tappa 1966, Wells 1970, Hutchinson 1971, Warshaw 1972, Nilsson & Pejler 1973, Nilsson 1978, Stenson 1979, Hamrin 1983, Andersson 1984). By negatively affecting the fitness of the predated zooplankton, fish predation acts as a selective force in their evolution. The vertical migration of zooplankton is one phenomenon which has been attributed to fish predation (Zaret & Suffern 1976, Wright et al. 1980, Stich & Lampert 1981).

Table 6.5. Food consumption (g ash-free dry wt) of the total perch population consisting of 2+ and 3+ perch in 1980 and 1+ and 2+ perch in 1982.

| Food item | 1980 | 1982 | Ratio 1982/1980 |
|---|---|---|---|
| Chironomidae | 68,700 | 124,200 | 1.8 |
| Hirudinea | 9,570 | 9,470 | 1.0 |
| Fish | 2,140 | 39,600 | 18.5 |
| Trichoptera | 6,050 | 12,100 | 2.0 |
| **Asellus aquaticus** | 2,840 | 6,750 | 2.4 |
| **Gammarus pulex** | 1,100 | 6,900 | 6.3 |
| Ephemeridae | 1,950 | 4,510 | 2.3 |
| **Chaoborus** | 1,980 | 4,460 | 2.3 |
| Cyclopoid copepods | 2,170 | 1,590 | 0.7 |
| Anisoptera | 960 | 1,180 | 1.2 |
| Ceratopogonidae | 1,060 | 2,050 | 1.9 |
| Total | 118,200 | 259,500 | 2.2 |

Effects by fish predation on the zooplankton community have mostly been documented for obligate planktivores e.g. alewife, *Alosa pseudoharengus* (Brooks & Dodson 1965, Wells 1970, Wells 1970, Hutchinson 1971, Warshaw 1972), blueback herring, *Alosa aestivalis* (Brooks & Dodson 1965) and smelt, *Osmerus mordax* (Reif & Tappa 1967). The effects are seen on both the size structure and species composition of the plankton. Galbraith (1967) showed similar effects on the zooplankton community after the introduction of three fish species, rainbow trout *(Salmo gairdneri)*, smelt and fathead minnow *(Pimephales promelas)* in a small lake. When studying the natural oscillations of a vendace population *(Coregonus albula)*, Hamrin (1983) found a negative relation betweeen the density of cladocerans and the abundance of vendace yearlings, indicating a strong predation effect on the cladoceran populations. Nilsson & Pejler (1973) compared the zooplankton communities and the fish fauna in 65 lakes in northern Sweden. Lakes inhabited by one or more species of whitefish *(Coregonus* spp) had a zooplankton fauna with a smaller mean size than lakes inhabited by trout *(Salmo trutta)* and/or arctic charr *(Salvelinus alpinus)*, indicating an impact of whitefish on the zooplankton community. There was also a difference in species composition with the larger zooplankton species being absent from the whitefish lakes.

The selective effect of fish predation on the zooplankton community can be due to an active choice by the fish to eat or not to eat an encountered zooplankton, but it can also be a result of the differences in visibility and catchability of different zooplankton. Generally the predation process consists of the initial encounter with the prey, the decision to attack, followed by pursuit and attack and finally, if successful, the swallowing of the prey. The encounter rate is a function of the reactive distance, the distance from which a predator detects a prey and initiates the attack, and the swimming speeds of the predator and prey (Gerritzen 1984). While the reactive distance increases with fish size (Breck & Gitter 1983), it decreases with increasing turbidity and decreasing light conditions (Ware 1972, Confer & Blades 1975, Confer et al. 1978). The reactive distance has been found to increase with body size of the prey (Werner & Hall 1974, Confer et al. 1978) (Fig. 6.10). The different visibilities of different zooplankton morphologies also have an effect on the reactive distance (Zaret & Keerfoot 1975). Thus, compared to the actual size distribution, the zooplankton size distribution encountered by the fish is biased towards the bigger zooplankton due to the differences in reactive distance for different plankton sizes.

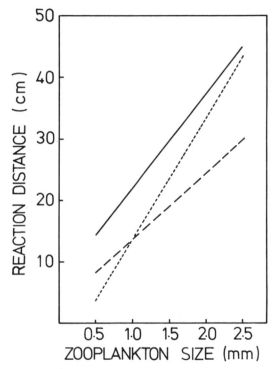

*Fig. 6.10.*
*Relation between reactive distance of bluegill and prey size for Daphnia magna (——), D. pulex (-----) and copepods (········) (data from Confer & Blades 1975).*

After encountering a prey, the fish decides whether to pursue the encountered prey (0'Brien 1979). If the fish encounters more than one prey at a time, it makes a choice among the encountered prey (Werner & Hall 1974, 0'Brien et al. 1976, Gardner 1981). The basis of this choice is, however, unclear (Janssen 1983). Cladocerans have a very poor capacity to escape an attack from a fish predator, and hence their capture probability is always close to 100% (Confer & Blades 1975, Winfield et al. 1983). Drenner et al. (1978) demonstrated the different abilities of zooplankton to escape by using a tube with a controlled suction as an artificial predator. The capture probability for cladocerans was not significantly different from air-bubbles and heat-killed zooplankters. In contrast, the capture probabilities for cyclopoid copepods and calanoid copepods was around or below 20% (Table 6.6).

Similar patterns was found by Winfield et al. (1983) for underyearling roach and by Confer & Blades (1975) for bluegill sunfish. The capture success of the fish may also be affected by the form and function of the mouthparts of the fish. Winfield et al. (1983) found that bream *(Abramis brama)* had a much higher capture success than roach on cyclopoid copepods. This was suggested to be an effect of the more protrusible mouth of bream which will result in a higher suction pressure.

Table 6.6. Capture success (%) for an artificial predator (siphon tube) on different zooplankton species (data from Drenner et al. 1978).

| | Distance from tube | |
| --- | --- | --- |
| Species | 2.5 mm | 7.5 mm |
| Heatkilled zooplankton | 100 | 100 |
| **Ceriodaphnia reticulata** | 100 | 100 |
| **Daphnia pulex** | 100 | 85 |
| **Cyclops scutifer** | 35 | 25 |
| **Diaptomus palladius** | 15 | 10 |

The difference in capture probability for different zooplankton will affect the impact of planktivorous fish on the zooplankton community. The largest effects are to be expected on cladocerans, while copepods are expected to be affected to a lesser extent. In agreement with this hypothesis, Hamrin (1983) found that cyclopoid copepods were unaffected or even favoured during years with a high abundance of vendace yerlings while the opposite was true for cladocerans.

### 6.4.2. The effect of cyprinid predation on zooplankton communities in eutrophic lakes

Experimental evidence for the effects of cyprinid predation on zooplankton comes mainly from experiments in ponds and enclosures. After the introduction of roach and bream into a formerly fish-free eutrophic pond, the dominant assemblage of *Daphnia magna* and *D. pulex* was replaced by one dominated by *Bosmina longirostris* and *Eudiaptomus* sp. (Andersson & Cronberg 1984). After two years the fish were killed with rotenone and the zooplankton community again became dominated by *Daphnia magna* and *D. pulex*. Andersson (1981, 1984) studied the effects of different fish species on the zooplankton community, primary production and water chemistry in enclosures in a eutrophic lake. The number of daphnids in an enclosure stocked with roach was low (4 ind/l) compared to that in enclosures with bream (20 ind/l) or perch (134 ind/l), again indicating the strong effect the cyprinids have on the zooplankton community.

Only a few projects have been carried out with the aim of studying the effects of cyprinid predation on the zooplankton community in whole lakes. Since cyprinids are present in most eutrophic lakes, it is difficult to make comparisons between zooplankton communities in lakes with cyprinids present and lakes with cyprinids absent. To obtain this information, pertubation experiments are necessary. Stenson (1973) carried out such an experiment in small oligotrophic lakes. He replaced the original fish fauna of roach, perch and pike *(Esox lucius)* with salmonids in four lakes. The zooplankton community responded with a marked increase in mean size and in size diversity compared to the control lakes (Fig. 6.11). There was also a change in the species composition of the zooplankton with *Daphnia longispina* replacing the smaller *Daphnia cristata* in the salmonid lakes, and *Bythotrephes longimanus* being found exclusively in the salmonid lakes.

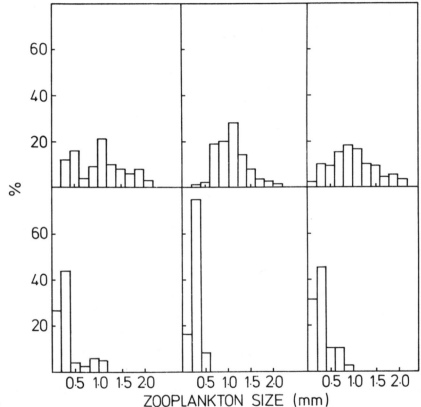

*Fig. 6.11.*
*Size class distributions of zooplankton in lakes inhabited by brook trout (Salvelinus fontinalis) or rainbow trout (Salmo gairdneri) (above) and perch and roach (below), respectively (data from Stenson 1972).*

To summarize, effects of fish predation on the zooplankton community have been documented in a vast number of studies. In eutrophic lakes, these effects are more pronounced for cyprinid species than for other species, and effects on primary production and water chemistry have also been documented (Andersson et al. 1978, Andersson 1984). The existence of these strong effects suggest that the interactions between different trophic levels are tightly interconnected and only by covering the whole trophic web including fish will it be possible to understand the dynamics of eutrophic lakes.

## 6.5. References

Aleksandrov, A. I. 1974. A morphological and ecological description of the ruffe of the middle reacher of the Dnieper. J. Ichtyol. 14: 53-59.

Ali, M. A., Ryder, R. A. & Anctil, M. 1977. Photoreceptors and visual pigments as related to behavioural responses and preferred habitats of perches (*Perca* spp.) and pikeperches (*Stizostedion* spp.). J. Fish. Res. Bd. Can. 34: 1475-1480.

Alm, G. 1946. Reasons for the occurrence of stunted fish populations with special reference to perch. Rep. Inst. Freshw. Res., Drottningholm 25.

Andersson, G. 1981. Influence of fish on waterfowl in lakes. Anser 20: 21-34. (In Swedish with English summary).

Andersson, G. 1984. The role of fish in lake ecosystems - and in Limnology. Nordisk limnologsymposium, Oslo 1984.

Andersson, G. & Cronberg, G. 1984. *Aphanizomenon* flos-aquae and large *Daphnia* - An interesting association in hypertrophic waters. Nordisk limnologsymposium, Oslo 1984.

Andersson, G. Bergren, H. & Hamrin, S. 1975. Lake Trummen restoration project III. Zooplankton, macrobenthos and fish. Verh. Int. Verein. Limnol. 19: 1007-1106.

Andersson, G., Bergren, H. Cronberg, G. & Gelin, C. 1978. Effects of planktivorous and benthivorous fish on organisms and water chemistry in eutrophic lakes. Hydrobiol. 59:9-15.

Bajkov, A. D. 1934. How to estimate the daily food consumption of fish under natural conditions. Trans. Am. Fish. Soc. 65: 288-289.

Bowen, S. H. 1976. Mechanism for digestion of detrital bacteria by the cichlid fish *Sarotherodon mossambicus* (Peters). Nature 260: 137-138.

Bowen, S. H. 1979a. A nutritional constraint in detrivory by fishes: the stunted population of *Sarotherodon mossambicus* in Lake Valencia (Venezuela). Ecol. Monogr. 49: 17-31.

Bowen, S. H. 1979 b. Determinants of the chemical composition of periphytic detrital aggregate in a tropical lake (Lake Valencia, Venezuela). Arch. Hydrobiol. 87: 166-177.

Brabrand, Å. 1983. Distribution of fish and food of roach *Rutilus rutilus* bleak *Alburnus alburnus* and ruffe *Acerina cernua* in Lake Vansjö south-east Norway. Fauna 36: 57-64.

Brandt, S. B. 1980. Spatial segregation of adult and young-of-the-year alewives across a thermocline in Lake Michigan. Trans. Am. Fish. Soc. 109: 469-478.

Breck, J. E. & Gitter, M. J. 1983. Effects of fish size on the reactive distance of bluegill *(Lepomis macrochirus)* sunfish. Can. J. Fish. Aquat. Sci. 40: 162-167.

Brooks, J. L. & Dodson, S. I. 1965. Predation, body size and composition of plankton. Science 150: 28-35.

Burrough, R. J., Bregazzi, P. R. & Kennedy, C. R. 1979. Interspecific dominance amongst three species of coarse fish in Slapton Ley, Devon. J. Fish Biol. 15: 535-544.

Chapham, D. W. 1978 a. Production p. 202-218. In: Bagenal, T. (ed.) Methods for the Assessment of fish production in Fresh Waters, Blackwell Scientific Publications, Oxford.

Chapham, D. W. 1978 b. Production in fish populations p. 5-25. In: Gerking, S. D. (ed.) Ecology of Freshwater Fish Production, Blackwell Scientific Publications, Oxford.

Confer, J. L. & Blades, P. I. 1975. Omnivorous zooplankton and planktivorous fish. Limnol. Oceanogr. 20: 571-579.

Confer, J. L., Hawick, G. L., Corzette, M. H., Kramer, S. L., Fitzgibbon, S. & Landesberg, R. 1978. Visual predation by planktivors. Oikos 31: 27-37.

Connell, J. H. 1980. Diversity and the coevolution of competitors, or the ghost of competition past. Oikos 35: 131-138.

Craig, J. F. 1978. A study of the food and feeding of perch, *Perca fluviatilis* in Windermere. Freshwat. Biol. 8: 59-68.

Disler, N. N. & Smirnow, S. A. 1977. Sensory organs of the lateral-line canal system in two percids and their importance in behaviour. J. Fish. Res. Bd. Can. 34: 1492-1503.

Drenner, R. W., Strickler, J. R. & O'Brien, W. J. 1978. Capture probability: The role of zooplankton escape in the selective feeding of planktivorous fish. J. Fish. Res. Bd. Can. 35: 1370-1373.

Eggers, D. M. 1979. Comments on some recent methods for estimating food consumption by fish. J. Fish. Res. Bd. Can. 36: 1018-1019.

Elliott, J. M. & Persson, L. 1978. The estimation of daily rates of food consumption for fish. J. Anim. Ecol. 47: 977-991.

Federova, G. V. & Verkasov, S. A. 1972. The biological characteristics and abundance of the Lake Umen ruffe. J. Ichtyol. 14: 836-841.

Galbraith Jr., M. G. 1967. Size-selective predation on Daphnia by rainbow trout and yellow perch. Trans. Am. Fish. Soc. 96: 1-10.

Gardner, M. B. 1981. Mechanisms of size selectivity by planktivorous fish; a test of hypotheses. Ecology 62: 571-578.

Gerritsen, J. 1984. Size efficiency reconsidered: a general foraging model for free-swimming aquatic animals. Am. Nat. 123: 450-467.

Giller, P. S. & McNeill, S. 1981. Predation strategies, resource partitioning and habitat selection in *Notonecta (Hemiptera/Heteroptera)*. J. Anim. Ecol. 50: 789-808.

Goldspink, C. R. 1979. The population density, growth rate and production of roach *Rutilus rutilus* (L.) in Tjeukemeer, the Netherlands. J. Fish. Biol. 15: 473-498.

Gumaá, S. A. 1978. The food and feeding habits of young perch, *Perca fluviatilis*, in Windermere. Freshwat. Biol. 8: 177-187.

Hamrin, S. F. 1973. Vertical distribution and relative composition of the fish populations in South and North Lake Bolmen 1970-1972. Mimeographed, Inst. of Limnology, Lund (In Swedish).

Hamrin, S. F. 1983. The food preference of vendace *(Coregonus albula)* in South Swedish forest lakes including the predation effect on zooplankton populations.

Hydrobiol. 101: 121-128.

Hamrin, S. F. 1984. The fish community and its food resources in Lake Vombsjön, 1983. Mimeographed, Inst. of Limnology, Lund (In Swedish).

Hartmann, J. & Nümann, W. 1977. Percids of Lake Constance, a lake undergoing eutrophication. J. Fish. Res. Bd. Can. 34: 1670-1677.

Hellawell, J. M. 1972. The growth, reproduction and food of the roach *Rutilus rutilus* of the river Lugg, Herefordshire. J. Fish. Biol. 4: 469-486.

Henriksen, L., Nyman, H. G., Oscarson, H. G. & Stenson, J. A. E. 1978. Trophic changes, without changes in the external nutrient loading. Hydrobiol. 68: 257-263.

Hickling, C. F. 1966. On the feeding process in the white amur, *Ctenopharyngedon idella*. J. Zool. 148: 408-419.

Holcik, J. 1977. Changes in the fish community of Klicava Reservoir with particular reference to Eurasian perch *(Perca fluviatilis)*, 1956-1972. J. Fish. Res. Bd. Can. 34: 1734-1747.

Hrbacek, J. 1962. Species composition and the amount of zooplankton in relation to the fish stock. Rozpr. CSAV, Ser. Mat. Nat. Sci. 72: 1-117.

Hutchinson, B. P. 1971. The effects of fish predation on the zooplankton of ten Adirondach lakes, with particular reference to the alewife, *Alosa pseudoharengus*. Trans. Am. Fish. Soc. 100: 325-335.

Janssen, J. 1982. Comparison of searching behaviour for zooplankton in an obligate planktivore, blueback herring *(Alosa aestivalis)* and a facultative planktivore, bluegill *(Lepomis macrochirus)*. Can. J. Fish. Aquat. Sci. 39: 1649-1654.

Janssen, J. 1983. How do bluegills "select" large *Daphnia* in turbid waters. Ecology 64: 403.

Kapoor, B. G., Smit, H. & Verighina, I. A. 1975. The alimentary canal and digestion in teleosts p. 109-239. In: Russel, F. & Younge, M. (eds.), Advances in Marine Biology Vol. 13. Academic Press, London.

Keast, A. 1977. Diet overlaps and feeding relationships between the year classes in the yellow perch *(Perca flavescens)*. Env. Biol. Fish. 2: 53-70.

Keast, A. 1978. Trophic and spatial interrelationships in the fish species of an Ontario temperate lake. Env. Biol. Fish. 3: 7-31.

Keast, A. & Harker, J. 1977. Fish distribution and benthic invertebrate biomass relative to depth in an Ontario lake. Env. Biol. Fish. 2: 235-240.

Kevern, N. R. 1966. Feeding rate of carp estimated by a radioisotopic method. Trans. Am. Fish. Soc. 95: 366-371.

Kitchell, J. F., Stewart, D. J. & Weininger, D. 1977a. Applications of a bioenergetics model to yellow perch *(Perca flavescens)* and walleye *(Stizostedion vitreum vitreum)*. J. Fish. Res. Bd. Can. 34: 1922-1935.

Kitchell, J. F., Johnson, M. G., Minns, C. K., Loftus, K. H., Greig, L. & Olver, C. H. 1977b. Percid habitat: the river analogy. J. Fish Res. Bd. Can. 34: 1959-1963.

Lammens, E. H. R. R. 1982. Growth, condition and gonad development of bream *(Abramis brama)* in relation to its feeding condition in Tjeukemeer. Hydrobiol. 95: 311-324.

Lammens, E. H. R. R. 1985. A test of a model for planktivorous filter feeding by bream *Abramis brama*. Env. Biol. Fish. 13: 289-296.

Laughlin, D. R. & Werner, E. E. 1980. Resource partitioning in two coexisting sunfish: pumpkinseed *(Lepomis gibbosus)* and northern longear sunfish *(Lepomis megalotis peltastes)*. Can. J. Fish. Aquat. Sci. 37: 1411-1420.

Leach, J. H., Johnson, M. G., Kelso, J. R. M., Hartmann, J., Nümann, W. & Entz, B. 1977. Responses of percid fishes and their habitats to eutrophication. J. Fish. Res. Bd. Can. 34: 1964-1971.

Lessmark, O. 1983. Competition between perch *(Perca fluviatilis)* and roach *(Rutilus rutilus)* in south Swedish lakes. Dissertation, University of Lund, Sweden.

Lobel, P. S. & Ogden, J. C. 19781. Trophic biology of herbivorous reef fishes: alimentary pH and digestive capabilities. J. Fish Biol. 19: 365-397.

Lyagina, T. N. 1972. The seasonal dynamics of the biological characteristics of the roach *(Rutilus rutilus* L.) under conditions of varying food availability. J. Ichtyol. 12: 210-226.

Menzel, D. W. 1959. Utilization of algae for growth by the angelfish. J. Cons. Explor. Mer. 24: 308-313.

Mittelbach, G. G. 1981a. Foraging efficiency and body size: a study of the optimal diet and habitat use by bluegills. Ecology 62: 1370-1386.

Mittelbach, G. G. 1981b. Patterns of invertebrate size and abundance in aquatic habitats. Can. J. Fish. Aquat. Sci. 38: 896-904.

Moriarty, D. J. W. 1973. The physiology of digestion of bluegreen algae in the cichlid fish, *Tilapia nilotica*. J. Zool. 171: 25-39.

Niederholzer, R. & Hofer, R. 1979. The adaptation of digestive enzymes to temperature, season and diet in roach, *Rutilus rutilus* L. and rudd, *Scardinius erythrophthalmus* L. Cellulase. J. Fish. Biol. 15: 411-416.

Niederholzer, R. & Hofer, R. 1980. The feeding of roach *(Rutilus rutilus L)* and rudd *(Scardinius erythrophthalmus* L.) 1. Studies on natural populations. Ekol. Pol. 28: 45-59.

Nilsson, N.-A. 1978. The role of size-biased predation in competition and interactive segregation in fish p. 279-302. In: Gerking, S.D. (ed.) Ecology of Freshwater Fish Production, Blackwell Scientific Publications, Oxford.

Nilsson, N.-A. & Pejler, B. 1973. On the relation between fish fauna and zooplankton composition in north Swedish lakes. Rep. Inst. Freshw. Res. Drottningholm 53: 51-77.

Nyberg, P. 1979. Production and food consumption of perch, *Perca fluviatilis* L., in two Swedish forest lakes. Rep. Inst. Freshw. Res., Drottningholm 58: 140-157.

Nümann, W. 1972. The Bodensee: effects of exploitation and eutrophication on the salmonid community. J. Fish. Res. Bd. Can. 29: 833-847.

O'Brien, J. W. 1979. The predator-prey interaction of planktivorous fish and zooplankton. Am. Sci. 67: 572-581.

O'Brien, Slade, N. A. & Vinyard, G. L. 1976. Apparent size as the determinant of prey selection by bluegill sunfish *(Lepomis macrochirus.* Ecology 57: 1304-1310.

Olson, R. J. & Mullen, A. J. 1986. Recent developments for making gastric evacuation and daily ration determinations. Env. Biol. Fish. 16: 183-191.

Persson, L. 1982. Rate of food evacuation in roach *(Rutilus rutilus)* in relation to temperature, and the application of evacuation rate estimates for studies on the reate of food evacuation. Freshwat. Biol. 12: 203-210.

Persson, L. 1983a. Food consumption and competition between age classes in a perch *Perca fluviatilis* population in a shallow eutrophic lake. Oikos 40: 197-207.

Persson, L. 1983b. Food consumption and the significance of detritus and algae to intraspecific competition in roach *Rutilus rutilus* in a shallow eutrophic lake. Oikos 41: 118-125.

Persson, L. 1983c. Effects of intra- and interspecific competition on dynamics and size structure of a perch *Perca fluativilis* and a roach *Rutilus rutilus* population. Oikos 41: 126-132.

Persson, L. 1984. Food evacuation and models for multiple meals in fishes. Env. Biol. Fish. 10: 305-309.

Persson, L. 1986. Patterns of food evacuation: a critical review. Env. Biol. Fish 16: 51-59.

Persson, L. 1986. Effects of reduced interspecific competition on resource utilization of a perch *(Perca fluviatilis)* population. Ecology 67: 355-364.

Prejs, A. 1984. Herbivory by temperate freshwater fishes and its consequences. Env. Biol. Fish. 10: 281-296.

Reif, C. B. & Tappa, D. W. 1966. Selective predation: smelt and cladocerans in Harveys lake. Limnol. Oceanogr. 11: 437-438.

Rice, J. A. & Cochran, P. A. 1984. Independent evaluation of a bioenergetics model for largemouth bass. Ecology 65: 732-739.

Rice, J. A., Breck, J. E., Bartell, S. M. & Kitchell, J. F. 1983. Evaluating the constraints of temperature, activity and consumption on growth of largemouth bass. Env. Biol. Fish. 9: 263-275.

Robson, D. S. 1970. On the relation between feeding rate and stomach content in fishes. Biometrics Unit Mimeo Series BU-328-M Cornell University, Itacha.

Robson, D. S. & Spangler, G. R. 1978. Estimation of population abundance and survival p. 26-51. In: Gerking, S. D. (ed.) Ecology of Freshwater Fish Production, Blackwell Scientific Publications, Oxford.

Rundberg, H. 1977. Trends in harvest of pike perch *(Stizostdion lucioperca)*, Eurasian perch *(Perca fluviatilis)*, and northern pike *(Esox lucius)* and associated environmental changes in lakes Mälaren and Hjälmaren, 1914-1974. J. Fish. Res. Bd. Can. 34: 1720-1724.

Sandheinrich, M. B. & Hubert, W. A. 1984. Intraspecific resource partitioning by yellow perch *(Perca flavescens)* in a stratified lake. Can. J. Fish. Aquat. Sci. 41: 1745-1752.

Sand-Jensen, K. 1979. Balancen mellem autotrofe komponenter i tempererede søer med forskellig næringsbelastning. Vatten 2/80: 104-115. (In Danish).

Schoener, T. W. 1983. Field experiments on interspecific competition. Am. Nat. 122: 240-285.

Sinha, G. M. & Moitra, S. K. 1975. Morpho-histology of the intestine in a freshwater major carp *Cirrhinius mrigala* (Hamilton) during the different life history stages in relation to food and feeding habits. Annat. Anz. Bd. 137: 395-407.

Sorokin, J. I. 1968. The use of $^{14}$C in the study of nutrition of aquatic animals. Mitt. Int. Verein. Limnol. 16.

Stenson, J. A. E. 1972. Fish predation effects on the species composition of the zooplankton community in eight small forest lakes. Rep. Inst. Freshw. Res., Drottningholm. 52: 132-148.

Stenson, J. A. E. 1979. Predatory-prey relations between fish and invertebrate prey in some forest lakes. Rep. Inst. Freshw. Res., Drottningholm 58: 166-183.

Stewart, D. J., Weininger, D., Rottiers, D. V. Edsall, T. A. 1983. An energetics model for lake trout, *Salvelinus namaycush:* Application to the Lake Michigan population. Can. J. Fish. Aquat. Sci. 40: 681-698.

Stich, H.-B. & Lampert, W. 1981. Predation evasion as an explanation of diurnal vertical migration by zooplankton. Nature 293: 396-398.

Stickney, R. R. & Shumway, S. E. 1975. Occurrence of cellulase activity in the stomachs of fishes. J. Fish. Biol. 6: 779-790.

Sumari, O. 1971. Structure of perch populations in some ponds in Finland. Ann. Zool. Fennici 8: 406-421.

Svärdson, G. 1976. Interspecific population dominance in fish communities of Scandinavian lakes. Rep. Inst. Freshwat. Res., Drottningholm 56: 144-171.

Svärdson, G. & Molin, G. 1973. The impact of climate on Scandinavian populations of sander *(Stizostedion lucioperca* (L.). Rep. Inst. Freshw. Res., Drottningholm 53: 112-139.

Thorpe, J. E. 1977. Daily ration of adult perch, *Perca fluviatilis* L. in Loch Leven, Scotland. J. Fish. Biol. 11: 55-68.

Ware, D. M. 1972. Predation by rainbow trout *(Salmo gairdneri):* the influence of hunger, prey density, and prey size. J. Fish. Res. Bd. Can. 29: 1193-1201.

Warshaw, S. J. 1972. Effects of alewives *(Alosa pseudoharengus)* on the zooplankton of lake Wononskopomuc, Connecticut. Limnol. Oceanogr. 17: 816-825.

Wells, L. 1970. Effects of alewife predation on zooplankton population in Lake Michigan. Limnol. Oceanogr. 15: 556-565.

Werner, E. E. 1977. Species packing and niche complementarity in three sunfishes. Am. Nat. 111: 553-578.

Werner, E. E. & Hall, D. J. 1974. Optimal foraging and the size selection of prey by the bluegill sunfish *(Lepomis macrochirus)*. Ecology 55: 1042-1052.

Werner, E. E., Mittelbach, G. G. & Hall, D. J. 1981. The role of foraging profitability and experience in habitat use by the bluegill sunfish. Ecology 62: 116-125.

Werner, E. E., Mittelbach, G. G., Hall, D. J. and Gilliam, J.F. 1983. Experimental tests of optimal habitat use in fish: the role of relative habitat profitability. Ecology 64: 1525-1539.

Werner, E. E. & Gilliam, J. F. 1984. The ontogenetic niche and species interactions in size-structured populations. Ann. Rev. Ecol. Syst. 15: 393-425.

Wetzel, R. G. 1979. The role of the littoral zone and detritus in lake metabolism.

Arch. Hydrobiol. Beih. Ergebn. Limnol. 13: 145-161.

Willemsen, J. 1969. Food and growth of pikeperch in Holland. Proc. 4th Brit. Coarse Fish. Conf. 72-78.

Willemsen, J. 1977. Population dynamics of percids in Lake Ijssel and some smaller lakes in the Netherlands. J. Fish. Res. Bd. Can. 34: 1710-1719.

Winberg, G. G. 1956. Rate of metabolism and food requirements of fishes. Minsk Belorussian State Univ. (J. Fish. Res. Bd. Can. Transl. Ser. No. 194).

Windell, J. T. 1978. Digestion and the daily ration of fishes p. 159-183. In: Gerking, S. D. (ed.), Ecology of freshwater fish production, Blackwell Scientific Publications, Oxford.

Winfield, I. J., Peirson, G., Cryer, M. & Townsend, C. R. 1983. The behavioral basis of prey selection by underyearling bream *(Abramis brama* (L.)) and roach *(Rutilus rutilus* (L.)). Freshwat. Biol. 13: 139-149.

Wright, D., O'Brien, W. J. & Vinyard, G. L. 1980. Adaptive value of vertical migration: a simulation model argument for the predation hypothesis p. 138-147. In: Kerfoot, W. C. (ed.), Evolution and ecology of zooplankton communities, University Press of New England, Hannover, New Hampshire.

Youngs, W. D. & Robson, D. S. 1978. Estimation of population numbers and mortality rates p. 137-164. In: Bagenal, T. (ed.), Methods for Assessment of Fish Production in Fresh Waters, Blackwell Scientific Publications, Oxford.

Zaret, T. M. & Kerfoot, W. C. 1975. Fish predation on *Bosmina longirostris:* body-size selection versus visibility selection. Ecology 56: 232-237.

Zaret, T. M. & Suffern, J. S. 1976. Vertical migration in zooplankton as a predator avoidance mechanism. Limnol. Oceanogr. 21: 804-813.

# Chapter 7. CARBON METABOLISM AND COMMUNITY REGULATION IN EUTROPHIC, TEMPERATE LAKES

by Bo Riemann, Morten Søndergaard, Lennart Persson and Lars Johansson.

## 7.1. Introduction

In the previous chapters we have outlined and discussed ecologically important features of carbon metabolism in the pelagic environment. In particular, we have attempted to emphasize the pathways and rates by which inorganic and organic carbon are transported between different pools and to emphasize the size and structure of the biotic community catalyzing the flow. The purpose of this concluding chapter is to synthesize the many details which have been presented on the roles of phytoplankton, bacteria, zooplankton and fish in freshwater environments by presenting and discussing two general carbon flow models. One model describes a moderately low-productive lake with phytoplankton production of 60 g C m$^{-2}$ year$^{-1}$; the other a productive lake with yearly phytoplankton production of 600 g C m$^{-2}$. The presentation of these two carbon budgets is used to illustrate changes in community carbon metabolism as a function of eutrophication.

Generalizing diagrams sometimes have a tendency to be viewed in a very rigid and definitive manner as if they contain the whole and final story. This is never the case as a diagram presents a "frozen" picture of a complex series of dynamic events. Furthermore, such diagrams are always biased by present knowledge. However, models - here presented as diagrams - can serve the purpose of challenging accepted theory. The following diagrams should be considered, therefore, as working hypotheses and we would like the reader to bear this in mind.

Although some specific diagrams describing carbon flow in lakes have been presented (Wetzel et al. 1972, Cole 1985), no counterpart for the general marine diagram of Williams (1981) has been presented for lakes. There are probably two major reasons for this: 1) Each lake has in limnology traditionally been considered unique and, therefore, different from other lakes and 2) quantitative information from all trophic levels has only become available during the past few years.

Basically, our model includes five compartments of standing stock: phytoplankton, bacteria, microzooplankton, macrozooplankton and fish (Fig. 7.1). The transport of carbon into and out of each compartment is indicated with arrows, where the arrow-symbols and their numbers indicate specific processes such as respiration, excretion, sedimentation and predation (grazing).

For the sake of simplicity, a specific pool of dissolved organic carbon is not included. However, excretion, cell lysis and other carbon release processes generate carbon which passes through the DOC pool. All particulate organic carbon is classified into specific biomass compartments, as it is not possible to distinguish

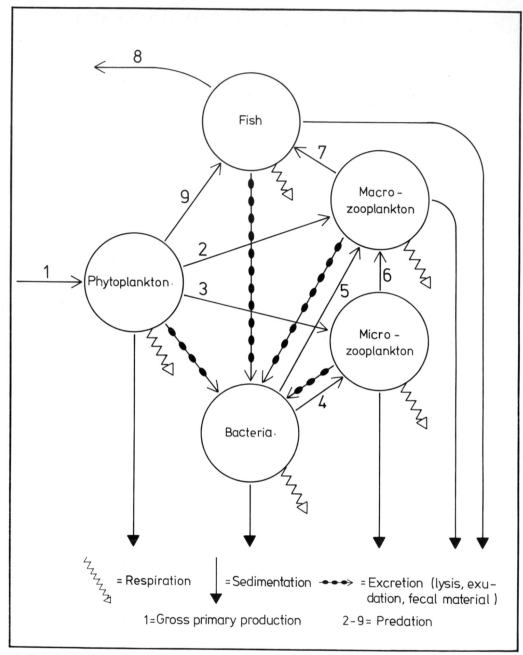

Fig. 7.1.
*The structure of the carbon budget. Circles represent the average annual biomass of the respective organisms.*

between between living and dead POC with a degree of certainty that would increase the quality of the model and information therein.

The most important assumptions relating to our budgets relate to time and space. The time is set by the general reference of one year and the depth is chosen as being a 10 m profile. Units of standing stocks are thus g C m$^{-2}$, and the carbon flows are given as g C m$^{-2}$ year$^{-1}$.

We have also been forced to compromise on matters where information and knowledge are limited, such as the effects on carbon flow of grazing by microheterotrophs and flagellated microalgae. Consequently, we have assumed the systems to be in steady state and adjusted the budgets to establish balance.

Finally, we have isolated the pelagic environment from the input from and influence of sediment contact (viz. hypolimnion), the littoral zone, reed swamps and external sources. The consequences of such an artificial isolation are not known. It is hoped that in spite of these limitations, the models and discussion presented will still prove relevant as an instrument and inspiration for further studies in pelagic limnological environments.

## 7.2. The Carbon budget in a low-productive lake
### 7.2.1. Phytoplankton

In general, phytoplankton primary production exhibits a more or less predictable seasonal pattern in temperate, low-altitude lakes. Virtually all primary production takes place during the spring, summer and autumn. The most conspicuous difference between the two lake types presented here is the normal lack of a distinct summer maximum in the low-productive lake.

In Fig. 7.2, 60 g C m$^{-2}$ gross production is generated by a phytoplankton biomass of 1 g C m$^{-2}$. Assuming that net production is 54 g C, this biomass turns over 54 times per year or, on average, once every 6.8 days.

Five pathways drain carbon from the phytoplankton: (1) High release of EOC is a common and expected feature of algae living close or fully exposed to a low nutrient regime. We assign a value of 20 g C m$^{-2}$ and include carbon loss via cell lysis in the release of EOC. Presumably, however, most of the 20 g carbon is lost via EOC. Only about 10% of primary production is lost via sedimentation and respiration, respectively. Both higher and lower values for respiration have been reported in the literature. In Forsberg's (1985) analysis of the loss processes from the phytoplankton in six lakes, respiratory processes were the major cause of loss of carbon from two of the lakes examined (Lanao and Bysjön). Inclusion of a higher respiratory loss in Figs. 7.2 and 7.3. would not change the validity of our conclusions, since the controlling feedback mechanisms and the carbon metabolism of the other compartments depend entirely on the net phytoplankton production which is transported throughout the food web. Zooplankton grazing is dominated by the macrozooplankton in that 24 g C m$^{-2}$ year$^{-1}$ or 44% of the net phytoplankton production is consumed by the macrozooplankton. This value pathway is markedly higher than that suggested by Forsberg (1985), who calculated the grazing from literature data or older studies carried out under laboratory conditions.

We believe that *in situ* studies are necessary to get reliable estimates of ingestion rates by natural populations of zooplankton feeding on natural populations of phytoplankton. Laboratory studies using animals in concentrations above those found in nature tend to lower the grazing estimates compared with those obtained from *in situ* experiments. In contrast to Forsberg's calculations, Gulati et al. (1982)

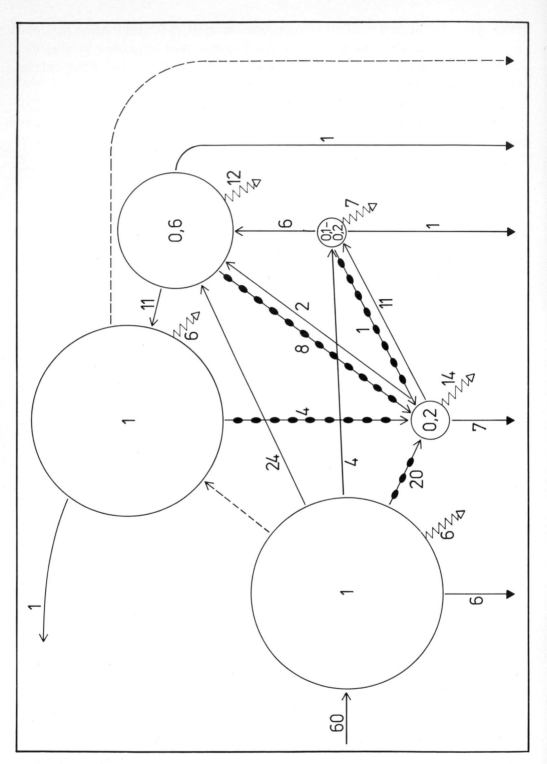

*Fig. 7.2.*
*Carbon budget from the low-productive lake with an annual phytoplankton primary production of 60 g*
*C m⁻². Pools are g C m⁻² and rates between the pools are g C m⁻² year⁻¹.*

reported that the annual mean ingestion rates of the zooplankton in Lake Vechten (annual primary production about 100 g C m$^{-2}$) were close to the mean particulate carbon fixation by the phytoplankton. However, large seasonal fluctuations in the grazing rates of zooplankton were reported by Gulati et al. (1982), with the lowest values occurring during the winter period. Thus, the highest values coincided with the period of maximum rates of primary production. Grazing rates from both macro- and micro-zooplankton on the phytoplankton need further verification. In view of existing knowledge, we suggest that about 70% of particulate phytoplankton primary production is consumed by macrozooplankton and about 7% by microzooplankton in our low-productive lake.

In a similar budget from Mirror Lake (annual phytoplankton net primary production of 40 g C m$^{-2}$), Cole (1985, Fig. V.D-1) estimated zooplankton ingestion to be at least 43% of particulate net primary production. He depicted a similar amount of particulate carbon as entering the sediment opposed to our 17% of the particulate primary production entering the sediment. In spite of this difference, however, there is a generally good agreement between the two budgets.

### 7.2.2. Zooplankton

The biomass of the macrozooplankton (0.6 g C m$^{-2}$) turns over 31.6 times per year or once every 11.5 days, whereas the corresponding values for the microzooplankton are highly variable. For microzooplankton we have chosen a range of 0.1-0.2 g C m$^{-2}$ with a turnover time of about 4-9.5 days.

Excretion of organic material from the zooplankton results mainly from "sloppy feeding" (Lampert 1978) and the production of fecal material, direct excretion is considered to be of minor importance. We suggest that about half of the 8 g C total excretion is generated via "sloppy feeding". This corresponds to about 17% of the annual ingestion rate.

The role of microzooplankton grazing on bacteria is highly speculative. Such grazing is assumed to be caused primarily by heterotrophic microflagellates. However, almost nothing is known about their ecological importance.

Recent evidence (Bird and Kalff 1986) has shown the chrysophyean *Dinobryon* to ingest a substantial amount of bacteria. In this way, part of the DOC is recycled to the phytoplankton compartment. The quantitative significance of this pathway is as yet unknown. However, as chrysophyceans are a dominant component of the phytoplankton in low-productive lakes, it is of great interest to determine if phytoplankton play a role in controlling bacterial biomass.

### 7.2.3. Bacteria

The bacterial biomass turns over 100 times per year or once every 3.7 days. Respiration is measured to be 40% of gross secondary production, although, as discussed in Chapter 4, this may be a serious underestimate.

Excretion of dissolved organic carbon from phytoplankton comprises 59% of the carbon requirement of the bacteria. The importance of the bacteria as decomposers of phytoplankton production is thus almost equal to the role of the herbivorous zooplankton.

### 7.2.4. Fish

From the discussion in Chapter 6, we see that the high proportion of piscivorous fish found in the low-productive lake, indicates that a relatively large amount of carbon is needed to sustain the fish biomass. Respiration for such organisms is about 55% of food consumption (Brett & Groves 1979, Elliott 1979). Fecal material is included in the excretion rate using values given in Elliott (1979) and Persson (1983b). Whether the excreted material is directly available to the bacteria is not known.

Direct predation on phytoplankton by fish is probably not important where piscivorous species dominate and planktivorous fish prefer animal food. Moreover, the concentration of the algae is so low that a direct carbon transport from algae to fish in the low-productive lake seems to be of little importance.

## 7.3. The carbon budget in the highly productive lake
### 7.3.1. Phytoplankton

The increased nutrient load and the reduced predation pressure from the zooplankton permit a marked increase in phytoplankton production rate and biomass in the highly productive lake. The turnover time of the phytoplankton biomass is 34.3 times per year, or once every 10.6 days, (Fig. 7.3) compared with once every 6.8 days in the low-productive lake (Fig. 7.2). This increase is caused by the successive development of large, mainly colonial, cyanobacteria.

The flow of carbon to the bacterial compartment is larger here than in the low-productive lake, both in absolute terms and relative to the net phytoplankton primary production. We assume that only 60 g C, or 10% of the gross primary production is EOC, and that the major part (240 g C m$^{-2}$) is released via cell lysis. The reasons for the decrease in the release of EOC are not clear. However, several explanations are suggested in Chapter 3. The ecological importance of cell lysis is not precisely known. It has been demonstrated, however, that only minor reductions in the phytoplankton biomass can release a substantial proportion of labile carbon compounds in a eutrophic lake (Hansen et al. 1986).

The absolute values of carbon transport from the phytoplankton to the zooplankton compartments are higher in the highly productive lake than in the low-productive lake, but lower relative to the net primary production. Thus, the major part of the phytoplankton carbon ( 70%) is channeled via the bacterioplankton.

Finally, in contrast to the low-productive lake, a significant proportion of the phytoplankton net production is grazed by planktivorous fish (see section 7.3.4) in the case of the highly productive lake.

### 7.3.2. Zooplankton

The increases in the biomasses of both macro- and micro-zooplankton reflect an increase in number and a decrease in size of the animals. Thus, the biomass of the

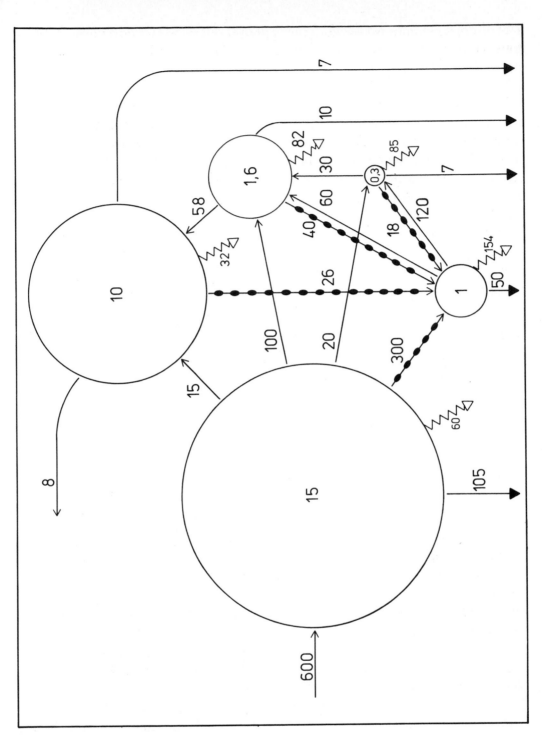

*Fig. 7.3.*
*Carbon budget from the highly productive lake with an annual phytoplankton primary production of*
*600 g C m⁻². All pools are g C m⁻² and all rates between the pools are g C m⁻² year⁻¹.*

macrozooplankton turns over 67.5 times per year or once every 5.4 days, whilst the biomass of the microzooplankton turns over 183 times per year, or once every 2 days. These turnover times are markedly lower than the corresponding values from the low-productive lake.

The high values of carbon transport from the bacteria to the macrozooplankton suggest that small species of *Daphnia* are particularly important as grazers of bacteria. They also suggest that a lot of bacteria attached to particles are transported to the macrozooplankton. Nevertheless, the microzooplankton still consume the majority of the bacteria, presumably due to a marked development in the number of heterotrophic microflagellates (Riemann 1985).

### 7.3.3. Bacteria

The bacterial biomass increases from 0.2 g C m$^{-2}$ in the low-productive lake to 1 g C in the highly productive lake (Fig. 7.3). This is caused predominantly by larger numbers and increased cell volumes. The bacterial biomass turns over 230 times per year, or once every 1.6 days, compared with 3.7 days in the low-productive lake.

### 7.3.4. Fish

A 10-fold increase of the fish biomass accompanies the 15-fold increase in the phytoplankton biomass and the 10-fold increase in the primary production in the highly productive lake. The size of the fish biomass are discussed by Persson (1983a, c) and Hamrin (1984). Changes in the community structure and the ecological consequences are discussed in Chapter 6 and below.

The direct grazing on the phytoplankton constitutes about 21% of the total consumption by fish and results from the numerical dominance of cyprinids.

## 7.4. Predator regulation and primary production in the trophic gradient.

Changes in fish communities in low- and highly productive lakes involve both quantitative and qualitative processes (see Chapter 6). Quantitatively, the numerical dominance of cyprinids in highly productive lakes can be related to the resource distribution and competitive interactions which favour cyprinids. Qualitatively, the size structure of perch populations varies with the trophic conditions. The size class diversity and the proportion of piscivorous perch is higher in low-productive lakes than in highly productive lakes. This suggests a difference in the proportion of piscivorous fish between low- and highly productive lakes (Fig. 7.4) see also Persson (1983a). While the proportion of piscivorsous fish of the total fish biomass in low-productive systems is generally above 40%, it is 10% or less in highly productive lakes.

These differences in the size and structure of the fish communities suggest that planktivorous fish are controlled by different factors in low and highly productive lakes. The fish populations in highly productive lakes, where a low proportion of

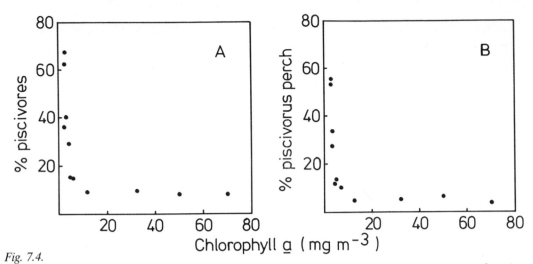

Fig. 7.4.

*The percentage of piscivorous fish in relation to phytoplankton biomass (A) and the percentage of pisci-vorous perch in relation to phytoplankton biomass (B).*

piscivores is found, can be explained by competitive interactions as outlined in Chapter 6.

In the previous chapters, we examined how planktivorous fish affect different pelagic communities and the physical-chemical conditions of the lakes. Considering the different composition of the fish communities, we would expect to observe different effects of the fish on lower trophic levels in low and highly productive lakes. Planktivorous fishes affect lower trophic levels by depressing zooplankton populations with a subsequent increase in the biomass of the phytoplankton (Brooks and Dodson 1965, Andersson et al. 1978, Henriksson et al. 1980, see also Chapters 5 and 6). This effect would be minimized, however, in the low-productive lakes where the fish community consists of two functional trophic levels (Fig. 7.4). To understand the dynamics of such a relatively complex system, a theoretical approach in exploring the type of interactions involved is useful.

An interesting theoretical model dealing with these types of questions has been developed by Oksanen et al. (1981). The ultimate regulator of the system in the model is the maximum gross primary productivity ($G$) allowed by the environment, which in lake systems can be assumed to be directly related to the nutrient input. It is assumed that the growth of phytoplankton agrees with the logistic growth equation and that herbivores and primary and secondary carnivores exhibit a Holling II functional response. The predictions of the model include that, in the least productive environments, phytoplankton production and biomass are too low to support any herbivores. In this region of the model, the phytoplankton biomass increases linearly with primary production ($G < g_o$) Fig.7.5. Once the phytoplankton biomass reaches a level which allows herbivores to enter the system, a stable herbivore-phytoplankton equilibrium is established. In this region, the phytoplankton biomass is expected to be constant and an increase in primary production will result in a grazing response by herbivores ($g_o < G < g_1$). In communities with high ecological efficiencies such as aquatic environments, an increase in $G$ may result in population oscillations ($g_1 < G < g_2$) (Oksanen et al. 1981). As $G$ increases further, primary carnivores (i.e. mainly planktivorous fish) will enter the system and the relation between phytoplankton biomass and $G$ is once again linear ($G > g_2$) (Fig. 7.5). In aquatic systems, a fourth trophic level - piscivorous fish - can be formed. The fourth trophic level will depress the level of primary carnivores. This, in turn results in increased herbivore density and a decreasing or constant phytoplankton biomass with increasing $G$ (Oksanen et al. 1981, Oksanen pers. comm.).

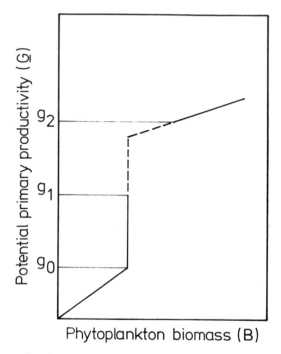

*Fig. 7.5.*
*The relationship between the phytoplankton maximum gross primary production and biomass (modified after Oksanen et al. 1981).*

Thus, the model by Oksanen et al. (1981) predicts that the relationship between primary production and phytoplankton biomass is not linear over the whole spectrum of $G$ (Fig. 7.5) but, rather, that zones exist where phytoplankton production and biomass are independent of each other. The first zone (phytoplankton and herbivores) is difficult to test, as most lakes include three trophic levels but experimental studies in enclosures support the prediction (Riemann and Søndergaard 1986). In low-productive lakes where more than 40% of the fish population is piscivores, the phytoplankton biomass is constant (2-4 mg chlorophyll a m$^{-3}$, see Fig. 7.4), while primary production may vary by a factor of 3 (from 20 to 70 mg C m$^{-2}$ year$^{-1}$, Lundqvist 1975). This supports the prediction that increased production does not result in an increasing phytoplankton biomass in a four level system.

In highly productive lakes, the ecosystem is functionally reduced to a three level system, as the low proportion of piscivorous fish offsets any regulating effects on planktivorous/herbivorous cyprinids. This situation is not predicted by Oksanen et al.'s (1981) model. It is not known whether the failure of Oksanen et al.'s model to arrive at this prediction results from the assumption of homogenous trophic levels being violated by the presence of size structure in the fish population or from the fact that there may not exist an equilibrium point for a four level lystem at high primary production. Evidence for the former explanation is that the recruitment of perch to the piscivorous class is affected by competition from planktivorous/benthivorous roach in highly productive systems (see Chapter 6).

Fretwell (1977) suggested that whether a certain trophic level was regulated by competition or predation should depend on the number of trophic levels in the system. For example, in a three level system planktivorous fish should be regulated by competition, while in a four level system they should be regulated by predation. In correspondence with this available data (see above) indicate that planktivorous fish are mainly regulated by competition in highly productive lakes while in low-productive lakes they are mainly regulated by predation. Based on Fretwell's idea it can further be predicted that zooplankton should be regulated to a greater extent by competition in low-productive lakes and by predation in highly productive. This results in the prediction that phytoplankton should be controlled more by competition in highly productive lakes than in low-productive lakes where they are to a greater degree regulated by grazing. The theory that different trophic levels are regulated by different factors in low- and highly productive lakes is supported by the biomass relationships between different trophic levels (Figs. 7.2 and 7.3).

## 7.5. Equilibrium and non-equilibrium conditions

The above comparison between low and highly productive lakes assumes that such systems are at or close to equilibrium conditions. Several factors will, however, act to create dis-equilibrium. One important factor, for example, might be death of fish during periods of long ice cover or during summer. Changes in nutrient input can also disturb equilibrium. Winter kills are most pronounced in shallow highly productive lakes (Andersson and Gelin 1970). Summer kills also occur in the same lakes and are presumably caused by high concentration of $NH_3$-N produced at high pH levels and/or by low oxygen conditions.

Since environmental factors together with size and physiology of the fish influences the outcome of several oxygen conditions, the effects on fish populations are difficult to predict (Hargeby 1985). However, any reduction in biomass of fish leads to a release of the competition intensity and species that are inferior competitors may increase in abundance. This type of density compensation has been described by Tonn (1985) in mudminnow *(Umbra limi)* -yellow perch assemblages of North America.

After a winter kill or summer kill with a reduction of the fish community, the fish populations enter a successional phase. In European fish assemblages, perch appears to have the most rapid numerical response to the competitive release as was shown by the dominance of perch during the initial successional phase in the Klicava reservoir (Fig. 7.6, Holcik 1977). After seven years, the roach population had increased in abundance, and the fish community showed a pattern typical of

278

eutrophic lakes. The same successional pattern was observed in eutrophic Lake Sövdeborg after a fish kill (Persson 1983a, c).

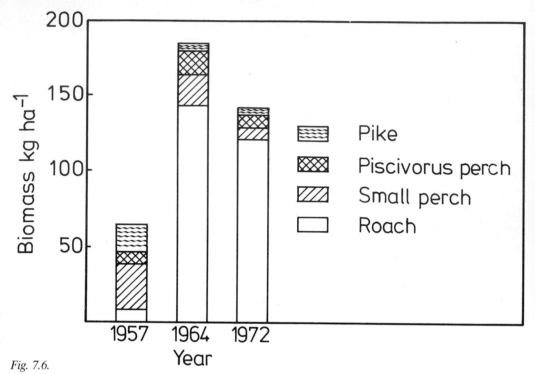

Fig. 7.6.
*The succession of various fish in the Klicava reservoir (after Holcik 1977).*

Environmental factors such as winter and summer kills may, thus, push the fish community away from equilibrium conditions and start a number of feedback mechanisms cascading down on lower trophic levels. Fish kills are examples of how stochastic environmental factors may affect the dynamics of lake ecosystems. Incorporating such factors into our analysis will further increase our understanding of the dynamics of freshwater systems.

## Acknowledgements

We wish to thank dr. Katherine Richardson for valuable linguistic suggestions and comments on Chapters 7 and 8.

# References

Andersson, G. & Gelin, C. 1970. Chemical conditions in icecovered lakes in South Sweden. Vatten 1970: 174-183. (In Swedish with English summary).

Andersson, G., Berggren, H., Cronberg, G. & Gelin, C. 1978. Effects of planktivorous fish on organisms and water chemistry in eutrophic lakes. Hydrobiologia 59: 9-15.

Bird, D. F. & Kalff, J. 1986. Bacterial grazing by planktonic lake algae. Science, 231: 493-495.

Brett, J. R. & Groves, T. D. D. 1979. Physiological energetics, p. 280-352. In: Hoar, W. S., Randall, D. J. and Brett. J. R. (Eds.), Fish Physiology Vol. VIII. Academic Press, New York.

Brooks, J. L. & Dodson, S. I. 1965. Predation, body size and composition of plankton. Science 150: 28-35.

Cole, J. J. 1985. Decomposition, p. 302-310. In: Likens, G. E. (Ed.), An ecosystem approach to aquatic ecology. Mirror Lake and its watershed. Springer Verlag. New York.

Elliott, J. M. 1979. Energetics of freshwater teleosts, p. 29-61. In: Miller, P. J. (Ed.), Fish phenology: anabolic adaptiveness in teleosts. Symp. Zool. Soc. London.

Forsberg, B. R. 1985. The fate of planktonic primary production. Limnol. Oceanogr. 30: 807-819.

Fretwell, S. D. 1977. The regulation of plant communities by food chains exploiting them. Perspect. Biol. Med. 20: 169-185.

Gulati, R. D., Siewertsen, K. & Postema, G. 1982. The zooplankton: its community structure, food and feeding, and role in the ecosystem of Lake Vechten. Hydrobiologia 95: 127-163.

Hamrin, S. F. 1984. The fish community and its food resources in Lake Vombsjön, 1983. Mimeographed, Inst. of Limnology, Lund. (In Swedish).

Hansen, L., Krog, G. F. & Søndergaard, M. 1986. Decomposition of lake phytoplankton. 1. Dynamics of short-term decomposition. Oikos 46: 37-44.

Hargeby, A. 1985. Fiskens överlevnad vid syrebrist vintertid - en fråga om att fly ille fläkta? Fauna och Flora 80:81-92 (in Swedish with English summary).

Henriksson, L., Nyman, H. G., Oscarson, H. G. & Stenson, J. A. E. 1980. Trophic changes, without changes in the external loading. Hydrobiologia 68: 257-263.

Holcík, J. 1977. Changes in fish community of Klícava reservoir with particular reference to Eurasian perch (Perca fluviatilis) 1957-1972. J. Fish. Res. Bd. Can. 34: 1734-1747.

Lampert, W. 1978. Release of dissolved organic carbon by grazing zooplankton. Limnol. Oceanogr. 23: 831-834.

Lundqvist, I. 1975. Fegen, Kalvsjö. Limnologisk undersökning. Mimeographed, Inst. of Limnology, Lund (in Swedish).

Oksanen, L., Fretwell, S. D., Arruda, J. & Niemalä, P. 1981. Exploitation ecosystems in gradients of primary productivity. Am. Nat. 118: 240-261.

Persson, L. 1983a. Food consumption and competition between age classes in a perch *Perca fluviatilis* population in a shallow eutrophic lake. Oikos 40: 197-207.

Persson, L. 1983b. Food consumption and the significance of detritus and algae to intraspecific competition in roach *Rutilus rutilus* in a shallow eutrophic lake. Oikos 41: 118-125.

Persson, L. 1983c. Effects of intra- and interspecific competition on dynamics and size structure of a perch *Perca fluviatilis* and a roach *Rutilus rutilus* population. Oikos 41: 126-132.

Petersson, B. J. 1984. Synthesis of carbon stocks and flows in the open ocean mixed layer, p. 547-554. In: Hobbie, J. E. & Williams, P. J. LeB. (Eds.), Heterotrophic activity in the sea. Plenum Press.

Riemann, B. 1985. Potential importance of fish predation and zooplankton grazing on natural populations of freshwater bacteria. Appl. Environ. Microbiol. 50: 187-193.

Riemann, B. & Søndergaard, M. 1986. Regulation of bacterial secondary production in two eutrophic lakes and in experimental enclosures. J. Plank. Res. 8: 519-536.

Tonn, W. M. 1985. Density compensation in *Umbra-Perca* fish assemblages of northern Wiconsin lakes. Ecology 66: 415-429.

Wetzel, R. G., Rich, P. H., Miller, M. C. & Allen, H. L. 1972. Metabolism of dissolved and particulate detrital carbon in a temperate hard-water lake. Mem. Ist. Ital. Idrobiol. Suppl. 29: 185-241.

Williams, P. J. LeB. 1981. Incorporation of microheterotrophic processes into the classical paradigm of the planktonic food web. Kieler Meeresforsch. Sonderh. 5: 1-28.

# Chapter 8. EPILOGUE

by Morten Søndergaard and Bo Riemann

The most recent scenario of pelagic carbon cycling is that given by Williams in his challenging paper on microheterotrophic processes in the planktonic food web in marine waters (Williams 1981). One major outcome of this work has been an accelerated interest in the role of the bacterioplankton and the mode by which DOC is made available to bacteria. Accordingly, there has been a dramatic increase in the available literature dealing with bacterial production and its relation to planktonic primary producers and other biological activities.

The few seasonal studies that have been carried out in freshwater systems have estimated the total bacterial carbon requirement to range from 20 to 60% of the annual phytoplankton primary production. In addition, the level of bacterial production and activity seems to be linked with the level of phytoplankton primary production and biomass. Seasonal studies integrating biological activity at various trophic levels, are however, scarce, both for marine and freshwater environments. It is our view that attempts to generalize such as that presented in Chapter 7, can be used to produce preliminary working hypotheses.

The carbon budget approach to the study of pelagic carbon cycling has the advantage of framing our ecological understanding and of providing a guideline for the evaluation of results which may be orders of magnitude in error. However, its predictive value is not better than the sum of erratic results. The accuracy of the budget is, in most cases, unknown.

New approaches to the study of pelagic carbon cycling can be identified by applying carbon budgets. For example, we can take the new evidence on bacterial growth yield (now 0.1-0.3, earlier 0.5-0.6) and, perhaps, the conversion factor used to calculate bacterial net carbon production from rates of $^3$H-thymidine incorporation. These factors, when applied to natural data, would, in a series of cases (e.g. Riemann and Søndergaard 1986), increase bacterial carbon demand to a level very close to, or even above, phytoplankton production. Very small, seemingly unrealisticly small, amounts of carbon would be left to herbivores and sedimentation. Thus, the carbon budget approach gives us cause to reconsider estimates of primary production. Such statements have been made before but primary production rates and their measurements still remain a controversial discussion subject at international meetings. One indirect way to evaluate primary production may be to give more attention to loss processes (i.e. respiration, lysis, grazing, sedimentation and EOC release).

In this book, we have emphasized major processes in the planktonic food web of eutrophic lakes and their possible control. In addition, we have attempted to critically evaluate the methods used to examine these processes. In only a few cases have we tried to evaluate data from a total carbon budget point of view.

The biological structures and processes in lakes are controlled from at least two different directions. The best known scenario is that an increased nutrient loading will immediately be followed by an increase in phytoplankton biomass and a concomitant change in the dominating species. However, community structure and function can also be greatly modified by biological events.

Changes in the composition of the fish population towards dominance by pelagic, planktivorous species constitute a distinct response of community structure to eutrophication. Small fishes preferentially prey on the macrozooplankton which, especially during summer periods, can control the development of algal standing stock. By escaping such grazer control, the phytoplankton can build up massive and persistent summer blooms. Furthermore, the diminishing, or even removal, of a selective grazing pressure favours specific algal species; mainly larger forms and colonies.

The removal of macrozooplankton affects not only algae but is rapidly followed by an increase in the number of flagellates and microzooplankton present in the community. Grazing on bacteria becomes dominated by these two groups. The entire structure of the community, then, is biologically controlled by the predators. In structural terms, eutrophication in temperate lakes can be described as a progressive change towards a community dominated by larger algae, microzooplankton, flagellates and small fish (e.g. roach).

Theoretically, eutrophicaction of a lake should result in a change in the dominating processes within the organic carbon cycle compared with more oligotrophic situations. The reduction in direct grazing pressure on algae in the eutrophic watermass increases the importance of the detrital pathway. Increased primary production and the resulting algal biomass can either be recycled in the pelagic zone or after sedimentation. The former leads to a higher ratio of bacterial production to primary production in eutrophic as compared to oligotrophic lakes.

Recently, Hobbie and Cole (1984), using large seawater enclosures (MERL), presented data collected to illustrate whether a development towards dominance by a detrital pathway could be induced in a eutrophication experiment. They concluded that this was not the case, inasmuch as the percentage of carbon passing through the bacterioplankton was nearly constant (31%) for all levels of primary production. In contrast, we conclude from studies in eutrophic lakes that such waters probably have a higher relative carbon flow through bacteria and other microheterotrophs than more oligotrophic waters. This discrepancy is most probably due to the structural and process- controlling role played by fish in lakes.

Admittedly, the data relating to pelagic carbon cycling and their interpretation from both marine and freshwater environments are not yet mature and there is a particular lack of seasonal studies which should be given high priority in the future. We feel that there is an urgent need to reach a comprehensive understanding of these fundamental ecosystem responses to eutrophication. Control and reduction of nutrient loading must still have priority but the "internal" ecosystem behaviour must be given greater attention than it is today.

Future research in the planktonic environment should, thus, concentrate on the dynamic aspects of carbon transport from algae via different pathways to bacteria, zooplankton and fish in order to identify the factors controlling and modulating each specific route of carbon flow. Is there a continuum of different situations moving from the low production to high production sites?

Recent years' studies have provided many new methods and results which can enlarge our understanding of various processes in the pelagic food web. However, we would also like to draw attention to some specific methodological problems which still have not been adequately dealt with.

At the primary production level, the release of EOC - its specific chemical composition and subsequent utilization by bacteria - must be studied in further detail. The uncertainty surrounding true release rates is mainly due to problems connected with the $^{14}$C-method. Solving the question concerning the specific activity of the EOC seems to be the single most important issue to be dealt with. However, bacterial preference for various EOC products, POC availability to bacteria, etc. are also interesting problems and important to an overall understanding of carbon flow.

The *in situ* biomass of bacteria is currently under study at a number of laboratories and new results point to a doubling of presently accepted estimates due to errors in the conversion of biovolumes to biomasses. The importance of picophytoplankton of bacterial size should probably be reconsidered. An understanding of overall bacterial growth yield is crucial to our concept of the detrital pathway. The recently proposed values of 0.1 to 0.3 should be verified as soon as possible.

Larger heterotrophic organisms - zooplankton and fish - are important components in the pelagic zone and their effects should always be considered. Their participation in DOC generating processes *in situ* is largely unknown compared with our knowledge of their effects on biological structures. The impact of fish on macrozooplankton communities and the cascade effects on phytoplankton and microheterotrophs, in particular, should be a focus for future research.

Many details and also some significant carbon budget aspects of pelagic structure and dynamics in lakes are still only partly understood. By identifying some of the gaps in our knowledge, we hope to have encouraged future research in these areas.

# References

Hobbie, J. E. & Cole, J. J. 1984. Response of a detrital foodweb to eutrophication. Bull. Mar. Sci. 35: 357-363.

Riemann, B. & Søndergaard, M. 1986. Growth of natural bacterial assemblages in relation to phytoplankton biomass and production in two eutrophic lakes and in experimental enclosures. J. Plank. Res. 8:. 519-536.

Williams, P. J. LeB. 1981. Incorporation of microheterotrophic processes into the classical paradigm of the planktonic food web. Kieler Meeresforsch. 5: 1-28.

# List of contributors

*Wayne Bell,* Center for Environmental and Estuarine Studies, University of Maryland, P.O. Box 775, Cambridge, MD 21613 U.S.A.

*Suzanne Bosselmann,* Freshwater Biological Laboratory, University of Copenhagen, Helsingørsgade 51, DK-3400 Hillerød, Denmark.

*Lars Møller Jensen,* Carbon 14 Centralen, The International Agency for [14]C determination, Agern Allé 5, DK-2970 Hørsholm, Denmark.

*Lars Johansson,* Fish Ecology Research Group, Institute of Limnology, P.O. Box 65, S-221 00 Lund, Sweden.

*Niels O. G. Jørgensen,* Department of Microbiology, Royal Veterinary and Agricultural University, Rolighedsvej 21, DK-1958 Frederiksberg C, Denmark.

*Jørgen Kristiansen,* Institute of Spore Plants, University of Copenhagen, Ø. Farimagsgade 2 D, DK-1353 Copenhagen K, Denmark.

*Lennart Persson,* Fish Ecology Research Group, Institute of Limnology, P.O. Box 65, S-221 00 Lund, Sweden.

*Bo Riemann,* Freshwater Biological Laboratory, University of Copenhagen, Helsingørsgade 51, DK-3400 Hillerød, Denmark.

*Morten Søndergaard,* Botanical Institute, University of Aarhus, Nordlandsvej 68, DK-8240 Risskov, Denmark.